PREFACE

W0231931

For a long time deterministic and statistical descriptions of physical phenomena existed in distinct mathematical environments. Topology, functional analysis, and differential geometry played the main role in deterministic mechanics, whereas measure theory and the theory of stochastic processes were the tools of statistical mechanics. Furthermore, statistical descriptions were usually associated with the lack of information on the precise dynamical specification of systems with a large number of degrees of freedom.

Although some of the ideas may be traced as far back as Poincaré and Hadamard, it was only recently that the physics community came to realize that systems with a small number of degrees of freedom and simple mathematical descriptions could display complex behaviour and, in some cases, have solutions which, although fully deterministic, are for all practical purposes indistinguishable from a random process. The simplicity of the mathematical description brought with it the possibility of abstracting, from the models, laws of complexity that are even found to hold universally, irrespective of the details of the system. Furthermore, whereas statistical descriptions mainly address equilibrium states and transitions between equilibrium states, deterministic descriptions handle more efficiently the dynamics of complexity. Hopefully some insight might also be gained into the problems of pattern formation and self-organization.

One of the purposes of the workshop was the exploration of the interface between the deterministic and stochastic points of view in the mathematical description of complex phenomena. Displaying and encouraging this interaction, one hopes that further insight will arise from the cross-breeding of the two cultures.

The design and construction of technological devices is always influenced by the theoretical understanding available at a given time. For example, most mechanical devices of engineering relevance are modelled as integrable Hamiltonian systems and the classical mechanics textbooks in many schools are still exclusively concerned with this very small class of systems. The improved theoretical understanding of complex systems is sure to have an impact on the design and control of new devices and the exploration of old ones in previously avoided regions of instability. Having this in mind, as the second aim of the workshop, we have attempted to explore the interface between theoretical models and the understanding of complex behaviour in engineering systems.

Lisbon, 1989

R. Lima

L. Streit

R. Vilela Mendes

Lecture Notes in Physics

Edited by H. Araki, Kyoto, J. Ehlers, München, K. Hepp, Zürich
R. Kippenhahn, München, D. Ruelle, Bures-sur-Yvette
H. A. Weidenmüller, Heidelberg, J. Wess, Karlsruhe and J. Zittartz, Köln
Managing Editor: W. Beiglböck

355

R. Lima   L. Streit
R. Vilela Mendes   (Eds.)

Dynamics and Stochastic Processes
Theory and Applications

Proceedings of a Workshop
Held in Lisbon, Portugal
October 24–29, 1988

Springer-Verlag Berlin Heidelberg GmbH

Editors

Ricardo Lima
Centre de Physique Théorique, C.N.R.S.
Luminy, Case 907, F-13288 Marseille Cédex 09

Ludwig Streit
Fakultät für Physik, Universität Bielefeld
Postfach 86 40, D-4800 Bielefeld 1, FRG

Rui Vilela Mendes
Centro de Física da Matéria Condensada
Av. Gama Pinto, 2, P-1699 Lisboa Codex, Portugal

ISBN 978-3-662-13797-0   ISBN 978-3-540-46969-8 (eBook)
DOI 10.1007/978-3-540-46969-8

This work is subject to copyright. All rights are reserved, whether the whole or part of the material
is concerned, specifically the rights of translation, reprinting, re-use of illustrations, recitation,
broadcasting, reproduction on microfilms or in other ways, and storage in data banks. Duplication
of this publication or parts thereof is only permitted under the provisions of the German Copyright
Law of September 9, 1965, in its version of June 24, 1985, and a copyright fee must always be
paid. Violations fall under the prosecution act of the German Copyright Law.

© Springer-Verlag Berlin Heidelberg 1990

Originally published by Springer-Verlag Berlin Heidelberg New York in 1990
Softcover reprint of the hardcover 1st edition 1990

CONTENTS

Some remarks on classical, quantum and stochastic dynamical systems

by

Sergio Albeverio*,****, **Teresa Arede#** ,
Astrid Hilbert*

* Fakultät für Mathematik, Ruhr-Universität, D-4630 Bochum 1
  (Fed. Rep. of Germany)
** BiBoS (Bielefeld), CERFIM (Locarno), SFB 237 (Essen – Bochum – Düsseldorf)
# Faculdade de Engenharia, Universidade do Porto, Porto (Portugal)

Abstract

We give a survey of some recent results on stochastic perturbation of classical dynamical systems of Hamiltonian type respectively of gradient type. We also discuss the latters as quantization of classical dynamical systems of the former type. Moreover we examine some relations between classical and quantum systems on manifolds, as well as infinite dimensional versions of these topics.

1. Introduction

The relations between classical and quantum dynamical systems are far from being understood. Stochastic dynamical systems bear relationships, of different types, with both subjects and besides presenting an interest on their own they can also serve as a bridge between the topics. In this paper we discuss a little bit these three types of systems, in the case of finite dimensional state space as well as infinite dimensional state space. In section 1 we study perturbations of classical finite dimensional Hamiltonian dynamical systems, by adding to the deterministic force a stochastic one. Such systems have been considered before, mainly in the case of 2-dimensional states space, by J. Potter, H. McKean, J. Goldstein, K. Narita, L. Marcus and A. Weeraninghe a.o. In addition there is a large heuristic literature having connections with this problem, where typically some moments of the phase space variables are computed in some approximation. We should also add that the case of linear deterministic force has also received great attention, see e.g. the references in [Hi]. We also like to mention the case of "multiplicative stochastic perturbation" (as opposite to the above additive one), studied in [Pi], [ArW], [ArK] (and references therein). We report essentially on recent results in [Hi], [AH], [AHZ].

As compared with general studies of perturbations of dynamical systems, see e.g. [Kh], [KrS], the systems we study present the difficulty of being degenerated (hypoelliptic rather than elliptic), in addition we do not assume smooth coefficients neither bounds on growth at infinity. An equivalence, for finite times, of the probability measures associated with a linear and a non linear case is discussed. Also results about the asymptotic behaviour of the systems for large time are reported. In section 3 we discuss stochastic perturbations of dynamical systems of gradient type and their relations with the quantization of classical Hamiltonian systems. This approach is related with stochastic mechanics, the theory of Dirichlet forms and the associated potential theory. It has its roots in work by L. Gross and by S. Albeverio and R. Høegh-Krohn, see [A-HK1], [A-HK2] (and references therein), and has been pursued systematically, particularly in the case of finite dimensional state space, since the work [AHKS], see also e.g. [ABR] and references therein. As a powerful technical tool it has at disposal the theory of Dirichlet forms, developed particularly by Fukushima and his school, see e.g. [Fu1], [Fu2] and references therein. After recalling briefly some main parts of the theory of Dirichlet forms, we discuss the relations between the study of (symmetric) Markov processes by "Dirichlet forms" and quantum mechanics. This points out the usefulness of the Dirichlet approach as a tool to handle strong singularities in the potentials involved as well as a suitable tool to pass to the case of infinite dimensional state space, connected with the theory of quantum fields.

In section 4 we study classical mechanics on manifolds and its quantization, particularly through heat kernel methods. In section 4.1 we give the main formulae of classical mechanics on manifolds, for complements see any of the modern books in analytical dynamics, e.g. [AMa]. In section 4.2 we discuss Schrödinger and heat equation on manifolds. Via a Feynman-Kac formula one can express the heat semigroup acting on functions as an expectation with respect to a Brownian motion on the manifold. The latter is a well studied subject starting with classical work by Hunt, probabilistic and potential theoretic, Yosida, analytic, and McKean, via stochastic equations, see e.g. [RoW] and references therein. In

Section 4.3 we describe a formula obtained by Elworthy and Truman [ElT], which gives an expression for the heat kernel on connected complete manifolds exhibiting a multiplicative factor analogous to the heat kernel in the flat case, with Euclidean distance and volume replaced by the corresponding Riemannian ones, times an expectation with respect to a suitable Brownian bridge with drift of an expression involving the Ruse invariant i.e. the determinant of the exponential mapping. We discuss several applications of the formula as obtained in [Ar]. These include simply harmonic manifolds, Cartan-Hadamard manifolds and in particular Clifford-Klein spaces. We also point out that Eskin's formulae for the heat kernel on compact semisimple Lie groups can be obtained from the probabilistic Elworthy-Truman expression, following [El] and [Ar2]. A new interpretation of Eskin's formula for the case of general semisimple compact Lie groups and certain symmetric spaces is also given, following [Ar2].

In all these cases we can say that in a sense the heat kernel is expressed exactly for all times by classical expressions, involving essentially only the length of geodesics. This observation should be put in relation with certain discussions going presently on in the physical literature concerning quantization of classical chaotic systems, cfr. e.g. [GrSt], and references therein.

We also briefly mention recursion formulae for hyperbolic spaces [DaM] and nilpotent Lie groups [AArH].

In section 5 we briefly discuss extensions of the topics of the previous sections to the infinite dimensional case, reporting particularly about some recent developments in the study of Dirichlet forms over infinite dimensional state spaces [ARö].

We illustrate the connection of these studies with the quantization of certain classical field theories. We also briefly mention on a new type of Markov fields and interacting quantum fields over 4-dimensional space-time obtained by solving a stochastic partial differential equation (based on work in [AHKI], [AIK] (and references therein)).

2. Perturbation of classical finite dimensional Hamiltonian dynamical systems

2.1 Existence and uniquenes

In this section we shall study essentially stochastic perturbations of classical Hamiltonian systems with phase space $\mathbb{R}^{2d}$, of the form

$$
\begin{aligned}
\dot{x} &= v \\
\dot{v} &= K(x,t),
\end{aligned}
\tag{2.1}
$$

where position x and velocity v run in $\mathbb{R}^d$, $\cdot$ means derivative with respect to time and $K(\cdot)$ is the (deterministic) force, which we assume to be either linear or such that the associated energy integral is bounded from below (which is assured by assuming the force is derived from a potential and is attracting towards a given point $x_0 \in \mathbb{R}^d$), and locally Lipschitz. Much is known about these and related systems, in particular about their asymptotics for large times, see e.g. [Ar]. To quote just a recent discussion [DiZ] we might mention that in the case where $d = 1$ and $K(x,t) = -x^{2k+1} - \sum_{j=0}^{2k} p_j(t)x^j$ we have that for $p_j \equiv 0$ all solutions are periodic with periods decreasing to zero as the energy is increased and that for $p_j \not\equiv 0$, periodic and smooth, all solutions are bounded. We might ask ourselves, following [AHZ], whether stochastic perturbations might change this picture drastically or not.

We shall report here basically on some answers obtained in work [AHZ] (in collaboration with E. Zehnder), and in [Hi] which extends in particular preceeding work by J. Potter [Po], H. Mc Kean [McK], J. Goldstein [Go], Narita [Na], L. Marcus – A. Weeransinghe [MaW].

In this work K is assumed not to depend explicitly on time and the stochastic perturbation replaces $K(x)$ by $K(x) + \dot{w}_t$, with $\dot{w}_t$ being Gaussian white noise, so that the equation of motion (2.1) becomes the stochastic system

$$\begin{aligned} dx(t) &= v(t)dt \\ dv(t) &= K(x(t))dt + dw_t, \end{aligned} \tag{2.2}$$

with w_t a Brownian motion in $\mathbb{R}^d$ started at time 0 from the origin. We assume $\binom{x(0)}{v(0)}$ be given (stochastic or deterministic) in $\mathbb{R}^{2d}$. We write (2.2) also in the form

$$dy(t) = \beta(y(t))dt + \sigma d\tilde{w}_t, \tag{2.3}$$

with

$$y \equiv \begin{pmatrix} x \\ v \end{pmatrix}, \quad \beta(y) \equiv \begin{pmatrix} v \\ K(x) \end{pmatrix}, \quad \sigma \equiv \begin{pmatrix} 0 & 0 \\ 0 & \mathbb{1} \end{pmatrix}$$

(with $0, \mathbb{1}$ being $d \times d$–matrices), $\tilde{w}_t = \binom{b_t}{w_t}$, with b_t a brownian motion independent of w_t. Here is the basic existence and uniqueness theorem:

Theorem 1: Each of the following conditions is sufficient for the existence and uniqueness of pathwise solutions (i.e. solutions $y(t,\omega)$ for a.e.w. in the probability space $(\Omega, \mathcal{A}, P)$:
a) $|K(\alpha)| \leq C(1 + |\alpha|) \; \forall |\alpha| \geq R$
(for some $R > 0$) ("sublinear growth condition")
or
b) K is a gradient field in the sense that $K = -\nabla V$ for some $V \in C^1(\mathbb{R}^d)$ and such that the energy functional $W(y) \equiv \frac{1}{2}|v|^2 + V(x) - V(0)$ is lower bounded (which is the case if $(\alpha - x_0)K(\alpha) \leq 0 \; \forall |\alpha - x_0| \geq R$
("attractiveness condition")).

Proof: The sufficiency of a) is proved by the usual Picard–Lindelöf iteration procedure. For b) we just remark that, assuming $x(0) = x_0 = 0$, $R = 0$ for simplicity, under the attractiveness condition one has $V(x) \geq V(0)$, hence $W(y) \geq \frac{1}{2}|v|^2$. $\hspace{2cm}$ (2.4)

Introducing the stochastic time $\tau(t) \equiv \int_0^t |v|^2 ds$ one has that $a(\tau(t)) \equiv \int_0^t v dw_s$ is a Brownian motion with respect to the filtration $\mathcal{F}_\tau$ associated with τ. This, together with the fact that Ito's formula yields

$$W(y(t)) = W(y(0)) + \int_o^t v dw_s + d\frac{t}{2} \hspace{2cm} (2.5)$$

implies that the explosion time $\zeta > 0$ for the equation (2.3) must be almost surely positive infinite (otherwise $\tau(t) < +\infty$ resp. $\tau(t) = +\infty$ imply $|y(\zeta)| = \infty$ resp. $|y(\zeta)| < \infty$, and in both situations we have a conflict with (2.4), (2.5)). See [AHZ], [Hi] for details and the proof of uniqueness.

∎

Having this basic result, we would like to discuss the behaviour of the solution y of (2.3) for finite times first and then for $t \to \infty$. As to the former question, a natural comparison is the one with a corresponding linear system.

2.2 Comparison with linear systems: a Girsanov type theorem

The relation with linear systems comes about due to existence of a Cameron–Martin–Girsanov–Maruyama–type theorem. More precisely this theorem permits to compare the probabilistic properties of our nonlinear system with those of any "corresponding" linear system (with first d components of the drift equal to the velocity and the latter d ones equal to a linear attracting force), in as much as almost sure statements can be transmitted from the latter to the former and viceversa. The comparison system is the following one

$$d\eta_t = \tilde{\beta}(\eta_t)dt + \sigma d\tilde{w}_t \hspace{2cm} (2.5)$$

with $\eta \equiv \binom{\eta_1}{\eta_2}$, $\eta_i \in \mathbb{R}^d$, $i = 1, 2$,

$$\tilde{\beta}(\eta) \equiv \begin{pmatrix} \eta_2 \\ -\gamma\eta_1 \end{pmatrix}, \quad \gamma > 0.$$

It turns out that the path space measures $P_{(y)}$ resp. $P_{(\eta)}$ belonging to the processes y resp. η are equivalent. This is proven in [AHZ], [Hi] (which extends a corresponding result for $d = 1$ and K sublinear of [MaW]).

The proof uses a limiting argument for $\tau_n \to +\infty$ after having checked that

$$E\left(e^{\frac{1}{2}\int_0^{t \wedge \tau_n} |+\gamma x(s) + K(x(s))|^2 ds}\right) < \infty$$

(where E means expectation and τ_n is a suitable sequence of stopping times) (whenever x solves (2.1)), which is essentially a consequence of the fact that $\int_0^t v_s dw_s$ is a martingale, implying $E\left(|v|^2\right) < \infty$, $E\left(|x|^2\right) < \infty$. As a consequence we have e.g.

1) $x(t)^2 + v(t)^2 > 0$ a.s. $\forall t > 0$ if it is so for $t = 0$.
2) for $d = 1$, $x(t)$ has a.s. infinitely many zeros which are all simple (proven as in [MaW])
3) "Winding" can be studied (following [MaW], see also [AGQ]).

2.3 Some additional results

a) We have already remarked that $\int_0^t vdw_s$ is a martingale. This is also the case for $W(y(t)) - \frac{d}{2}t$, see [AHZ], [Hi] for proofs. As a consequence of these facts, essentially following [Po], one obtains various inequalities controlling the phase space behaviour of the process. E.g.

$$\left[W(y(0)) + \frac{d}{2}t\right]^n \leq E\left[W(y(t))^n\right]$$

$$\leq d^n \sum_{k=1}^n \prod_{l=1}^n (2l-1)\binom{2l}{2k} W(y(0))^{n-k}\left(d\frac{t}{2}\right)^k.$$

See [AHZ], [Hi] for other inequalities.

b) $y(t)$ is a diffusion. The associated Feller Markov semigroups (which is non symmetric) has a density with respect to Lebesgue measure. All σ-finite invariant measures for y are of the form of a constant times Lebesgue measure in $\mathbb{R}^{2d}$ (at least when K is smooth).That Lebesgue measure is an invariant measure follows easily from $L^*1 = 0$, where L^* is the formal adjoint of the infinitesimal generator L of y, given on smooth functions by

$$L = \frac{1}{2}\Delta_v + v\cdot\nabla_x + K(x)\cdot\nabla_v.$$

The proof of uniqueness is however more involved, see [Hi].

c) As known since Potter [Po], for $d = 1$ and K linear, y is recurrent if $K \not\equiv 0$ and non recurrent if $K \equiv 0$, y is null recurrent, for K non necessarily linear but of the form

$$K(\alpha) = -\frac{d}{d\alpha}V(\alpha), \text{ if } \int \frac{d\alpha}{\sqrt{1+V(\alpha)}} < \infty.$$

For $d \geq 3$ it is proven in [AHZ], [Hi] that $y(t)$ is transient. This is a consequence of estimates on the transition probabilities and above results under a). The relevant estimate on transition probabilities is

$$\int P_t(y_0, y)\chi_A(y)dy \leq (ct+1)^{-\frac{d}{2}} \exp\left[-\frac{cW(y_0)}{ct+1}\right],$$

with c a constant depending on the bounded Borel set A.

Remark: In the proof of the above results, and in particular of c), the difficulty of the problem, comes from the <u>degeneracy</u> of L (with respect to x) (and σ). If σ and L were not degenerated one could use the powerful Khasminskij theory for non degenerated elliptic generators. The vectors fields in L span $\mathbb{R}^{2d}$ (hence if K is smooth we have the so called weak hypoellipticity): this can be exploited to study the weak noise limit [AHK].

3. Stochastic perturbation of dynamical systems of gradient type and quantization of classical Hamiltonian systems

3.1 Drifts which are gradients, diffusion generators, Dirichlet forms

By introducing a damping term $-\gamma v$, $\gamma > 0$ and replacing the forces $K, \dot{w}_t$ by $\gamma K, \gamma \dot{w}_t$ in the second equation of (2.1) one can show (at least when K is globally Lipschitz) that x converges a.s. $\gamma \to +\infty$ to the solution of the equation

$$dx(t) = K\left(x(t)\right) dt + dw(t) \tag{3.1}$$

(with the same initial condition $x(0) = x_0$ as for the component x of the solution process y of (2.3)), see [Ne, Th. 10.1].

Let us more generally consider thus an equation of the type

$$dx(t) = \beta\left(x(t)\right) dt + dw(t)$$

with β a gradient field i.e. $\beta = \nabla\varphi$ for some $\varphi \in C^1(\mathbb{R}^d)$.

Remark: The drift coefficient $\tilde{\beta}$ of the process y in (2.3) is not of the gradient type (but rather a "symplectic gradient"). This makes the dynamics of (3.2) (or (3.1)) very different from the one of (2.3).

The theory of stochastic differential equations of the type (3.2) is well developed. The most powerful technique is the one by which one looks upon (3.2) as the stochastic equation satisfied by a diffusion process associated with a Dirichlet form of the type

$$E_\mu(f, f) \equiv \frac{1}{2} \int |\nabla f|^2 d\mu \quad \text{in } L^2(\mathbb{R}^d, d\mu), \tag{3.2}$$

with $d\mu \equiv e^{2\varphi} dx$. This is so in the sense that for any bounded continuous f on $\mathbb{R}^d$ and quasi every x (i.e. every x outside a capacity zero set)

$$E^x\left(f(x(t))\right) = \left(P_t^\mu f\right)(x);$$

with P_t^μ the (symmetric Markov) semigroup on $L^2(\mathbb{R}^d, \mu)$ uniquely associated with the above Dirichlet form in the sense that $P_t^\mu = e^{-tH_\mu}$, $t \geq 0$, with the generator H_μ satisfying

$$\left((H^\mu)^{\frac{1}{2}} f, (H^\mu)^{\frac{1}{2}} f \right) = E_\mu(f, f) \tag{3.3}$$

(with $(,)$ on the left hand side being the $L^2(\mu)$-scalar product). On smooth functions we have $H^\mu = -\frac{1}{2}\Delta - \beta \cdot \nabla$ ("diffusion generator"). For these concepts and results see e.g. [Fu1], [Fu2], [ABR], [A2]. The asymptotics of (3.2) for $t \to +\infty$ is then reduced to the study of spectral properties of H_μ. These in turn are connected with the spectral properties of an associated "generalized Schrödinger" operator, of which we come now to speak.

We finally remark that under some weak smoothess assumptions on φ one can reduce, via a Girsanov formula, expectations with respect to the solution process of (3.1) to expectations with respect to Brownian motion:

$$E^x(f(x_t)) = E\left[e^{\int_0^t \beta(x_s) dw_s} e^{-\frac{1}{2}\int_0^t \beta(x_s)^2 ds} f(x_t) \right], \quad x_t \equiv x(t).$$

3.2 Relation with Schrödinger operators and quantization of classical Hamiltonian systems.

The relation with Schrödinger operators (as well as with both the Schrödinger and heat equations) relies on the following basic observation: if $\varphi > -\infty$ a.e. then $L^2(\mathbb{R}^d, \mu)$ is unitary equivalent $L^2(\mathbb{R}^d, dx)$ by the map $f \in L^2(\mathbb{R}^d, \mu) \to e^\varphi f \in L^2(\mathbb{R}^d, dx)$. By this equivalence the symmetric semigroup $P_t^\mu = e^{-tH_\mu}$ is mapped into a unitarily equivalent symmetric semigroup e^{-tH}, $t \geq 0$, acting in $L^2(\mathbb{R}^d, dx)$, with generator H, called (generalized) Schrödinger operator. In fact in case $\varphi \in C^2(\mathbb{R})$ and setting $V \equiv \frac{1}{2}(\Delta e^\varphi) e^{-\varphi}$ we have (e.g. on smooth functions of compact support)

$$H = -\frac{1}{2}\Delta + V,$$

hence V appears, in quantum mechanical terms, as a "potential" (multiplication by a function). Note that $He^\varphi = 0$, and $H \geq 0$, hence e^φ is the ground state for H (positive eigenfunction to the lowest end zero of the spectrum of H, a simple eigenvalue). The above transformation from H_μ to H is called <u>Darboux</u> or <u>ground state transformation</u>. The solution of Schrödinger equation

$$i\frac{\partial}{\partial t}\psi_t = \left(-\frac{1}{2}\Delta + V \right) \psi_t \tag{3.4}$$

with initial condition $\psi_0 \in L^2(\mathbb{R}^d, dx)$ is then given by $\psi_t = e^{-itH}\psi_0$, $t \in \mathbb{R}$, and correspondingly the solution of the heat equation

$$\frac{\partial}{\partial t}f_t = \left(\frac{1}{2}\Delta - V \right) f_t \tag{3.5}$$

with initial condition f_0 is given by

$$f_t = e^{-tH} f_0, \quad t \geq 0.$$

Feynman–Kac formula, valid if V is "not too negative" (see e.g. [Si], [AMa1,2,3] for some sufficient assumptions) expresses $f_t(x)$ (for a.e. x) as an integral with respect to Wiener measure: $f_t(x) = E^0 \left[e^{-\int_0^t V(w_s + x) ds} f_0(w_t + x) \right]$, E^0 being expectation with respect to a brownian motion w_s, $0 \leq s \leq t$ in $\mathbb{R}^d$, started at the origin. We can express this in the form, for $f, g \in L^2(\mathbb{R}^d, dx)$:

$$(f, e^{-tH} g) = \int f(w_0) g(w_t) d\mu_{E,t}(w),$$

where on the l.h.s. $(,)$ means scalar product in $L^2(\mathbb{R}^d, dx)$ and

$$d\mu_{E,t}(w) \equiv Z_t^{-1} e^{-\int_0^t V(w_s) ds} dP(w),$$

with $Z_t \equiv \int e^{-\int_0^t V(w_s) ds} dP(w_s)$ the "Euclidean" path space measure (for the time–space region $[0, t] \times \mathbb{R}^d$), P being Wiener measure. $\int_0^t V(w_s) ds$ is the Feynman–Kac functional. In order to get the connection with the case of infinite dimensional state space to be discussed below, let us point out that $d\mu_{E,t}$ can be looked upon as the restriction to the σ–algebra generated by the paths w_s, $0 \leq s \leq t$ of an "Euclidean" path space measure μ_E on $C(\mathbb{R} \times \mathbb{R}^d)$ given heuristically by

$$d\mu_E(w) \equiv \text{``} Z^{-1} e^{-\int_\mathbb{R} W(w_s) ds} dw \text{''} \tag{3.6}$$

with

$$\text{``} Z \equiv \int e^{-\int W(w_s) ds} dw \text{''}, \quad dw = \prod_{0 \leq s \leq t} dw_s,$$

$$W(w) \equiv \frac{1}{2} \int \dot{w}_s^2 ds + \int V(w_s) ds. \tag{3.7}$$

Let us finally make a simple but basic remark concerning the connection between classical mechanics and quantum mechanics. Let us consider the classical mechanical system described by (2.1) with $K(\alpha) = -\nabla V(\alpha)$, V being the classical potential. Quantization consists in replacing the dynamics given by solving (2.1) by the one given by the Schrödinger equation (3.4). By the fact that, for "nice" V, (see e.g [Re–Si] for sufficient conditions), e^{-itH}, $t \in \mathbb{R}$ is the analytic continuation of $e^{-tH} = e^{\varphi} e^{-tH_\mu} e^{-\varphi}$, $t \geq 0$, and e^{-tH_μ} is the semigroup uniquely associated with the solution process of (3.2), with $\beta = \nabla\varphi$, we see that we can look at the passage from the solution $\binom{x(t)}{v(t)}$ of the deterministic Hamiltonian equation (2.1) to the solution of the stochastic gradient type equation (3.2) (with β not given by the force but rather by $\nabla\varphi$!), as a way of quantizing the Hamiltonian equation (2.1). This has been called the "Dirichlet (form) approach to quantum mechanics". Another way is to form the Euclidean path space measure μ_E given by (3.6), (3.7) (which might be called the Feynman–Kac approach).

Remark1:   A third approach ("Euclidean quantum mechanics") has been pursued recently by J.C. Zambrini extending earlier work by Schrödinger, see e.g. [AYZ] and references therein.

Remark2:   The Dirichlet approach to quantum mechanics works in situations where the potential V is very singular (and in fact does not make sense directly neither in the classical nor in the quantum mechanical equation of motion). A well known example, discussed in details e.g. in [AGHKH], [AFHKL] is the case where $d = 3$, $e^\varphi = \frac{e^{4\pi\alpha|x|}}{|x|}$ with $\alpha \in I\!\!R$, in which case e^{-tH_μ} as well as e^{-tH} are well defined, but $H = -\frac{1}{2}\Delta + V$ holds only heuristically, since V is of the type $\frac{1}{2}(4\pi\alpha)^2 + \text{``}\frac{\pi}{3}\left[-\frac{1}{4}\pi^2 + 4\pi\alpha\varepsilon\right]\varepsilon\delta_\varepsilon(x)\text{''}$ for $\varepsilon \downarrow 0$ and δ_ε a δ–sequence converging to Dirac's δ measure at zero).

Remark3:   The above Dirichlet resp. Feynman–Kac approach to quantum mechanics has lead to many important developments both in the study of quantum mechanical problems as well as diffusions with gradient type drift (e.g. estimates on eigenvalues and eigenfunctions, see e.g. [Jo], [Car], study of irreducibility, confinement and ergodic problems e.g. [AFKS], [Fu2]; most recently, a solution of the inverse problem of finding β and V given spatial observation of the solution process outside a given bounded region, see [ABKS]).

Remark4:   The above way of quantizing (2.1) should not be confused with "stochastic quantization" in the sense of Parisi–Wu, Jona–Lasino–Mitter and others [Jo–Mi]. In fact the latter procedure consists in looking at the heuristic stochastic differential equation

$$dX(t,\tau) = \beta_{st}\left(X(t,\tau)\right)d\tau + \dot{w}(t,\tau) \tag{3.8}$$

where $\beta_{st} = \nabla W$ (with W as in (3.7)) and $\dot{w}$ is white noise over $I\!\!R^2$. A heuristic limit $\tau \to \infty$ yields then the Euclidean measure μ_E as equilibrium measure. This has been, at least partly, carried through also mathematically, see [Jo–Mi], [ARö] and references in [AZ].

Remark5:   We like also to mention stochastic mechanics, see e.g. [Ne], [BCZ], [Jo], [Gu], [Ca] and references therein, which is a stochastic way of quantizing different from all above ways in the non stationary case, with real time instead of imaginary time.

4. Classical mechanical motions on manifolds, their quantisation and the heat kernel

4.1 Classical mechanics on manifolds

Let us first recall how classical (Hamiltonian) mechanics is formulated on (pseudo) Riemannian manifolds. Let M be a (pseudo) Riemannian d–dimensional manifold (configuration space). Let $TM \equiv \cup_{x \in M} T_x M$ be its tangent bundle and let $L : TM \to I\!R$ be defined by $L(v) = \frac{1}{2}\langle v, v \rangle$, where $\langle , \rangle$ is the (pseudo) Riemannian metric on M. In coordinates $q^1, \ldots, q^d$ we have $L(v) = \frac{1}{2} g_{ij} v^i v^j$, where g_{ij} is the matrix of the metric and v^i are the components of the vector v.

L is the "free Lagrangian" (describing free motion on M). Let ω_L be the Lagrange closed 2–form on TM defined by $\omega_L = (FL)^* \omega_0$, ω_0 being the canonical symplectic form on $T^* M$, FL being the flat map associated with the metric i.e. $FL(\omega) \cdot v \equiv < w, v >, v, w \in TM$, in local coordinates $\omega_0 = \sum_{i=1}^d dq^i \wedge dp_i$, $\omega_L = \sum_{i=1}^d dq^i \wedge d\dot{q}^i$.

Let X_E be the unique Lagrangian vector field on TM s.t.

$$\omega_L(X_E, X_1) = dE ,$$

X_1 any vector field on M, with $E \equiv A - L$, A being the action: $TM \to I\!R$ given by $A(v_x) = FL(v_x) \cdot v_x =< v_x, v_x >, v_x \in T_x M$. In local coordinates $A = \sum_{i=1}^d (\dot{q}^i)^2 = 2L$, $E = \sum_{i=1}^d (\dot{q}^i)^2 - L = L$.

One shows that $c_0(t)$ is a base integral curve of X_E iff $c_0(t)$ is a geodesic (for the (pseudo) Riemannian manifold M). In local coordinates if $c_0(t) = \big(q^1(t), \ldots, q^d(t)\big)$ we have

$$\ddot{q}^i + \sum_{j,k} \Gamma^i_{jk} \dot{q}^j \dot{q}^k = 0 ,$$

i.e. the geodesic equations, with $\Gamma^i_{jk} \equiv \frac{1}{2} g^{il}(g_{lk,j} + g_{lj,k} - g_{jk,l})$ the Christoffel symbols of the affine (Levi–Civita) connection given by the metric (torsion zero and parallel translation being an isometry) (with $g_{lk,j} \equiv \frac{\partial}{\partial q^j} g_{lk}$).

Now let a function $V : M \to I\!R$ (potential) be given on M. Define the perturbed Lagrangian L_V (incorporating the potential) by

$$L_V \equiv \frac{1}{2}\langle v, v \rangle - V(\tau_M v) ,$$

with τ_M the tangent bundle projection for TM to M.

Let again A: $A(v) = \langle v, v \rangle$ be the action and E the energy. Then

$$E(v) = \frac{1}{2}\langle v, v \rangle + V(\tau_M v) \ .$$

$c_0(t)$ is a base integral curve of the Lagrangian vector field X_E i.e. satisfies Lagrange's equation of motion iff

$$\nabla_{\dot{c}_0}\dot{c}_0 = -\nabla V(c_0(t)) \ ,$$

with ∇ the covariant derivative (Levi–Civita connection of the metric). It is well known that Hamilton's principle holds: a curve $c_0 : [a, b] \to M$ joining $c_0(a)$ to $c_0(b)$ satisfies Lagrange equation iff c_0 is a critical point of the real–valued function J defined on C^2 curves, c starting at $c_0(a)$ and ending at $c_0(b)$, by

$$J(c) \equiv \int_a^b L\big(c(t), \dot{c}(t)\big)\, dt \ .$$

T^*M is the phase space of the classical dynamical system. (T^*M, ω_0) is a symplectic manifold.

The Hamiltonian H associated with the Lagrangian L is given by the map $T^*M \to \mathbb{R}$ defined by

$$H = \frac{1}{2}\langle \alpha, \alpha \rangle_{\tau_M^*(\alpha)} + V_0\tau_M^* \ , \tag{4.1}$$

where $\langle, \rangle_x$ is the metric on T_x^*M given by $\langle \alpha, \beta \rangle_x = \langle \gamma^\sharp(x)(\alpha), \gamma^\sharp(x)(\beta) \rangle_x$ for $\alpha, \beta \in T_x^*M$, and $\gamma^\sharp : T^*M \to TM$ the isomorphism of vector bundles $\gamma^\sharp = (\gamma^\flat)^{-1}$, $\gamma^\flat(v_x) \equiv \langle \cdot, v_x \rangle_x$. τ_M^* is the projection $T^*M \to M$. The Hamiltonian vector field X_H associated with H is defined by $\omega_0(X_H, Y) = dH \cdot Y$. The integral curves $(q(t), p(t))$ of X_H are called the Hamiltonian flow. They satisfy $H(q(t), p(t)) = const.$. In local coordinates we have $\dot{q}^i = H_{p_i}(u, v)$, $\dot{p}_i = -H_{q^i}(u, v)$.

In the next subsection we shall see how the quantisation of these classical flows can be done.

4.2 Schrödinger or heat operators on manifolds

Let M be an oriented Riemannian d–dimensional manifold, with metric $\langle, \rangle$, locally given by the matrix g_{ij}, $i, j = 1, \ldots, d$.

The Laplace–Beltrami operator Δ on functions is defined by $\Delta = \text{div} \cdot \nabla$, where ∇ is as before the gradient (s.t. $\langle \nabla f(x), v_x \rangle = df(x) \cdot v_x$, $\forall v_x \in T_x M$; for any $f \in C^1(M, \mathbb{R})$) and div is the divergence operator (i.e. the unique function div X on M s.t. $L_X \rho = (\text{div } X)\rho$, where ρ is the volume form on M and L_X the Lie derivative given by X).

In local coordinates we have

$$(\Delta f)(x) = \frac{1}{\sqrt{\det g_{ij}}} \frac{\partial}{\partial x^k}\left(g^{ik}\sqrt{\det g_{ij}}\,\frac{\partial f}{\partial x^i} \right) \ ,$$

with g^{ik} the inverse matrix to g_{ik}, and $\det(\cdot)$ being the determinant.

Let $L^2(M, \rho)$ be the natural L^2 space given by the Riemannian metric. We can regard Δ as a symmetric operator with domain $C_0^\infty(M, \mathbb{R})$ in $L^2(M, \rho)$. It is known that Δ is essentially self–adjoint if M is complete relative to the metric g (cfr. [Fu2]).

The closure of $-\Delta$, denoted again by $-\Delta$, is then a positive self–adjoint operator in $L^2(M, \rho)$, uniquely associated with the Dirichlet form obtained by closure from

$$E^\circ(u, v) = \frac{1}{2} \int \langle du, dv \rangle_x \rho(dx) , \qquad (4.2)$$

where $u, v \in C_0^\infty(M, \mathbb{R})$ and $\langle \ , \ \rangle_x$ is the scalar product in the space $T_x^* M$ of 1–forms.

Remark For results on Markov uniqueness and essential self–adjointness of the infinitesimal generators associated with Dirichlet forms obtained by closure from (4.2) with $\rho(dx)$ replaced by $\varrho(x)\rho(dx)$, with a suitable density ϱ, see [Fu2], [ABR], and references therein.

Having that $-\Delta$ is essentially self–adjoint it is not difficult to find criteria for

$$-\frac{1}{2}\Delta + Q_V ,$$

with Q_V the operator multiplication in $L^2(M, \rho)$ by the function V defined on M, to define by closure a unique self–adjoint operator in $L^2(M, \rho)$ (it suffices e.g. that Q_V be small with respect to $-\frac{1}{2}\Delta$ in the sense of quadratic forms, see e.g. [Re-Si]).

The closure H_V of $-\frac{1}{2}\Delta + Q_V$ in $L^2(M, \rho)$ can be looked upon as the quantisation of the operator H given by (4.1).

It is then natural to look, in analogy with the flat case, to both the Schrödinger equation

$$i\frac{\partial \psi}{\partial t} = H_V \psi$$

and heat equation

$$\frac{\partial f}{\partial t} = -H_V f$$

on M.

Let us start with the latter.

We report here essentially on recent results contained in [El1], [El2], [ElT], [Ar], [AArH], see also [AAr] for further references. We consider uniquely the cases $V = 0$ and the initial condition concentrated at a point, so that the solution of the above equation is the fundamental solution $p(t, x, y)$ of the heat equation on M (C^2 in x, C^1 in t) i.e.

$$\frac{\partial}{\partial t}p = \frac{1}{2}\Delta_x p, \quad t > 0$$

with Δ the Laplace-Beltrami operator on M, and

$$\lim_{t \downarrow 0} \int p(t, x, y)f(y)dy = f(x)$$

$\forall f \in C_c^2(M), dy$ being the Riemann-Lebesgue volume on M. The case $V \not\equiv 0$ can be essentially reduced to the case $V \equiv 0$ using a Feynman-Kac formula. The case of other initial conditions can be handled by superpositions. Assume M is connected and complete and such that there exists a point $y_0 \in M$ s.t. y_0 has no conjugate points (i.e. such that there exists no point along the geodesic $\exp_{y_0}(tX), X \in T_{y_0}M$ which is critical for the map $\exp_{y_0}$). Then the fundamental solution p in y_0 is given by

$$p(t, x, y_0) = \sum_{X_i \in \exp_{y_0}^{-1} x} (2\pi t)^{-\frac{d}{2}} \tilde{\Theta}^{-\frac{1}{2}}(X_i) \exp\left(-\frac{\tilde{d}(X_i, 0)}{2t}\right)$$

$$E(\exp(\frac{1}{2}\int_0^t \tilde{\Theta}^{\frac{1}{2}}(X_s^i)\tilde{\Delta}\tilde{\Theta}^{-\frac{1}{2}}(X_s^i)ds)),$$

where d is the dimension of M, $X_s^i, 0 \le s \le t$ is a Brownian bridge between X_i and the origin 0 in $T_{y_0}M$, for any $X_i \in \exp_{y_0}^{-1} x$, associated with the drift $-\frac{X}{t-s} - \frac{1}{2}\tilde{\nabla}\log\tilde{\Theta}(X)$, where $\tilde{\Delta}$ is the Laplace-Beltrami operator over $T_{y_0}M$ associated with the Riemannian metric induced by $\exp_{y_0}$ form that of M, $\tilde{\Theta}$ is the Ruse invariant on $T_{y_0}M$ (i.e. the modulus of the determinant of the derivative map of the corresponding $\exp_0$ mentional map), $\tilde{\nabla}$ is the gradient.

The expectation is with respect to the above bridge. $\tilde{d}$ is the Riemannian distance on $T_{y_0}M$.

For a proof of this fundamental formula see [ElT]. The effectivity of the application of the formula depends essentially on the properties of the Ruse invariant. If M is harmonic the Ruse invariant on M, $\Theta_{y_0}(x)$ only depends on the Riemannian distance of y_0 and x; if M is simply harmonic then $\Theta_{y_0}(\cdot) = 1$: examples of such spaces are the spaces of zero sectional curvature i.e. Euclidean or locally isometric to Euclidean spaces (for $d \ge 4$ examples of curvature $\neq 0$ are known).

Another case where the formulae for p simplifies considerably is the one of the complete simply connected Riemannian manifolds with non positive sectional curvature, so called Cartan-Hadamard manifolds. In this case for every point $y_0 \in M$ the above formula holds with the X_i reducing to only one X.

In the case where the sectional curvature is constant negative one has isometry with the hyperbolic space $H^n \equiv \{x \in \mathbb{R}^{n+1} | x_1^2 + ... + x_n^2 - x_{n+1}^2 = -1, x_{n+1} > 0\}$, (with pseudo riemannian metric $ds^2 = dx_1^2 + ...dx_n^2 - dx_{n+1}^2$). Here one has $\Theta_y(x) = \left(\frac{shd(x,y)}{d(x,y)}\right)^{n-1}$. For $d = 3$ one gets the special formula (cfr. [El])

$$p(t, x, y) = (2\pi t)^{-\frac{3}{2}} \frac{d(x, y)}{shd(x, y)} \exp\left(-\frac{d^2(x, y)}{2t} + \frac{R}{12}t\right)$$

with $R = -6$ the scalar curvature. In the general case of Cartan-Hadamard manifolds which are not simply connected one can still apply the above formula, since no two points are conjugates. In particular one gets explicit formulae for the Clifford-Klein spaces $\mathbb{R}^n|\Gamma_0$

and $H^n|\Gamma_1$ (where Γ_i are acting proper discontinuously and freely on $I\!\!R^n$ resp. H^n and Γ_0 is a subgroup of the group of isometries of H^n and Γ_1 is a subgroup of the Lorentz group $O(n,1)$)

In these cases one has e.g.

$$p(t,\tilde{x},\tilde{y}) = \sum_{\gamma\in\Gamma_0} (2\pi t)^{-\frac{d}{2}} e^{-\frac{d^2(\gamma x, y)}{2t}}$$

where $\tilde{x},\tilde{y}$ are two classes of equivalence modulo Γ_0 of x,y (as an example we get the torus $I\!\!R^n/2\pi Z\!\!\!Z^n$). Similar formulae hold for $H^n|\Gamma_1$.

The case where M is the manifold of a compact Lie group or is a symmetric space can essentially be handled using an extension of the above formula to the case, treated in [Ndu], where $y_0 \in M$, the cut locus Cut (y_0) of y_0 has codimension 2 and the formula is restricted to $x \in M -$ Cut (y_0). The formula holds then in the form

$$p(t,x,y_0) = (2\pi t)^{-\frac{d}{2}} \Theta_{y_0}^{-\frac{1}{2}}(x) e^{-\frac{d^2(x,y_0)}{2t}}$$

$$E(\chi_{\tau>t} \exp \frac{1}{2} \int_0^t \Theta_{y_0}^{\frac{1}{2}}(X_s) \Delta\Theta_{y_0}^{-\frac{1}{2}}(X_s) dx) \ ,$$

where τ is the exit time of a Brownian bridge X_s from M-Cut (y_0).

Let thus, following [Ar2], $M = G$ be a compact semisimple Lie group. It is proven in [Ar2] that one can reduce the discussion to the case of a compact semisimple simply connected Lie group (by going if necessary to the universal covering).

For the fundamental solution $p^U(t,x,e)$ of the heat equation in an open neighbourhood U of the identity e with Dirichlet boundary conditions on the border ∂U of U one has

$$p^U(t,x,e) = (2\pi t)^{-\frac{d}{2}} \prod_{\alpha(H)>0} \frac{i\alpha(H)}{2\sin(\frac{i\alpha(H)}{2})}$$

$$e^{-\frac{\|H\|^2}{2t}+<\rho,\rho>\frac{t}{2}} E_x(\chi_{\tau^0>t}) \ ,$$

where $x \in U$, H is given by $H = \exp^{-1} h_x$ with h_x in the maximal torus T of G. α is the root of G (relative to T),

$$\|H\| \equiv <Adu_x H, Adu_x H >^{\frac{1}{2}} \ ,$$

with $u_x \in G$ s.t.

$$x = u_x^{-1} h_x u_x \ .$$

Moreover

$$\rho \equiv \frac{1}{2}\sum_{\alpha>0} \alpha \ .$$

On the other hand one has, see [Ar2]:

Cut $(e) = \{\exp Ad(g)H | H$ such that $i\alpha(H) = \pm 2\pi$ for some $\alpha > 0$ and all $g \in G\}$,

which then yields the fundamental solution of the heat equation on G, for $x \in G-$ Cut (e), as given by the r.h.s. of the above formula, with τ the exit time of the Brownian bridge from $M -$ Cut (e) and H is a regular element s.t. $x = \exp Ad(g)H$.

The expectation has been computed for $G = SU(2)$ in [Ar2], by a direct method obtaining for all $u \neq -e$:

$$p(t, u, e) = \sum_{j \in \mathbb{Z}} (2\pi t)^{-\frac{3}{2}} \frac{4\sqrt{2}j\pi + |\lambda|}{2\sqrt{2}\sin\left(\frac{[4\sqrt{2}j\pi + |\lambda|]}{2\sqrt{2}}\right)}$$

$$e^{-\frac{(4\sqrt{2}j\pi + |\lambda|^2)}{2t}} e^{\frac{t}{16}}$$

for $|\lambda| < 2\sqrt{2}\pi$ (λ being the length in g corresponding to $d(u, e)$).

This expression can also be obtained by adapting a method of images from [El2], and possibly the latter can be extended to $SU(n)$, see [Ar2].

In [Ar2] it is shown on the other hand that Eskin's formula for the restriction to the maximal torus of the fundamental solution of the heat equation on a compact semisimple Lie group can be written in the form

$$p(t, h, e) = \sum_{A} (2\pi t)^{-\frac{\ell}{2}} \prod_{\alpha > 0} \frac{i\alpha(H + A)}{2\sin\left(\frac{i\alpha(H+A)}{2}\right)} e^{-\frac{\|H+A\|^2}{2t} + R\frac{t}{12}},$$

where $h = \exp H$, the sum being over the elements of $\exp^{-1} e$.

An extension to a formula for the fundamental solution of the heat equation on a symmetric (non compact type) space G_1/K, dual to a compact semisimple Lie group G, G_1 being a connected semisimple non compact Lie group and K a compact connected maximal subgroup, is also given in [Ar2]. It should be pointed out that in all expressions obtained in this way the heat kernel is always expressed essentially by lengths of geodesics, hence in terms of quantities associated with the classical free motion of a particle on the manifold (thus these group manifolds give examples of systems for which quantum mechanics can be written in terms of classical mechanics).

Finally let us mention that "explicit formulae", at least in the sense of a recursion in the dimension, for heat kernels can be obtained for all hyperbolic spaces, see [DaM] and nilpotent Lie groups [AArH]. See also [Ar2] for more details.

Of course there are a variety of other results (estimates e.g., small time expansions) on heat kernels on manifolds and associated diffusions. For some recent discussions we refer to existing excellent surveys like e.g. [Pi], [El3], [Da].

5. Stochastic perturbations in infinite dimensions and quantization of classical field theories

In this section we expose shortly some new developments in the study of stochastic perturbations of certain infinite dimensional dynamical systems, which are an infinite dimensional version of the gradient type systems discussed in Section 3, and can also be looked upon as providing a quantization of certain classical field theories. We also comment briefly on infinite dimensional versions of the systems considered in sections 2 and 4. Let us start from the basic observation that a classical relativistic equation as Klein-Gordon's equation

$$\left(\Box - m^2\right) X(t, \vec{x}) = 0 \tag{5.1}$$

(with $t \in \mathbb{R}$ "time", $\vec{x} \in \mathbb{R}^{d-1}$ "space", m a constant, the "mass", $\Box = \frac{\partial^2}{\partial t^2} - \Delta_{\vec{x}}$, the D'Alembert operator) can be looked upon as a "Newton equation" for the variable $t \longrightarrow X(t, \vec{x})$ taking values in a space of functions of $\vec{x}$:

$$\ddot{X}_{\vec{x}}(t) = (\Delta_{\vec{x}} + m^2) X_{\vec{x}}(t). \tag{5.2}$$

Heuristically this is a Newton equation for a "degree of freedom" with configuration space a space of functions and linear "force" $(\Delta_{\vec{x}} + m^2) X_{\vec{x}}(t)$. In analogy with what we discussed in Sect. 3, a quantization of (5.1) can be obtained heuristically by looking at the stochastic equation

$$dX_{\vec{x}}(t) = -\sqrt{-\Delta_{\vec{x}} + m^2} X_{\vec{x}}(t) dt + dw_{\vec{x}}(t) \tag{5.3}$$

where $dw_{\vec{x}}(t)$ is the natural Brownian motion associated with the Hilbert space $L^2(\mathbb{R}^{d-1}, dx)$.

In this picture $\beta(X_{\vec{x}}(t)) = -\sqrt{-\Delta_{\vec{x}} + m^2} X_{\vec{x}}(t)$ is the drift associated with the ground state of the harmonic oscillator Hamiltonian

$$H = \text{``} -\frac{1}{2} \int \frac{\delta^2}{\delta X(\vec{x}, 0)^2} d\vec{x} + \frac{1}{2} \int X(\vec{x}, 0)(-\Delta_{\vec{x}} + m^2) X(\vec{x}, 0) d\vec{x} + \text{ const. ''} \tag{5.4}$$

acting in L^2 with respect to a heuristic flat measure on the space of all functions $X(\vec{x}, 0)$. Obviously H is a purely heuristic, non existing object, cfr. also the discussion in [AHK 1,4,5].

However we may ask the question whether the Dirichlet form picture, sketched in Sect. 3 for the finite dimensional situation, would not help us in this case.

By the analogy with the case of a finite dimensional harmonic oscillator Hamiltonian $H = -\frac{1}{2}\Delta + \frac{1}{2} x A^2 x$ in $L^2(\mathbb{R}^d, dx)$ (with A^2 a $d \times d$ positive matrix), where H is unitary equivalent to H_μ, with H_μ the self-adjoint operator associated with the Dirichlet form (3.2) with $\mu^0 = N(0; (-\Delta_{d-1} + m^2)^{-\frac{1}{2}})$, we guess that H in our present infinite dimensional situation may be realised as H_μ, with H_μ heuristically defined as the self-adjoint operator associated with a Dirichlet form given heuristically by

$$\frac{1}{2} \int \int \left| \frac{\delta f}{\delta X(o, \vec{x})} \right|^2 d\mu^0(X(o, \vec{x})) d\vec{x} \tag{5.5}$$

with f in some domain of functionals of the variables $X(o, \vec{x})$.

Another way to formulate this heuristics is to look for path-space measures over $\mathbb{R}^d$ which can play a corresponding role as Wiener measure over $\mathbb{R}$, heuristically given by "$e^{-\frac{1}{2} \int \dot{w}(t)^2 dt} dw$". Such measures are heuristically given by

$$\text{``} e^{-\frac{1}{2} \int \dot{X}(t,\vec{x})^2 dt d\vec{x}} e^{-\frac{1}{2} \int |\nabla_{\vec{x}} X(t,\vec{x})|^2 dt d\vec{x}} e^{-\frac{1}{2} \int m^2 X(t,\vec{x})^2 dt d\vec{x}} dX\text{''} , \qquad (5.6)$$

with m a positive constant, rigorously as $N(0; (-\Delta_d + m^2)^{-1})$, the normal (i.e. Gaussian) distribution with mean zero and covariance $(-\Delta_d + m^2)^{-1}$.

The introduction of interaction corresponds heuristically to adding a term $-\int V(X(t, \vec{x})) dt d\vec{x}$ under exp in (5.6), obtaining an "interacting measure" μ, whose restriction to "the subspace generated by time zero fields $X(o, \vec{x})$" yields an "interacting measure" to be inserted for μ^0 in (5.5).

In this way we see that it is interesting to study quadratic forms given heuristically by (5.5), with μ^0 some probability measure. The systematic study of quadratic forms of this type has been initiated in [AHK1] and pursued in [AH4] in a rigged Hilbert space setting. Further work was done by Paclet and Kusuoka, the latter in a Banach space setting, with a detailed construction of an associated process. Incorporating also a more abstract setting by Bouleau-Hirsch, in 1988 Röckner and one of the authors [ARö] started a systematic treatment of forms (5.5) in a Souslin setting. More precisely let E be a Souslin topological, Hausdorff vector space (Souslin means continuous image of a completely separable metric space: examples are $E = $ Banach or $E = \mathcal{S}'(\mathbb{R}^d)$). Let FC_b^∞ be the smooth finitely based functions over E (smooth cylinder functions). Let μ be a probability measure on E. In [ARö1] necessary and sufficient conditions for closability of

$$\mathcal{E}_k^0(f, g) \equiv \frac{1}{2} \int \frac{\partial f}{\partial k} \frac{\partial g}{\partial k} d\mu , \qquad (5.7)$$

with $\frac{\partial}{\partial k}$ the derivative in the direction k, are given. The conditions depend roughly speaking on the way μ "has zeros along finite dimensional subspaces (the conditions in [ARö1] are actually weak and new even in for Dirichlet forms over finite dimensional spaces. Closability of (5.7) for all k forming a base of a subspace K of E implies closability of

$$\mathcal{E}^0(f, g) \equiv \sum_{k \in K} \mathcal{E}_k^0(f, g). \qquad (5.8)$$

If one assumes that there exists a separable real Hilbert space H densely and continuously contained in E, s.t. K is a dense linear subspace of H, then

$$\mathcal{E}^0(f, g) = \frac{1}{2} \int \nabla f \cdot \nabla g d\mu, \qquad (5.9)$$

with a natural definition of the gradient ∇. Closability of all $\mathcal{E}_K^0$ implies the closability of $\mathcal{E}^0$. Under certain further assumptions it is possible to give a more direct interpretation of the closure $\mathcal{E}$ of $\mathcal{E}^0$, see [AK], [AKRö].

It is also possible to associate a process with the Dirichlet form $\mathcal{E}$. In fact a compactification $\hat{E}$ (complete separable metric compact) of E can be constructed s.t. $E \subset \hat{E}$ continuously and densely and s.t. the image $\hat{\mathcal{E}}$ of $\mathcal{E}$ in $L^2(\hat{E}, \hat{\mu})$ (μ being the measure μ lifted to $\hat{E}$) is a regular Dirichlet form on $L^2(\hat{\mu})$ (in the sense of [Fu]). Under some additional assumptions e.g. E Hilbert with $H \subset E$ compactly or E conuclear and μ s.t. the linear functions are in $L^1(\mu)$, there exists a diffusion process X_t associated with $\mathcal{E}$ in the sense that, for any bounded continuous function f on E, $E^z(u(X_t))$ is a quasi-continuous version of $e^{-tH}f(z)$, for quasi-every $z \in E$, e^{-tH} being the semigroup associated with $\mathcal{E}$. If there exists a dense linear subspace K of the topological dual E' of E s.t. the linear functions on E are in $L^2(\mu)$ and a certain finite dimensional smoothness condition on μ is satisfied, then

$$Hf = -\frac{1}{2}\Delta f - \beta \cdot \nabla f$$

for a certain vector field β computable from μ, for all $f \in FC_b^\infty$, with

$$\Delta f \equiv \sum_i \frac{\partial}{\partial e_i}\left(\frac{\partial}{\partial e_i}f\right)$$
$$\beta \cdot \nabla \equiv \sum \beta_{e_i}\frac{\partial}{\partial e_i}f$$

with e_i a complete orthonormal system in H. X_t satisfies then in a weak sense the stochastic equation

$$dX_t = \beta(X_t)dt + dw_t, \tag{5.10}$$

with $X_0 = z$, for quasi every $z \in E$. This is an equation of the same type as (3.2). Similarly as we looked in 3.2 at such an equation as a quantization of the classical system with finitely many degrees of freedom (through the finding of the corresponding semigroup P_t^μ) described by (2.1) with force $K = -\nabla V$, provided V resp. β are related to the measure $d\mu = \varphi(x)^2 dx$ by $V = \frac{1}{2}\Delta\varphi/\varphi$ resp. $\beta = \nabla \ln \varphi$, we can now find examples of classical mechanical systems with infinitely many degrees of freedom whose quantization can be expressed by the above infinite dimensional stochastic differential equation. In fact this is heuristically easily understood as follows. Let us reconsider the arguments (5.1) – (5.5), adding an interaction term. I.e. we consider a classical field (mechanical system of infinitely many degrees of freedom) $\mathcal{X}(t, \vec{x})$, $t \in \mathbb{R}$, $\vec{x} \in \mathbb{R}^s$ (s space–dimension, t time), satisfying a relativistic equation of motion of the form

$$\Box \mathcal{X}(t, \vec{x}) = -\tilde{V}'\left(\mathcal{X}(t, \vec{x})\right), \tag{5.11}$$

with $\Box \equiv \frac{\partial^2}{\partial t^2} - \Delta_{\vec{x}}$, the D'Alembert operator, $V : \mathbb{R} \longrightarrow \mathbb{R}$ a (smooth) nonlinearity. As in (5.2), we can look upon the above equation as

$$\ddot{\mathcal{X}}(t, \vec{x}) = \Delta_{\vec{x}}\mathcal{X}(t, \vec{x}) - \tilde{V}'\left(\mathcal{X}(t, \vec{x})\right)$$

which formally is of the same type as (2.1) with $-V'\left(\mathcal{X}(t)\right)$ replaced by $\Delta_{\vec{x}}\mathcal{X}(t, \vec{x}) - \tilde{V}'\left(\mathcal{X}(t, \vec{x})\right)$. Thus the quantization of this equation should be provided, in analogy with the one of (2.1), by a stochastic equation of the form (5.10), with associated transition semigroup $P_t^\mu = e^{-tH_\mu}$, with P_t^μ the Markov semigroup in $L^2(\mu)$ associated with the pre – Dirichlet form

$$\mathcal{E}_{(\mu)}^0(f,g) = \frac{1}{2} \int \int_{\mathbb{R}^s} \frac{\delta}{\delta\mathcal{X}(0,\vec{x})} f\left(\mathcal{X}(0,\vec{x})\right) \frac{\delta}{\delta\mathcal{X}(0,\vec{x})} g\left(\mathcal{X}(0,\vec{x})\right) d\vec{x} d\mu\left(\mathcal{X}(0,\vec{x})\right)$$

with μ the probability measure on the space of $\mathcal{X}(0,\vec{x})$ – fields which plays the role of the "ground state measure" $d\mu = \varphi(x)^2 dx$ with $-\frac{1}{2}\Delta\varphi + V\varphi = 0$ in the finite dimensional case.

Heuristically μ should be, as in the finite dimensional case, the restriction to the σ–algebra generated by the time zero fields $\mathcal{X}(0,\vec{x})$ of a Euclidean path space measure μ_E obtained from (3.6) by replacing dw by $d\mathcal{X}(t,\vec{x})$ and $\int_{\mathbb{R}} W(w_s) ds$ by

$$\int_{\mathbb{R}^d} W\left(\mathcal{X}(t,\vec{x})\right) dt d\vec{x}$$

with

$$W\left(\mathcal{X}(t,\vec{x})\right) = \frac{1}{2} \int_{\mathbb{R}^d} |\nabla_{t,\vec{x}}\mathcal{X}(t,\vec{x})|^2 dt d\vec{x} + \int_{\mathbb{R}^d} \tilde{V}\left(\mathcal{X}(t,\vec{x})\right) dt d\vec{x}$$

(with $d = s + 1$ the space–time dimension).

It turns out that it is possible to construct μ_E and thus μ for $\tilde{V}(\alpha) = m^2\alpha^2 + \tilde{\tilde{V}}(\alpha)$, $m^2 \geq 0$ for all d if $\tilde{\tilde{V}} = 0$ if $m > 0$ (for $d \geq 3$ also $m = 0$ is allowed), and for $d = 2$ if $\tilde{\tilde{V}}$ is a polynomial of even degree and strictly positive highest order coefficient (the so called $P(\varphi)_2$ model), or $\tilde{\tilde{V}}$ is a trigonometric function (Sine–Gordon type model) or $\tilde{\tilde{V}}$ an exponential function (Høegh–Krohn's model, also called Liouville model if $m = 0$). See [AH5], [AHKZ] and references therein. The probability measures μ_E resp. μ can be realized on $\mathcal{S}'(\mathbb{R}^d)$ resp. $\mathcal{S}'(\mathbb{R}^{d-1})$. The processes $X(t,\vec{x})$ (associated with the classical field $\mathcal{X}(t,\vec{x})$) can then be realized as coordinate process $< f, X >$, $f \in \mathcal{S}(\mathbb{R}^d)$, on $E = \mathcal{S}'(\mathbb{R}^d)$. μ_E is the Euclidean measure of the quantum field, it is the path space measure for the process $< g, X(t) >$, $g \in \mathcal{S}(\mathbb{R}^{d-1})$ in as much as the analogue of the Feynman–Kac formula holds

$$E_{\mu_E}\left(F(< g_1, X(t) >)G(< g_2, X(0) >)\right) = E_\mu\left(F(< g_1, X(0) >)\tilde{P}_t^\mu G(< g_2, X(0) >)\right),$$

for any $g_1, g_2 \in \mathcal{S}(\mathbb{R}^{d-1})$ and for any F, G real–valued bounded continuous on $\mathbb{R}$. $\tilde{P}_t^\mu$ is the semigroup given by the fact that the random field $< f, X >$ has the global Markov property, in particular with respect to the hyperplane $t = 0$ in $\mathbb{R}^d$, cfr. [AHKZ] (by the way, the proof of the global Markov of μ_E for all models took nearly two decades to be performed). $\tilde{P}_t^\mu$ has infinitesimal generator coinciding with the one associated with the classical Dirichlet form given by μ, at least on some dense subset [ARö], [AHPRS] (the full identification of the generators is still open, except for $\tilde{\tilde{V}} = 0$).

That μ defines a classical Dirichlet form with associated diffusion satisfying (5.10) is part of the recent developments in [ARö] – [AHPRS]. The analytic continuation of $\tilde{P}_t^\mu$ yields then the unitary evolution for the relativistic quantum fields given by the interaction. In this sense then the above stochastic techniques give a quantization of the classical nonlinear wave equation (5.11).

Remark: It is also possible to use μ_E to define another classical Dirichlet form. The associated diffusion solves in the weak sense an equation of the form (3.8), with $X(t,\tau)$ replaced by $X(t,\vec{x},t)$ and $\beta_{st}(X(t,\vec{x},\tau)) = - : \tilde{V}'(X(t,\vec{x},\tau)) : -\Delta_{t,\vec{x}}X(t,\vec{x},\tau)$, where : : means "Wick ordering" (the limit of a certain regularization procedure applied to the heuristic expression $\tilde{V}'(X(t,\vec{x},\tau))$, see [ARö]).

Finally let us report shortly on work contained in [AHKI], [AIK] (and references therein). Probabilistically it amounts to the solution of the problem of showing existence of global Markov fields which are stationary (homogeneous) with respect to the Euclidean group over $I\!R^4$. The Euclidean fields are defined as solutions of a stochastic partial differential equation (of a type similar to the one we discussed before), best formulated by using the isomorphism of $I\!R^4$ (as a vector space with scalar product) with the field H of quaternions. The isomorphism is obtained by identifying a vector x with components $x_\mu, \mu = 0, 1, 2, 3$, with the quaternion with the same components, denoted by the same symbol. The form of the equation alluded above also depends on the fact that $SO(4)$ is isomorphic $(\mathrm{Sp}(1) \times \mathrm{Sp}(1))/Z\!\!\!Z_2$, with $\mathrm{Sp}(1) \cong SU(2)$, identifiable with the quaternions of unit norm.

Let ∂ be the natural quaternionic Cauchy-Riemann operators, and $\bar{\partial}$ its quaternionic adjoint. Then $\partial\bar{\partial} = \bar{\partial}\partial = \Delta$, with Δ the Laplacian in $I\!R^4$ (note that this product decomposition of the Laplacian is possible only in 1,2,4,8 dimensions, by a theorem of Hurwitz; incidentally it is also related with the introduction of Dirac's operator). Now let F be a generalized (non necessarily Gaussian) vector white noise over $I\!R^4$, with values in $H \cong I\!R^4$. The distribution of F looked upon as a generalized random field with a suitable function space $\mathcal{T}$ is given by

$$E(e^{i<f,F>}) = e^{-\int_{R^4} \psi(f(x))dx} , \ f \in \mathcal{T} ,$$

with $\psi = \psi_p + \psi_G$ a given function consisting of a "Gaussian part" ψ_G and a "Poisson part" ψ_p, see [AHKI],[AIK] for details.

The first order stochastic partial differential equation $\partial A = F$ admits a Euclidean generalized vector random field A (over $I\!R^4$ with values in $I\!R^4 \cong H$) as solution. A has Markov properties, in particular with respect to a time zero hyperplane. Its Schwinger functions can be computed explicitly (and are not free for $\psi_p \neq 0$).

In the case $\psi_p = 0$ A is a realization of the free Euclidean electromagnetic potential field.

In suitable gauges A is also time-reflection invariant; in the case $\psi_p \neq 0$, A is only time-reflection invariant, for a natural definition of time-reflection, when the Lévy measure on $I\!R^4$ associated with ψ_p has support only in $I\!R\backslash\{0\}$.

But also in the general case one can find directly an analytic continuation of the Schwinger functions to relativistic functions satisfying the usual properties of spectrum, invariance and locality [AIK]. An extension of our results to 2 and 8 dimensions has been obtained by E. Osipov (see [O] and references therein) (in 2 dimensions reflection positivity has been proven). We also like to mention a recent result by Tamura on "confinement" in the 4-space-time dimensional model [T].

In conclusion it seems that the method of stochastic perturbations of classical equations is a very useful tool for quantization as well as for intrinsic studies of dynamical systems.

Acknowledgements

It is a pleasure to thank the organizers, for their kind invitation to a very stimulating conference. The joy of collaboration with Z. Haba, T. Hida, K. Iwata, T. Kolsrud, Ma Zhiming, J. Potthoff, M. Röckner, L. Streit, B. Zegarlinski, E. Zehnder on topics in this lecture is also gratefully acknowledged, as well as the help of Eva Aich, Martin Jarrath, Regina Kirchhoff, Heike Nierenheim, Carsten Welge in the setting of the manuskript.

References

[A1] S. Albeverio: *Some new developments concerning Dirichlet forms, Markov fields and quantum fields*, SFB-Preprint, pp. 250–259 in "IXth International Congress on Mathematical Physics", July 1988, Swansea '88, Edts., B. Simon, A. Truman, I.M. Davis, Adam Higer, Bristol and New York (1989)

[A2] S. Albeverio: *Some points of interaction between stochastic analysis and quantum theory*, pp. 1–26 in "Stochastic Differential Systems", Proc. Bad Honnef Conference 1985, Ed. N. Christopeit, K. Helmes, M. Kohlmann, Lect. Notes Control Inform. Sciences **78**, Springer, Berlin (1986)

[AAr] S. Albeverio, T. Arede: *The relation between quantum mechanics and classical mechanics: a survey of some mathematical aspects*, pp 37–76 in "Chaotic behaviour in Quantum Systems, Theory and Applications", (Proc. Como 1983) Ed. G. Casati, Plenum Press, New York (1985)

[AArH] S. Albeverio, T. Arede, Z. Haba: *On left invariant Brownian motions and heat kernels of nilpotent Lie groups*, Bochum Preprint (1988), to appear in J. Math. Phys.

[ABKS] S. Albeverio, Ph. Blanchard, S.Kusuoka, L. Streit: *An inverse problem for stochastic differential equations*, J. Stat. Phys. (1989)

[ABR] S. Albeverio, J. Brasche, M. Röckner: *Dirichlet forms and generalized Schrödinger operators*, Edts. H. Holden, A. Jensen, Lect. Notes Maths., Springer, Berlin (1989)

[AFHKL] S. Albeverio, J.E. Fenstad, R. Høegh-Krohn, T. Lindstrøm: *Non standard methods in stochastic analysis and mathematical physics*, Academic Press (1986)

[AFKS] S. Albeverio, M. Fukushima, W. Karwowski, L. Streit: *Capacity and quantum mechanical tunneling*, Commun. Math. Phys. **80**, 301-342 (1981)

[AGHKH] S. Albeverio, F. Gesztesy, R. Høegh-Krohn, H. Holden: *Solvable models in quantum mechanics*, Springer Verlag, Berlin (1988)

[AGQ] S. Albeverio, Guanglu Gong, Minping Quian, in preparation

[AH1] S. Albeverio, A. Hilbert: *Some remarks on stochastically perturbed (Hamiltonian) systems*, BiBoS - Preprint, to appear in Nonlinear Fields Proc. Bielefeld Conf., Edts. Ph. Blanchard, J. Stubbe, Lect. Notes Phys., Springer, Berlin

[AH2] S. Albeverio, A. Hilbert: *Some results on Newton equation with an additional stochastic force*, to appear in Proc. Bad Honnef Conference "Stochastic Systems", Edts. Christopeit et al., Lect. Notes Control and Inform., Springer, Berlin (1989)

[AH3] S. Albeverio, R. Høegh-Krohn: *Dirichlet forms and diffusion processes on rigged Hilbert spaces*, Zeitschrift für Wahrscheinlichkeitstheorie und verwandte Gebiete **40**, 1–57 (1977)

[AH4] S. Albeverio, R. Høegh-Krohn: *Hunt processes and analytic potential theory on rigged Hilbert spaces*, Ann. Inst. H. Poincaré (Probability Theory) **B13**, 269–291 (1977)

[AH5] S. Albeverio, R. Høegh-Krohn: *Diffusion fields, quantum fields and fields with values in groups*, pp. 1–98 in "Stochastic Analysis and Applications", Edts. M. Pinsky, M. Dekker, New York (1984)

[AHK] S. Albeverio, A. Hilbert, A. Klar: work in preparation

[AHK1] S. Albeverio, R. Høegh-Krohn: *Quasi invariant measures, symmetric diffusion processes and quantum fields*, pp. 11-59 in "Proceedings of the International Colloquium on Mathematical Methods of Quantum Field Theory", Editions du CNRS. 1976, (Colloques Internationaux du Centre National de la Recherche Scientifique, No. 248)

[AHK2] S. Albeverio, R. Høegh-Krohn: *A remark on the connection between stochastic mechanics and the heat equation*, J. Math. Phys. **15**. 1745–1747 (1974)

[AHKI] S. Albeverio, R. Høegh-Krohn, K. Iwata: *Covariant Markovian random fields in four space-time dimensions with nonlinear electromagnetic interaction*, pp. 69–83 in "Applications of Self- Adjoint Extensions in Quantum Physics", in Proc. Dubna Conf. 1987, Edts. P. Exner, P. Seba, Lec. Notes in Physics **324**, Springer, Berlin (1989)

[AHKS] S. Albeverio, R. Høegh-Krohn, L. Streit: *Energy forms, Hamiltonians and distorted Brownian paths*, J. Math. Phys. **18**, 907–917 (1977)

[AHKZ] 1) S. Albeverio, R. Høegh-Krohn, B. Zegarlinski: *Uniqueness and global Markov property for Euclidean fields. The case of general polynomial interactions*, Commun. Math. Phys.

2) S. Albeverio, R. Høegh-Krohn, B. Zegarlinski: *Uniqueness of Gibbs states for general $P(\varphi)_2$-weak coupling models by cluster expansion*, Commun. Math. Phys.

[AHPRS] S. Albeverio, T. Hida, J. Potthoff, M. Röckner, L. Streit: *Dirichlet forms in terms of white noise analysis I - Construction and QFT examples* and *Dirichlet forms in terms of white noise analysis II - Closability and Diffusion Processes*, BiBoS Preprint (1989)

[AHPS] S. Albeverio, R. Høegh-Krohn, S. Paycha, S. Scarlatti: *A probability measure for random surfaces of arbitrary genus and bosonic strings in 4 dimensions*, Nucl. Phys. B (Proc. Suppl.), **6**, 180–182 (1989) (Proc. Eugene Wigner Symposium on Space - Time Symmetries, Washington 1988).

[AHZ] S. Albeverio, A. Hilbert, E. Zehnder: *Hamiltonian systems with a stochastic force: nonlinear versus linear, and a Girsanov formula*, in preparation

[AIK] S. Albeverio, K. Iwata, T. Kolsrud: *Random fields as solutions of the inhomogeneous quaternionic Cauchy-Riemann equation I. Invariance and Analytic Continuation*, SFB 237 - Preprint (1989)

[AK]   S. Albeverio, S. Kusuoka: *Maximality of infinite dimensional Dirichlet forms and R. Høegh-Krohn's model of quantum fields*, to appear in Memorial Volume for Raphael Høegh- Krohn

[AKRö]   S. Albeverio, S. Kusuoka, M. Röckner: *On partial integration in infinite dimensional space and applications to Dirichlet forms*, to appear in J. London Math. Soc. (1989)

[AMa]   R. Abraham, J. Marsden: *Foundations of Mechanics*, Benjamia/Cummings, Reading (1978)

[AMa1]   S. Albeverio, Ma Zhiming: *Additive functionals, nowhere Radon and Kato class smooth measures associated with Dirichlet forms*, in preparation

[AMa2]   S. Albeverio, Ma Zhiming: *Perturbation of Dirichlet forms - lower semiboundedness, closability and form cores*, in preparation

[AMa3]   S. Albeverio, Ma Zhiming: *Nowhere Radon smooth measures, perturbations of Dirichlet forms and singular quadratic forms*, in Proc. Bad Honnef Conf. 1988, ed. Christopeit, Lect. Notes Control and Inform. Sciences, Springer, Berlin (1989)

[Ar1]   T. Arede: *La géometrie du noyau de la chaleur sur les variétés*, Thèse 3^{éme} Cycle, Université d' Aix Marseille II (Luminy) (1983)

[Ar2]   T. Arede: *Equação do calor em grupos de Lie e alguns espaços simétricos*, Lisboa Thesis (1989)

[ArK]   L. Arnold, W. Kliemann: *On unique ergodicity for degenerate diffusions*, Bremen Rept. <u>147</u> (1986)

[Arn]   V. Arnold: Ed. *Dynamical Systems III*. Springer Berlin (1988)

[ARö 1] S. Albeverio, M. Röckner: *Classical Dirichlet forms on topological vector spaces - closability and a Cameron- Martin formula*, BiBoS - SFB 237 - Preprint (1988), to appear in J. Funct. Anal.

[ARö 2] S. Albeverio, M. Röckner: *Dirichlet forms, quantum fields and stochastic quantisation*, pp. 1–21 in "Stochastic Analysis, path integration and dynamics", Emanations from "Summer Stochastics", Warwick 1987, Edts., D. Elworthy, J.C. Zambrini, Pitman Res. Notes, Longman, Harlow (1989)

[ARö 3] S. Albeverio, M. Röckner: *Classical Dirichlet forms on topological vector spaces - the construction of the associated diffusion process*, BiBoS - SFB 237 Preprint, to appear in Prob. Theory and Rel. Fields.

[ARö 4] S. Albeverio, M. Röckner: *New developments in theory and applications of Dirichlet forms*, to appear in "Stochastic Processes, Physics and Geometry", Proc. 2nd Int. Conf. Ascona - Locarno - Como 1988, Ed. S. Albeverio, G. Casati, U. Cattaneo, D. Merlini, R. Moresi, World Scient. (1989)

[ARö 5] S. Albeverio, M. Röckner: *On Dirichlet forms on topological vector spaces: Existence and maximality*, SFB 237 - Preprint, to appear in Proc. Bad Honnef '88 Conf., Edts. N. Christopeit et al., Lect. Notes Inform. Control, Springer Verlag (1989)

[ARö 6] S. Albeverio, M. Röckner: *On partial integration in infinite dimensional space and applications to Dirichlet forms*, SFB 237 - Preprint, to appear in J. London Math. Soc. (1989)

[ARö 7] S. Albeverio, M. Röckner: *Infinite dimensional stochastic equations: solutions via Dirichlet forms*, in preparation

[ArW] L. Arnold, V. Wihstutz: Edts. Lyapunov Exponents, LN Math., Springer, New York (1986)

[AYZ] S. Albeverio, K.Yasue, J.C, Zambrini: *Euclidean quantum mechanics: analytic approach*, in Ann. Inst. H. Poincaré $\underline{49}$, 259–308 (1989)

[AZ] S. Albeverio, B. Zegarlinski: *Contribution to the proceedings of the 1989 Cargèse Conf.*, Edts. Damgaard, Hüffel, Plenum Press in preparation

[BCZ] Ph. Blanchard, Ph. Combe, W. Zheng: *Mathematical aspects and physical aspects of stochastic mechanics*, Lect. Notes Phys. 281, Springer, Berlin (1987)

[BeG] A. M. Berthier, B. Gaveau: *Critère de convergence des fonctionnelles de Kac et application en mécanique quantique et en géométrie*, J. Funct. Anal. $\underline{29}$, 416 (1978)

[Car] R. Carmona: *Regularity properties of Schrödinger and Dirichlet semigroups*, J. Funct. Anal. $\underline{29}$, 227-237 (1974)

[Ca] E. Carlen: *Stochastic mechanics of free scalar fields*, pp. 40–60 in "Stochastic mechanics and Stochastic processes", Edts. A. Truman, J.M. Davies, Lect. Notes Maths. $\underline{1325}$, Springer, Berlin (1988)

[Da] B. Davies: *Heat kernels on manifolds*, Cambridge Univ. Press (1988)

[DaM] E. B. Davies, N. Mandouvalos: *Heat kernels and spectral theory*, Proc. London Math. Soc. (3) **57**, 182-208 (1988)

[Di-Z] R. Dieckerhoff, E. Zehnder: *An a priori estimate for non linear oscillatory differential equations*, Ann. Scuola Norm. Pisa, Bd. 14, S. 79-95 (1987)

[El1] D. Elworthy:*Stochastic Differential Equations on Manifolds*, Cambridge University press (1982)

[El2] D. Elworthy: to appear in Proc. Ascona Conf. 1988, Edts. S. Albeverio, G. Casati, U. Cattaneo, R. Moresi, D. Merlini, World Scient., Singapore (1990)

[El3] D. Elworthy: to appear in Proc. Evanston Conf. 1989, Ed. M. Pinsky

[ElT] K.D. Elworthy, A. Truman The diffusion equation and classical mechanics: an elementary formula in "Stochastic Processes in Quantum Physics" ed. S. Albeverio et al., pp.136–146, Lecture Notes in Physics **173** Springer (1982)

[Fu1] M. Fukushima: *Dirichlet forms and Markov processes*, North Holland, Amsterdam (1980)

[Fu2] M. Fukushima: *Energy forms and diffusion processes*, "Mathematics and Physics", Vol. 1, Ed. L. Streit, World Scient., Singapore (1985)

[Go] J.A. Goldstein: **Second Order Ito Processes**, Nagoya Math. J. **36** (1969), 27–63

[GrSt] C. Grosche, F. Steiner: *The path integral on the pseudosphere*, Ann. Phys. $\underline{282}$, 120 (1988)

[Gu] F. Guerra: *Quantum field theory and probability theory. Outlook on new possible developments*, pp. 214–243 in "Trends and Developments in the Eighties", Edts. S. Albeverio, Ph. Blanchard, World Scient., Singapore (1985)

[Hi] A. Hilbert: *Stochastic perturbations of Hamiltonian systems*, Ph.D. Thesis, Bochum (1989)

[Jo] G. Jona - Lasinio: *Stochastic processes and quantum mechanics*, Astérisque $\underline{132}$, 203-216 (1985)

[Jo-Mi] G. Jona - Lasinio, P.K. Mitter: *On the stochastic quantization of field theory*, Comm. Math. Phys. $\underline{101}$, 409-436 (1985)

[Kh] R.Z. Khas'minskii: *Stochastic Stability of Differential Equations*, Sifthoof, Alplen aan den Rijn (1980)

[KrS] P. Krée, C. Soize: *Mécanique aléatoire*, Dunod (1983)

[MaW] L. Markus, A. Weerasinghe: *Stochastic Oscillators*, J. Diff. Equ. **21**, 288-314 (1988)

[McK] H.P. McKean, **Stochastic Integrals**, Academic Press New York 1969

[Na1] K. Narita: *No Explosion Criteria for Stochastic Differential Equations*, J. Math. Soc. Japan **34**, 192–203 (1982)

[Na2] K. Narita:*Explosion Time of Second-Order Ito Processes*, J. Math. Anal. Appl. **104**, 418–427 (1984)

[Na3] K. Narita:*On explosion and growth order of inhomogeneous diffusion processes*, Yokohama Math. J. **28**, 45–57 (1980)

[Ndu] M. Ndumu: *An elementary formula for the Dirichlet heat kernel on Riemannian manifolds* in "From local times to global geometry, control and physics". Ed. K.D. Elworthy, Pitman Research Notes in Mathematical series, 150, Longman, Scientific and Technical (1986)

[Ne] 1) E. Nelson: *Dynamical Theories of Brownian Motion*, Princeton University Press, Princeton (1967)
2) E. Nelson: *Quantum fluctuations*, Princeton University Press (1985)

[O] E. P. Osipov: *Two-dimensional random fields as solution of stochastic differential equations*, Bochum Preprint

[Pi] 1) M.A. Pinsky: *Instability of the harmonic oscillator under small noise*, SIAM J. Appl. Math. **46**, 451–463 (1980)
2) M.A. Pinsky: *Inverse questions in stochastic differential geometry*, Singapore Workshop Lectures, in Proc. Singapore Probability Conf., Walter de Gruyter, (1989)

[Po] J.Potter: *Some Statistical Properties of the Motion of a Nonlinear Oscillator Driven by White Noise*, Ph.D. Thesis, M.I.T. (1962)

[ReSi] M. Reed, B. Simon: *Methods of Modern Mathematical Physics*, I-IV, Academic Press New York (1975)

[RoW] L.C.G. Rogers, D. Williams: *Diffusions, Markov Processes, and Martingales*, J. Wiley, Chichester (1987)

[Si] B. Simon: *Functional Integration and Quantum Physics*, Academie Press, New York (1979)

[T] H. Tamura: *Nonlinear electromagnetic fields confine charges*, Kanazawa University Press, Preprint (1989)

NEURAL NETWORKS:
DETERMINISTIC AND STOCHASTIC DYNAMICS

*Bruno Apolloni
*Alberto Bertoni
°Paola Campadelli
*Diego de Falco

*Dipartimento di Scienze dell'Informazione, Università di Milano, I-20133 Milano, Italy
°Istituto di Fisiologia dei Centri Nervosi del C.N.R., Milano I-20133, Milano, Italy

Abstract: Problems of combinatorial optimization, beyond their interest in applied research, play a crucial role in fundamental issues of theoretical computer science, for their inherent computational complexity. Here we use them as test bed on which to gauge the many perspectives and problems offered by neural networks.

The realization that optimization problems for quadratic functions of many Boolean variables which are, in a technical sense to be made precise, as difficult as they can be, are conveniently dealt with by neural networks contributes to the interest of such dynamical systems: the parameters controlling their evolution can indeed be assigned in such a way that they have precisely the function to be minimized as a Lyapunov function. The recognition that such an evolution will, in general, stop in a local minimum of this Lyapunov function, as opposed to the global minima one is searching for, motivates the idea of endowing the dynamics of a neural network with a stochastic transition rule leading to a stationary distribution strongly peaked around global minima.

Here we discuss several problems related to the dynamics of both deterministic and stochastic networks with an emphasis on the problem of quantitatively assessing their computational capabilities .

1. Computational Complexity

In this section we review a few notions and facts of life pertinent to the theory of computational complexity, without any pretense of rigour or self-containedness. The interested reader is referred to the monographs by Garey and Johnson [Ga79] or by Stockmeyer [St87] for more precise statements.

Solving a problem can mean providing a Yes/No answer (decision problem) or, more generally, evaluating a function (functional problem). In this paper we are interested in a particular class of functional problems, i.e. combinatorial optimization problems.

A combinatorial optimization problem Π (say, the one informally stated as "finding the minimum energy of a two dimensional antiferromagnetic array of Ising spins in a magnetic field") is determined by:

(1) A set D_Π of instances (for the example alluded to, an instance $I \in D_\Pi$ would be specified by an integer n, a set $\Lambda \subseteq Z^2$ with $|\Lambda| = n$, a subset E of $\Lambda \times \Lambda$ specifying which pairs of spins are actually coupled);

(2) For each instance $I \in D_\Pi$, a finite set S_Π (I) of candidate solutions for this instance (in our example S_Π (I) would be the set of functions $s : i \in \Lambda \rightarrow s_i \in \{-1,1\}$ giving the value of the spin sitting at each site $i \in \Lambda$);

(3) A funtion H_Π that assigns to each instance $I \in D_\Pi$ and to each candidate solution $s \in S_\Pi(I)$ a rational number $H_\Pi(I,s)$, called the solution value for s (in our case, say, $H_\Pi(I,s) = \sum_{(i,j) \in E} s_i s_j + \sum_{i \in \Lambda} s_i$).

In a minimization problem an optimal solution for an instance $I \in D_\Pi$ is a candidate solution $s' \in S_\Pi(I)$ such that, for all $s \in S_\Pi(I)$, $H_\Pi(I,s') \leq H_\Pi(I,s)$.

To every optimization problem a decision problem can be associated in a natural way; it can be posed in the following way: given the instance (I,k), where $I \in D_\Pi$, and k is a rational number, does there exist $s \in S_\Pi(I)$ such that $H_\Pi(I,s) \leq k$?

In order to solve a problem with a machine, one must be able to estimate the amount of resources (say time or memory space) which must be spent to obtain the answer.

The fact that a problem is "technically" solvable, namely that there is an algorithm which for each instance provides the answer, does not necessarily mean that it is "practically" solvable: one often faces situations in which any exact solving algorithm requires an amount of resources rising so sharply with the size of the instance as to make it practically unfeasible to search for exact solutions. It may be wise to realize a priori that this is the situation for a given optimization problem and concentrate, instead, every effort on the more realistic task of searching for good approximate solutions (say, look for low local minima instead of looking for global minima by an exhaustive enumeration which might require many times the age of the universe)

We sketch below a few notions relevant to a quantitative measure of the notion of "practically solvable or unsolvable" vaguely given above. For definiteness sake we focus on decision problems (thus giving at least lower bounds on the "difficulty" of the optimization problems) and refer to the computational model provided by deterministic Turing machines (referring to your PC or to a state of the art mainframe would not change the picture in any essential respect).

To state the decision problem Π in such a way that Turing machines can work on it, it is necessary first of all to codify the instances over some "suitable" alphabet Σ: for our prototype Ising decision problem any fixed reasonable binary description of the numbers n and k and of the incidence matrix of the graph (Λ,E) will do, so that in such a case each instance (n,Λ,E,k) is easily encoded by a finite sequence x of elements of (a word over) the finite alphabet $\Sigma = \{0,1\}$. The dimension of an instance is the length $l(x)$ of the word x associated to it.

Solving the problem Π is equivalent to recognizing the distinguished subset L_Π of the set Σ^* of all the words over Σ which is made of those words which encode "yes" instances: in our model example the issue is to recognize the "language" containing the binary words encoding those arrays of Ising spins for which, corresponding to the geometric setup (n,Λ,E) and to the threshold value k there is a spin configuration of energy below the given threshold.

For a given deterministic Turing machine M, call L_M the language recognized by M, namely the set of words which, given as input to M, lead it to the distinguished final state "yes". For given input $x \in L_M$ call $T_M(x)$ the number of computation steps performed (the time on input x) and $S_M(x)$ the number of memory cells visited (the space on input x) before stopping.

The following quantities can be usefully associated to a Turing machine M as a measure of the dependence on the dimension n of the input of the efficiency of M in recognizing elements of its own L_M:

$$t_M(n) = \max_{\substack{x \in L_M \\ l(x)=n}} T_M(x)$$

$$s_M(n) = \max_{\substack{x \in L_M \\ l(x)=n}} S_M(x)$$

Having formalized "solving problem Π" into "recognizing language L_Π" an intrinsic measure of how the difficulty of Π scales with the size n of the instance will be given by the n dependence of $t_M(n)$ and $s_M(n)$ for the smartest M having $L_M = L_\Pi$.

More precisely, having fixed some time or space bound, a complexity class is the class of all languages which can be recognized (of all problems which can be solved) by at least one Turing machine running within the prescribed bound. For a given function $f: \mathbb{N} \to \mathbb{N}$ prescribing the bound, one singles out the two classes:

$$\text{TIME}(f) = \{L : \exists M \text{ such that } L = L_M \text{ and } t_M(n) \leq f(n) \text{ for every } n\}$$

$$\text{SPACE}(f) = \{L : \exists M \text{ such that } L = L_M \text{ and } s_M(n) \leq f(n) \text{ for every } n\}.$$

"Easy", "practically solvable" problems are those which can be solved in polynomial time by deterministic Turing machines; they form the Polynomial class

$$P = \cup_{k \geq 1} \text{TIME}(n^k).$$

Unfortunately, there is an embarrassingly large number of practically relevant problems whose membership in P is unresolved, and among these there are most combinatorial optimization problems. Let us return to our standard example: it is hard to find, among the 2^n configurations of n Ising spins those, if any, which have energy below a given threshold k. Given, however, a configuration it is easy (takes a polynomial number of algebraic operations) to check on it whether this is the case. Problems with such a feature, namely such that a candidate solution can be verified in polynomial time, form the class NP.

NP can be formally defined as the class of languages L which can be specified by

$$x \in L \quad \text{iff} \quad \exists y \, [\, l(y) \leq p(l(x)) \ \text{and} \ (x,y) \in R \,]$$

for some polynomial p and language $R \in P$. In other words, NP is the class of languages L such that

$x \in L$ iff there is a "short certificate" y of the fact that $x \in L$, where "short" means that the length of y is bounded by a polynomial in the length of x and that the validity of the certificate can be checked in polynomial time.

It is pretty obvious that $P \subseteq NP$, while the problem whether the inclusion is proper is one of the major problems in theoretical computer science. Work initiated by Cook [Co71] and independently by Levin [Le73], has shown that there is at least one problem in NP such that if this problem is in P then P=NP. Doing some violence to the historical development of the subject we can state that such is the case for the 2-dimensional Ising system adopted as reference example in our previous discussion [Ba82] or for more general Ising spin glass Hamiltomians

$$H = -\frac{1}{2} \sum_{i,j} J_{i,j} s_i s_j + \sum_i h_i s_i$$

Problems like these (NP-complete problems) are the hardest problems in the class NP: the defining property of this class NPc is that any problem in NP can be reduced to one in NPc in polynomial time

(a language $L_\Pi \subseteq \Sigma_1^*$ is polynomially reducible to a language $L_{\Pi'} \subseteq \Sigma_2^*$ iff there is a polynomial time

computable function $\phi: \Sigma_1^* \to \Sigma_2^*$ such that $x \in L_\Pi$ iff $\phi(x) \in L_{\Pi'}$)

Other complexity classes we shall have to consider in the following are:

$\qquad$ PSPACE = $\cup_{k \geq 1}$SPACE (n^k) and Co-NP, the class of complements of the languages in NP.

The definitions of complete problems and polynomial reduction apply on these classes as well.

The following inclusions are easily proven:

$\qquad$ P $\subseteq$NP $\subseteq$PSPACE

$\qquad$ P $\subseteq$Co-NP $\subseteq$PSPACE

It is an open problem whether they are proper.

In Section 2 we will try to examine some dynamical problems relative to deterministic networks of binary threshold "neurons" [Mc43,Ca61] in terms of the complexity classes just defined. In Section 3 we will consider stochastic neural networks [Ho82;Hi84]. It is an interesting open research problem to discuss dynamical problems presented by such stochastic systems in terms of an analogous classification scheme made possible by the consideration of Probabilistic Turing Machines

[Gi77,Za82], namely of (ideal) Turing machines with three possible outputs ("yes",no","?") and such that for some computational steps a fair coin needs to be tossed in order to decide between two predetermined alternative next moves. As an example one can define the class RP of Random Polynomial problems as the class of languages L recognized by polynomially bounded PTM's , such that :

$$x \in L \Rightarrow P(M(x)="yes") \geq 1/2 \text{ and } x \notin L \Rightarrow P(M(x)="no") = 1,$$

the random variable M(x) being the ouput of the machine M on input x.

2. Deterministic symmetric neural networks.

A network of binary threshold neurons is a system of interconnected elementary processors which operate at discrete time steps through mutual interaction. The structure of the network and the iteration mode, i.e. the rule specifying the order in which neurons are updated, completely specify a discrete dynamical system.

The structure N of a network of n neurons is defined by the pair (W, T) where $W = \|w_{ij}\|$ is a nxn matrix of reals called the connection matrix , w_{ij} being called the weight of the connection from node i to node j, and $T = (t_1,..., t_n)$ is a n-component vector of reals called threshold vector , t_i being called the threshold of neuron i.

If W is a symmetric matrix, the network N is called a symmetric network.

Non symmmetric networks and their stochastic counterparts are interesting dynamical systems on their own right, presenting intriguing dynamical phenomena [Cl88,Br89]; in this article, however, we will concentrate exclusively on symmetric networks, for their relationship with combinatorial optimization.

At each time step $t \in \{0,1,...\}$ each neuron i=1,...,n is in a state $\tau_i(t) \in \{0,1\}$ The state of the network is identified by the vector $\underline{\tau}(t) = (\tau_1(t),..., \tau_n(t))$ or, equivalently by the corresponding vertex of the unit cube $C_n = \{0 \leq x_i \leq 1; i=1.....n\}$ in R^n. Given the structure (W,T), to each neuron there is associated a Boolean linearly separable function [Hu65] $f_i : \{0,1\}^n \rightarrow \{0,1\}$ given by

$$f_i(\tau_1,..., \tau_n) = H(\sum_{j=1}^{n} w_{ij} \tau_j - t_i)$$

where
$$H(x) = \begin{cases} 1 \text{ if } x \geq 0 \\ 0 \text{ if } x < 0 \end{cases}$$

Of the many iteration modes which have been defined for a network [Mc43, Ho82, Fo85, Ro87, Be89], here we mention only the parallel and the sequential one.

If a network operates according to the parallel mode its evolution from the initial state

$\underline{\tau}(0) = (\tau_1(0),...,\tau_n(0))$ is given by:

$$\tau_i(t+1) = f_i(\tau_1(t),..., \tau_n(t)) \qquad i=1,...,n.$$

If it evolves according to the sequential mode, the updating rule is given by

$$\tau_i(t+1) = \begin{cases} f_i(\tau_1(t),...,\tau_n(t)) & \text{if } t = i \bmod n \\ \tau_i(t) & \text{otherwise.} \end{cases}$$

The idea is that, in the sequential iteration, an external controller cyclically scans (1,2, ...,n) (or any assigned permutation) calling one neuron at a time to update on the basis of the present values of all the others.

Call $\text{tr}_N(t,\underline{\tau}_0)$ the trajectory starting from $\underline{\tau}_0$. As we are dealing with a dynamical system with a finite number of states, whatever iteration mode is adopted each trajectory will ultimately be periodic, entering, after a transient, an equilibrium state or a limit cycle.

More specifically, it can be shown , for a symmetric W, that:

(1) Under a sequential mode, if W has non negative diagonal elements, every trajectory will end into an equilibrium state [Ho82];

(2) Under the parallel mode, every trajectory will end into an equilibrium state or will enter a limit cycle of length 2 [Fo85].

Both statements have been originally proved by exhibiting for each mode a different Lyapunov function monotonically non increasing along each trajectory. A subsequent paper [BrGo88] has unified the proof of both results, showing that a network evolving in parallel mode can be reduced to a "dynamically isomorphic" network with vanishing diagonal elements evolving in a sequential mode. This is done by a simple duplication trick, which explains why cycles of length two can appear in parallel activation: before updating $\underline{\tau}$ copy it into $\underline{\tau}'$, sequentially scan $\underline{\tau}$, taking the data for its update from $\underline{\tau}'$; exchange the role of $\underline{\tau}$ and $\underline{\tau}'$.

The fact that, under the above hypothesis on the symmetric structure (W,T) and under sequential iteration, the function

$$V(\underline{\tau}) = -\frac{1}{2} \sum_{i,j=1}^{n} w_{ij}\tau_i\tau_j + \sum_{i=1}^{n} t_i\tau_i$$

satisfies

$$V(\underline{\tau}(t+1)) \leq V(\underline{\tau}(t)),$$

equilibrium points being "local" minima of V, suggests to use such networks as devices for performing a local search for minima of V. This is the reason why they have been proposed as a "general hardware" for both associative memories [Ho82] and optimization problems [Ho85]. This point of view and its stochastic improvements will be examined in the next section.

Here we wish to pursue the following perspective: a triple (W,T,m), where (W,T) is a network structure for a set of n neurons, and m is an iteration mode, singles out the distinguished domain $D \subseteq \{0,1\}^n$ of initial points which end into equilibrium points and the function F which to each initial point in D associates the corresponding equilibrium point; how well does the "machine" (W,T,m) perform in computing F ?

The question of "robustness" with respect to change in the parameters (W,T), or otherwise stated of the number of significant digits with which the parameters must be assigned is conveniently discussed in terms of the notion of equivalent network structures: two network structures of n neurons $N_1=(W_1,T_1)$ and $N_2=(W_2,T_2)$ are equivalent if for each iteration mode it is for every t and τ_0 $tr_{N_1}(t,\tau_0)=tr_{N_2}(t,\tau_0)$.

As the number $\Phi(n)$ of distinct threshold functions of n Boolean variables is finite, the set of network structures gets partitioned into a finite number of equivalence classes: the point is that the dynamically relevant fact about $f_i(\tau_1,\ldots,\tau_n) = H(\sum_{j=1}^{n} w_{ij}\tau_j - t_i)$ is not the actual numerical value of the parameters but the partition they determine of the set of vertices of C_n into two subsets lying on opposite sides of the plane $\Sigma_j w_{ij}\tau_j - t_i = 0$.

In particular, one can show that of each class one can take a representative N=(W,T) with integer parameters each of which of absolute value not larger than $(n+1)^{(n+1)/2}$ [Mu71,Hong87] and therefore representable by a binary number of length O(n lg).

For short: given a network structure of n neurons there is an equivalent one whose description takes no more than $O(n^3 \lg n)$ bits. Notice that one cannot improve much on this estimate. There are, indeed, examples of symmetric networks such that with less than $O(n^3)$ significant bits in the assignment of the parameters it is impossible to preserve the dynamics. This is shown, again, by a duplication trick: let f be the vector of threshold functions determined by this structure. Set $M = \Sigma_{ij} |w_{ij}|$ and consider the network structure of 2n neurons given by

$$W' = \begin{pmatrix} \begin{matrix} 4M & & 0 \\ & \ddots & \\ 0 & & 4M \end{matrix} & W^T \\ W & 0 \end{pmatrix} \qquad T' = (\underbrace{2M,\ldots,2M}_{n\ times} ; T)$$

In one parallel iteration step under (W',T'), (τ,σ) goes into $(\tau,f(\tau))$, so that the number S_{2n} of equivalence classes of network structures for 2n neurons is at least $\Phi(n)^n$. As $\Phi(n) \geq 2^{n(n-1)/2}$ [Mu71] we have a bound of the form $S_n \geq$ const. 2^{n^3}. At least one of these equivalence classes requires therefore at least $O(n^3)$ significant bits to be described.

The problem of the size of the weights touched above has something to say about the problem of the speed of (W,T,m) in computing F; this problem can be stated as the problem of giving bounds on the transient length. The usual estimates present in the literature are based on upper bounds on the initial absolute value of the Lyapunov function V and on lower bounds on the change of V in each step, by which one estimates from above the number of steps it takes to spend an initial $V(\underline{\tau}_0)$.

The above estimates of S_n indirectly show that there is at least one class of symmetric networks (with large integer weights) for which the transient length scales exponentially with n.

Some hypothesis of small coefficients is needed to prevent this phenomenon.

We single out the class R_k of network structures for which $\max_{1\leq i,j,k\leq n} (|w_{ij}|,|t_k|) \leq kn^k$, w_{ij} and t_k being integers.

For the symmetric structures in this class one can show that the transient lengths are polynomially bounded in n. Consequently the "Reachability" decision problem stated below

"Instance: $(N,\underline{\tau},\underline{\sigma},m)$, where N is a symmetric structure, $\underline{\tau}\in\{0,1\}^n$, $\underline{\sigma}\in\{0,1\}^n$, m is the parallel or a sequential iteration mode.

Question: Starting from $\underline{\tau}$, will the state of the network ever reach $\underline{\sigma}$? "

is in P if $N\in R_k$. The idea of the proof is very simple: start the network from $\underline{\tau}$ and wait (at most a polynomial time if $N\in R_k$) for the end of the transient.

As hinted by the previous analysis of the size of the weights, the hypothesis $N\in R_k$, or some similar hypothesis of small weights, cannot be released. Without such a hypothesis, in fact, "Reachability" is PSPACE complete.

The symmetry requirement is also essential: without this hypothesis "Reachability" for R_1 is PSPACE complete.

The peculiarity of symmetric networks is confirmed by the following results on the computational complexity of the "Stabilization" problem "is $D=\{0,1\}^n$?" (namely: "does every initial condition lead to an equilibrium state?"): for symmetric networks with parallel mode "Stabilization" is Co-Np complete; for arbitrary networks it is PSPACE complete [Be88].

3.Stochastic combinatorial optimization and stochastic networks

The fact, observed above, that the "typical" combinatorial optimization problem is "hard" leads to the pragmatic attitude, explored here, to give up any ambition of finding the exact solution and to settle for a reasonable approximate solution or for a reasonably high probability of finding a good one.

Stochastic augmentations of neural networks, to be discussed in this section, seem particularly well fit to the task of finding reasonable estimates of global minima for hard optimization problems. This largely experimental fact, witnessed for instance by the numerical success of the simulated annealing algorithm (for an exhaustive review see [vL87]) may seem surprising, in view of the fact that the deterministic counterparts of the stochastic networks we are going to consider in this section are at most local optimizers and are, in any case, systems whose deterministic dynamics is, as quantitatively shown above, difficult to control.

An optimistic point of view is that the numerical success of random search of minima is determined precisely by the complexity of the underlying deterministic dynamics which, in the presence of any amount of noise, causes a fast spreading of the search in the space of candidate solutions. Moreover, the fact that the convergence to the equilibrium distribution of the Markov chains used to implement the random search is monotonic in relative entropy [Ku59], [Sc76] is, as briefly discussed below, an indication that, in a simulated annealing search, time is efficiently spent in monotonically improving the average situation.

In this section we just wish to set the stage for a critical assessment of the above issues, mostly posing questions for future research.

For definiteness sake, we consider the problem of minimizing a given real valued function V of v

Boolean variables $\tau_1,..., \tau_v$,

$$V: \underline{\tau} \in \{0,1\}^v \rightarrow V(\underline{\tau}) \in R$$

Without any loss of generality, setting $\Lambda_v = \{1,2,..,v\}$ we can write it as:

$$V(\underline{\tau}) = - \sum_{M \subseteq \Lambda_v} \alpha_M \prod_{i \in M} \tau_i$$

and dispose of an additive constant and of a positive scale factor, both inessential for the problem of

minimizing V, by requiring that $\alpha_\varnothing = 0$ and $\inf (\{| V(\underline{\tau}) - V(\underline{\sigma})|\} - \{0\}) = 1$.

For k in Λ_v, we define $I_k: \{0,1\}^v \rightarrow \{0,1\}^v$ by

$$I_k(\tau_1,..., \tau_k,... \tau_v) = (\tau_1,..., 1-\tau_k,... \tau_v)$$

and easily compute, by the linearity of V in each of its variables that

$$V(I_k\underline{\tau}) - V(\underline{\tau}) = (2\tau_k - 1) \sum_{\substack{N \subseteq \Lambda_v \\ N \ni k}} \alpha_N \prod_{i \in N-\{k\}} \tau_i \equiv (2\tau_k - 1) X_k$$

We assume in what follows that, for every k and $\underline{\tau}$, $V(I_k\underline{\tau}) - V(\underline{\tau}) \neq 0$ or, equivalently

$$X_k(\underline{\tau}) \equiv \sum_{\substack{N \subseteq \Lambda_v \\ N \ni k}} \alpha_N \prod_{i \in N-\{k\}} \tau_i \neq 0$$

Under this hypothesis

$$V(I_k\underline{\tau}) - V(\underline{\tau}) > 0 \Leftrightarrow \begin{cases} X_k > 0 \text{ and } \tau_k = 1 \\ \text{or} \\ X_k < 0 \text{ and } \tau_k = 0 \end{cases} \Leftrightarrow \tau_k = H(x_k)$$

and all the solutions of the fixed point problem

$$\tau_k = H(X_k(\underline{\tau})) \quad k=1,2,....v$$

and only these are one step local minima of V in the sense that changing the value of any given single coordinate of such a point the value of V increases.

It is convenient therefore to improve any, however given, candidate solution $\mathfrak{x}_0$ to the minimization problem for V by an iterative search for a nearby fixed point.

One simple, far from unique, way of doing this is sketched below.

For $k \in \Lambda_v$, define $H_k : \{0,1\}^v \to \{0,1\}^v$ by

$$H_k(\tau_1,....\tau_k,...\tau_v) = (\tau_1,....H(X_k(\mathfrak{x})),...\tau_v).$$

One easily computes that

$$V(H_k\mathfrak{x}) - V(\mathfrak{x}) = X_k(\mathfrak{x}) (\tau_k - H(X_k)) \leq 0$$

with equality holding iff $\tau_k = H(X_k(\mathfrak{x}))$. Under sequential application of the above rule, say

$$\mathfrak{x}(0) = \mathfrak{x}_0$$
$$\mathfrak{x}(1) = H_1\mathfrak{x}_0$$
$$\cdots\cdots$$
$$\mathfrak{x}(hv+m) = H_m...H_1(H_v...H_1)^h\mathfrak{x}_0$$

$\mathfrak{x}_0$ evolves therefore toward a one step local minimum of V.

Interesting phenomena take place if one softens the rigid, sequentially deterministic evolution discussed above under two respects:

1. choose randomly, say with uniform probability $1/v$, the k-th coordinate to be updated;

2. being in $\mathfrak{x}$ at a given step, and having chosen the k-th component as the one to be updated, accept the proposed transition $\mathfrak{x} \to I_k \mathfrak{x}$ with large probability (say bounded below by $(1+e^{-\beta})^{-1}$ for some positive β) if $I_k\mathfrak{x}$ turns out to be equal to $H_k\mathfrak{x}$, with small (say bounded above by $(1+e^{\beta})^{-1}$ but not vanishing) probability otherwise.

Both requisites are satisfied for example by the Markovian dynamics defined by a transition matrix of the form

$$P_{\mathfrak{x},\mathfrak{x}'} = \begin{cases} \dfrac{1}{v}\dfrac{1}{1+e^{-\beta(V(\mathfrak{x})-V(I_k\mathfrak{x}))}} & \text{if } \mathfrak{x}'=I_k\mathfrak{x} \quad \text{for some } k \\[2mm] 1-\displaystyle\sum_{k=1}^{v} P_{\mathfrak{x},I_k\mathfrak{x}} & \text{if } \mathfrak{x}=\mathfrak{x}' \\[2mm] 0 & \text{otherwise} \end{cases}$$

The point is that, as the detailed balance condition

$$e^{-\beta V(\mathfrak{x})} P_{\mathfrak{x},\mathfrak{x}'} = e^{-\beta V(\mathfrak{x}')} P_{\mathfrak{x}',\mathfrak{x}}$$

holds, the unique stationary distribution for the above dynamics is

$$\rho_\beta(\underline{x}) = \frac{e^{-\beta V(\underline{x})}}{\sum_{\underline{x}' \in \{0,1\}^v} e^{-\beta V(\underline{x}')}} \equiv \frac{e^{-\beta V(\underline{x})}}{Z(\beta)}$$

Of course, as compared with the deterministic sequential evolution $\underline{x} \to H_k \underline{x}$, the nice fact of V being a Lyapunov function, $V(H_k\underline{x}) \le V(\underline{x})$, is lost (this is the price one pays in order to climb over local maxima in simulated annealing, or to tunnel beyond them in the "quantum annealing" approach we have proposed in [Ap88, Ap89], in search of low lying values of V).

This loss is however compensated by the fact that in the stochastic evolution described above the Helmoltz free energy is monotonically decreasing in time.

In order to make this statement precise, we review first of all a few elementary facts about Markov chains.

Let $P = \{p_{ij}, \, i,j \in \{1,...,N\}\}$ be a stochastic matrix, satisfying one of the usual hypotheses under which the existence of a unique stochastic vector $\underline{\phi} = (\phi_1,..., \phi_N)$ invariant under P is assured (suppose, say that $p_{ii} > 0$ for some i, and that for each h,k there exists r such that $(P^r)_{hk} > 0$).

Let $\underline{\pi}(t)$ be the stochastic vector reached, from an initial $\underline{\pi}(0)$ after evolution for t time steps under P:

$\underline{\pi}(t) = \underline{\pi}(0)) P^t$

For t large enough the components of $\underline{\pi}(t)$ are strictly positive, as those of $\underline{\phi}$ are, so that the definition

$$I(t) = \sum_{i=1}^{N} \pi_i(t) \, lg \frac{\pi_i(t)}{\phi_i} \qquad \text{(relative entropy from } \underline{\pi}(t) \text{ to } \underline{\phi} \text{ [Ku59])}$$

is well posed.

By Jensen's inequality $I(t) \ge 0$ and, obviously, $\lim_{t \to \infty} I(t) = 0$.

Relevant to our considerations is the fact that I(t) is monotonically decreasing, as can be seen by the following adaptation to discrete time of an argument of [Sc76]: by insertion of

$$\pi_i(t+1) - \pi_i(t) = \sum_{j=1}^{N} (\pi_j(t)p_{ji} - \pi_i(t)p_{ij})$$

and by Jensen's inequality

$$I(t+1) - I(t) = \sum_{i,j=1}^{N} \pi_j(t)p_{ji} lg \frac{\pi_i(t+1)\phi_j}{\pi_j(t)\phi_i} \le lg \sum_{i,j=1}^{N} \pi_j(t)p_{ji} \frac{\pi_i(t+1)\phi_j}{\pi_j(t)\phi_i} = 0,$$

because of the equilibrium condition $\phi_i = \sum_{j=1}^{N} \phi_j p_{ji}$ and of the normalization $\sum_{i=1}^{N} \pi_i(t+1) = 1$.

Specializing to our context the observation that $I(t+1) - I(t) \le 0$, and calling now $\rho(t; \underline{x})$ the probability distribution of τ at time t one concludes that, as t→∞

$$\sum_{\tau \in \{0,1\}^v}^{N} \rho(t;\tau)V(\tau) + \frac{1}{\beta} \sum_{\tau \in \{0,1\}^v}^{N} \rho(t;\tau)\, \lg\rho(t;\tau)$$

decreases monotonically to its equilibrium value $-\frac{1}{\beta}\lg Z(\beta)$.

A rather intuitive picture emerges, from the above considerations, of the underlying strategy for the search of minima of V: uphill climbs, corresponding to an increase of the internal energy $U=\Sigma\rho V$ are not excluded, but they are accompanied by an increase of the entropy $S=-\Sigma\rho\lg\rho$ corresponding to a larger spreading of the search; viceversa, if an iteration step decreases S, as it typically would happen for the focusing effect of the attractors of the deterministic dynamics, U decreases as a remnant of the fact that V does in the deterministic ($\beta\to\infty$) evolution.

Once equilibrium is reached, or reasonably approximated, many interesting conclusions can be drawn. As some of these conclusions seem to affect the complexity landscape outlined in Section 1, we wish to warn that this would be the case only if equilibrium could be reached or reasonably approximated in "reasonable time". This is a big if: it involves such issues (beyond the fundamental one of the computational meaning of drawing random numbers) as having quantitative control on the number of time steps over which to take the ergodic averages which only give operational meaning to the ensemble quantities named below, and relating the rate of approach to equilibrium with the complexity of decision problems related to V.

The first conclusion refers to the fact that, under the stochastic dynamics described above, one would have a global optimizer, as opposed to the local optimizers provided by deterministic evolution.

Calling $\mathcal{M}_V$ the set of points where V reaches its absolute minimum V_{min} it is indeed

$$P_\beta(\tau \in \mathcal{M}_V) \equiv \sum_{\tau \in \mathcal{M}_V} \rho_\beta(\tau) \geq \frac{1}{1+e^{-\beta}\left(\frac{2^v}{|\mathcal{M}_V|}-1\right)}$$

so that for each $0<\epsilon\leq1$ there exists a β such that:

$$P_\beta(\tau \in \mathcal{M}_V) \geq 1-\epsilon.$$

The second conclusion refers to the implementability of the above dynamics on a neural architecture. Notice that no hypothesis is made here about V being a polynomial of second degree, though some hypothesis on V is necessary in order to prevent the need of assigning an exponentially large number of coefficients (say that the degree of V is bounded above by a constant independent of v, or more generally that V is sparse, in the sense that at most a polynomially bounded number of coefficients is different from zero). The key to the network implementability of the stochastic optimization problem for such a V is the following simple observation [Ros 75]: suppose in V there is a term of degree >2, say for example $\tau_1\tau_2\tau_3$; one can formally write this as a quadratic term $\tau_1\tau_{2,3}$, introducing the additional

Boolean variable $\tau_{2,3}$ and adding to V a quadratic "penalty" term $\lambda(\tau_2\tau_3 + (3 - 2\tau_2 - 2\tau_3)\tau_{2,3})$ with $\lambda > 0$. The idea is that this term is zero iff $\tau_{2,3} = \tau_2\tau_3$, otherwise it is not smaller than λ. For λ large enough with respect to the coefficients of V, the function $W_\lambda(\underline{\tau},\tau_{2,3})$ constructed this way can have absolute minima only where $\tau_{2,3} = \tau_2\tau_3$, coinciding there with V. The above simple idea can be formalized into the proposition [Ros 75]:

Given V: $\underline{\tau} \in \{0,1\}^v \to V(\underline{\tau}) \in \mathbb{R}$

there exists an integer h and there exists a quadratic function

$$W: (\underline{\tau}, \underline{\sigma}) \in \{0,1\}^v \times \{0,1\}^h \to W(\underline{\tau}, \underline{\sigma}) \in \mathbb{R}$$

(a sketch of an idea of whose construction has been given above) such that

i. If $(\underline{\tau}_0, \underline{\sigma}_0)$ belongs to the set $\mathcal{M}_W$ of absolute minima of W, then $\underline{\tau}_0$ belongs to the set $\mathcal{M}_V$ of absolute

minima of V, and $W(\underline{\tau}_0, \underline{\sigma}_0) = V(\underline{\tau}_0)$.

ii. If $\underline{\tau}_0 \in \mathcal{M}_V$, there exists a unique $\underline{\sigma}_0 \in \{0,1\}^h$ such that $(\underline{\tau}_0, \underline{\sigma}_0) \in \mathcal{M}_W$, and it is $W(\underline{\tau}_0, \underline{\sigma}_0) = V(\underline{\tau}_0)$.

No claim of uniqueness of W holds, and in fact the nice problem emerges of the optimal choice of the number h of "hidden" nodes and of the penality parameters λ (additional thresholds and weights) in terms of the number v of "visible" nodes and the number and size of "bare" parameters α_M. Suffice to say here that both quantities are easily bounded for a sparse V.

Once the extended network on which W can be mapped is supposed to have been driven to equilibrium at $\rho_\beta(\underline{\tau},\underline{\sigma}) \sim e^{-\beta W(\underline{\tau},\underline{\sigma})}$, the probability that the configuration of the visible nodes is in $\mathcal{M}_V$ is

$$P_\beta(\underline{\tau} \in \mathcal{M}_V) = \sum_{\underline{\tau} \in \mathcal{M}_V} \sum_{\underline{\sigma} \in \{0,1\}^h} \rho_\beta(\underline{\tau},\underline{\sigma}) \geq \sum_{(\underline{\tau},\underline{\sigma}) \in \mathcal{M}_W} \rho_\beta(\underline{\tau},\underline{\sigma}) \geq \frac{1}{1 + e^{-\beta}\left(\frac{2^{h+v}}{|\mathcal{M}_V|} - 1\right)}$$

(No analogous claim is possible for deterministic dynamics, as there is no clear cut relationship between the local minima of the functions V and W).

The reducibility to the quadratic case just discussed adds a realistic touch to the hardware implementability of the stochastic dynamics of the general purpose global optimizer architecture resulting from the above considerations, namely:

a. a geometrically orderly and strongly hierarchical organization of the hidden nodes, emerging from a systematic application of Rosenberg's rule;

b. a fast activation of the transition rule on each single node based on linear operations on the signals coming to it from nodes with which it communicates;

c. (in the approximation in which the time needed for step b is negligible) a random asynchronous activation of each node, in a continuous time approach, upon calls of a Poisson clock independent of the Poisson clocks residing in every other node.

Proving, disproving or just analytically or experimentally determining the range of validity of the conjecture that such a machine works has, no doubt, much to say about the feasibility of some goals of artificial intelligence .

The problem whether resources (hidden nodes and weights) can be, on the basis of algorithmically predetermined choices, allocated in such a way as to generate on the "environmental" visible nodes a

probability distribution concentrated on a set analytically described as the critical set $\mathcal{M}_V$ of an assigned function V is in fact preliminary to the following fascinating problem: is it possible to allocate the hidden resources, on the basis of the statistical analysis of a large enough sample exhibited to the "sensorial" visible nodes, in such a way that the ensuing marginal environmental distribution reproduces the "significant" features of the population from which the sample was drawn?

There is encouraging but extremely preliminary experimental evidence [Ac85,Bo87] that the above learning skills are present in the so called Boltzmann machines [Hi84], of which we have tried above to give a constructive presentation aimed at stressing the hypotheses hidden under the assumption that they "work". Of paramount importance in the study of the Boltzmann machines is the question: are they in any sense a model of the learning paradigm stringently defined by Valiant in [Va84, Va88] ?

REFERENCES:

[Ac85]   Ackley D.H., Hinton G.E., Sejnowski T.J. :"A learning algorithm for Boltzmann machines". Cognitive Science 9, 147-169 (1985)

[Ap88]   Apolloni B., Carvalho C., de Falco D. :"Quantum stochastic optimization" to appear in Stochastic Processes and their Applications.

[Ap89]   Apolloni B., Cesa-Bianchi N., de Falco D. : "Quantum tunnelling in stochastic mechanics and combinatorial optimization" in [Ca89]

[Ba82]   Barahona F. :"On computational complexity of Ising spin glass models". Journal of Physics A 15, 3241-3253 (1982)

[Be88]   Bertoni A., Campadelli P., Morpurgo A. :"Total stabilization in symmetric networks". Proceedings of the international workshop Neural Networks and their Applications, Nimes (1988)

[Be89]   Bertoni A., Campadelli P. : "Neural networks and non uniform circuits", in [Ca89]

[Bo87]   Bounds D.G. : "A statistical mechanical study of Boltzmann machines". Journal of Physics A 20, 2133-2145 (1987)

[Br89]   Bruschi D., Campadelli P. "Reachability and stabilization in antisymmetric networks" in[Ca89]

[BrGo88] Brook J. , Goodman J.W. : "A generalized convergence theorem for neural networks" Stanford preprint (1988)

[Ca61]   Caianiello E.R. :"Outline of a theory of thought processes and thinking machines". Journal of Theoretical Biology 1, 204-235 (1961)

[Ca89]   Caianiello E.R., ed :"Parallel architectures and neural networks". World Scientific (1989)

[Cl88]   Clark J.W. : "Statistical Mechanics of neural networks". Physics Reports 158, 91-157 (1988)

[Co71]   Cook S.A. :"The complexity of theorem proving procedure". Proceedings of the third ACM symposium on the theory of computing.

[Fo85]   Fogelman F., Goles E., Pellegrin D. :"Decreasing energy functions as a tool for studying threshold networks". Discrete and Applied Mathematics 12, 261-277 (1985)

[Ga79]   Garey M.R., Johnson D.S. :"Computers and intractability" Freeman (1979)

[Gi77]   Gill J. :"Computational complexity of probabilistic Turing machines". SIAM Journal of Computing 6, 675-695 (1977)

[Hi84]   Hinton G.E., Sejnowski J.J., Ackley D.H. :"Boltzmann Machines: constraint satisfaction networks that learn". Technical Report CMU-CS-119 Carnegie-Mellon University (1984)

[Ho82]   Hopfield J.J. : "Neural networks and physical systems with emergent collective computational abilities". Proceedings of the National Academy of Science, 79, 2554-2558 (1982)

[Ho85]   Hopfield J.J., Tank D. : "Neural computation of decisions in optimization problems". Biological Cybernetics 52, 141-152 (1985)

[Hong87] Hong J. : "On connectionist model". Beijing Computer Institute preprint (1988)

[Hu]   Hu S.T. "Threshold Logic" University of California Press (1965)

[Ku59]   Kullback S. : "Information theory and statistics". Wiley (1959)

[Le73]   Levin L.A. :"Universal sorting problem" Problemy Peredachi Informatsii, 9, 115-116; English translation in: Problems of Information Transmission 9, 255-256 (1973)

[Mc43]   McCulloch W.S., Pitts W.A. : "A logical calculus of ideas immanent in nervous activity". Bulletin of Mathematical Biophysics 5, 115-133 (1943)

[Mu71]   Muroga S. "Threshold logic and its application". Wiley (1971)

[Ro87]   Robert F. :"An introduction to discrete iterations". in: "Automata networks in computer science", Fogelman, Robert, Tchuente eds. Manchester University Press (1987)

[Ros75] Rosenberg I.G. : "Reduction of bivalent maximization to the quadratic case" Cahiers Centre Etudes Rech.Oper. 17, 71-74 (1975)

[Sc76]   Schnakenberg J. :"Network theory of microscopic and macroscopic behavior of master equation systems". Reviews of Modern Physics 48, 571-585 (1976)

[St87]   Stockmeyer L.:"Classifying the computational complexity of problems". The Journal of Symbolic Logic, 52, 1-43, (1987)

[Va84]   Valiant L.G. :"A theory of the learnable". Communications of the ACM 27, 1134-1142 (1984)

[Va88]   Valiant L.G. :"Functionality in neural networks". Harvard preprint (1988)

[vL87]   van Laarhoven P.J.M., Aarts E.H.L. :"Simulated annealing". Reidel (1987).

[Za82]   Zachos S. :"Robustness of probabilistic computational complexity classes under definitional perturbations"; Information and Control, 54, 143-154 (1982)

AKNOWLEDGEMENTS:

This research was supported in part by Consiglio Nazionale delle Ricerche as part of the project Sistemi Informatici e Calcolo Parallelo - Parallel Computing on Neural networks.

LEARNING IN NEURAL NETWORKS

J. Bernasconi

Asea Brown Boveri Corporate Research

CH-5405 Baden, Switzerland

Abstract

Learning is one of the most important aspects of neural networks, and there exist many different learning paradigms. In this article, we concentrate on supervised learning from examples and provide a brief introduction to two of the most widely used learning procedures, "Error Backpropagation" and "Boltzmann Machine Learning". Both procedures can be viewed as strategies to minimize a suitably chosen error measure, and their performance depends on a number of parameters and implementation details. A simple model problem is used to illustrate how these dependences can affect the learning behavior.

1. Introduction

An artificial neural network consists of a set of units (formal neurons), each connected to some number of other units in the system. The state of the i-th unit is described by a scalar variable S_i, and each connection $j \rightarrow i$ carries a weight W_{ij} which can be positive, zero, or negative. Depending on the type of network considered, the weights are chosen symmetric ($W_{ij} = W_{ji}$) or asymmetric ($W_{ij} \neq W_{ji}$), and the S_i either assume only a discrete set of values (e.g., $\{0,1\}$ or $\{-1,+1\}$) or vary continuously (e.g., between 0 and 1, or between -1 and $+1$).

Artificial neural networks can be considered as grossly simplified models of the human brain. The units represent the neurons whose state of activity is measured by the variables S_i (e.g., $S_i = 1$ if neuron i is firing, and $S_i = 0$ if neuron i is quiescent), and the W_{ij} denote the strengths of the synapses. These can be excitatory ($W_{ij} > 0$) or inhibitory ($W_{ij} < 0$). Neural networks are also closely related to spin systems in statistical physics (S_i= spin variable, W_{ij}= interaction strength), and this analogy has recently led to considerable advances in the analysis of neural network properties [1-3].

If we consider neural networks as computing architectures, the units represent simple processing elements which update their states in a synchronous or asynchronous manner. The update rule is local and uniform, and usually taken to be of the form

$$S_i = f(\Sigma_j W_{ij} S_j - \theta_i) \tag{1}$$

where f is a nonlinear activation function, e.g., a threshold function or a sigmoid-type function such as $f(x) = 1/(1 + exp(-x))$. The updated value of S_i thus only depends on the total weighted input to unit i and on a threshold θ_i which can be regarded as an extra weight (associated with the connection to a unit whose value is always equal to -1). Certain types of neural networks (e.g., the Boltzmann machine [4,5]) employ stochastic units. In these cases, Eq.(1) is replaced by a probabilistic rule, i.e., $f(\Sigma_j W_{ij} S_j - \theta_i)$ represents the probability that S_i takes one of two possible values.

In neural networks, input and output are represented by the S_i-configurations of certain groups of units, and Eq.(1) defines a dynamical process which associates each input configuration with an output configuration. The resulting output configurations, of course, depend on the chosen weights W_{ij}, i.e., information or knowledge is stored in the pattern of weights and not in the processing units. In a learning phase, these weights therefore have to be adjusted in such a way that the network performs a given task as well as possible.

Quite generally, a neural network is characterized by its topology, by the type of units used, by the form of the update rule, and by the learning procedure. In this paper, we are primarily concerned with the learning behavior of neural networks, and we restrict ourselves to supervised learning from examples. In section 2, we introduce two of the most widely used learning procedures, "Error Backpropagation" and "Boltzmann Machine Learning", and section 3 is devoted to a discussion of some implementation issues. In section 4, we briefly review some recent results concerning the performance of these learning algorithms. The efficiency of a given algorithm depends on a number of parameters and implementation details, and in section 5 we use a simple model problem to illustrate how these dependences can affect the learning behavior.

2. Supervised Learning from Examples

We shall be concerned with neural networks in which the units are divided into input units, output units, and so-called hidden units. If an explicit distinction is required, the state variables S_i of the input units will be denoted by I_i, and those of the output units by O_i. The networks are supposed to perform a given pattern association task (classification, diagnosis, etc.) which can be expressed in terms of a specific input/output relation, $\{I_i\} \rightarrow \{O_i\} = \{D_i\}$, where $\{D_i\}$ denotes the desired output configuration.

A learning example is thus represented by an input/output pair, $\{I_i\}/\{D_i\}$. If the input pattern $\{I_i\}$ is held fixed, the network produces an output pattern $\{O_i\}$ which depends on the weights W_{ij}. Learning then consists in an adaptation of these weights, such that the discrepancy between $\{O_i\}$ and $\{D_i\}$, averaged over all inputs, is as small as possible.

More precisely, a supervised learning procedure can be described as follows:

a) Choose a set of learning examples, i.e., a set of input/output pairs $\{I_i^\mu\}/\{D_i^\mu\}, \mu = 1, ..., N$, where N may be much smaller than the number of possible inputs.

b) Define an error measure for the discrepancy between the actual output, $\{O_i^\mu(\{W_{ij}\})\}$, and the desired output, $\{D_i^\mu\}$,

$$F^\mu = F(\{O_i^\mu(\{W_{ij}\})\}, \{D_i^\mu\}). \qquad (2)$$

c) Adjust the W_{ij}'s such that the total error, $\Sigma_\mu F^\mu$, is minimized. Usually, this is done by a stochastic gradient descent procedure, i.e., after each presentation of a learning example, W_{ij} is changed by an amount proportional to the respective negative gradient of the error measure,

$$\Delta W_{ij} = -\eta \frac{\partial F}{\partial W_{ij}}, \qquad (3)$$

where we have dropped the index μ which labels the learning examples.

Learning is thus nothing else than the (stochastic) minimization of a suitably chosen error measure, and in the following we shall briefly introduce two specific and widely used procedures for supervised learning from examples.

(A) Error Backpropagation

In its original form, error backpropagation learning [5-7] is restricted to networks with feedforward connections only. Such networks consist of one layer of input units, one or more layers of hidden units, and one layer of output units, and there exist no backward connections from output to hidden, or from hidden to input units. (A simple example of a feedforward network is shown in Figure 1 below). The dynamics is assumed to be deterministic, i.e., the update rules are given by Eq.(1), and we note that in a feedforward network the units are updated in a single pass.

The standard version of error backpropagation uses gradient descent to minimize the mean squared error of the output signal,

$$F = \frac{1}{2}\Sigma_i(D_i - O_i)^2. \tag{4}$$

For a weight W_{ij} which is associated with a connection from an (input or hidden) unit j to an output unit i, we then simply obtain

$$-\frac{\partial F}{\partial W_{ij}} = (D_i - O_i)f'(\Sigma_k W_{ik}S_k)S_j \equiv \delta_i S_j, \tag{5}$$

where f' is the derivative of the nonlinear activation function introduced in Eq.(1). By applying the chain rule of differentiation, and by taking advantage of the previously calculated error signals δ_i, the error gradient can then easily be "backpropagated" through the network, i.e., recursively computed for the weights which do not connect to an output unit.

The occurrence of f' in the expressions for $-\partial F/\partial W_{ij}$ implies that backpropagation learning is only defined for continuous activation functions f. The backpropagation scheme can, however, be used with alternative error measures. For units with $0 \leq O_i, D_i \leq 1$, a recently proposed function [8,9] takes the form

$$F = \Sigma_i[D_i ln\frac{D_i}{O_i} + (1 - D_i)ln\frac{1 - D_i}{1 - O_i}]. \tag{6}$$

If O_i and D_i are interpreted as probabilities, F represents the information difference between the desired and the obtained outputs.

(B) Boltzmann Machine Learning

Boltzmann machines [4,5] are stochastic neural networks with discrete units (e.g., $S_i = \pm 1$) and with symmetric weights ($W_{ij} = W_{ji}$). They can thus be regarded as spin systems, and an energy,

$$E = -\Sigma_{(i,j)}W_{ij}S_iS_j, \tag{7}$$

can be associated with each configuration $\{S_i\}$. The stochastic update rules are chosen in such a way that the resulting dynamics is equal to the equilibrium dynamics of the spin system at some temperature T.

In the learning phase, one determines the equilibrium Boltzmann distributions for the states (configurations) of the network, first for the situation where both the input and output units are clamped at their desired values ("+" mode), and secondly for the situation where only the input units are clamped ("−" mode). The error function F is an information theoretic measure [4,5] for the difference between the two equilibrium distributions, and it can be shown [10] that an equivalent choice is the free energy difference, $F = \mathcal{F}^+ - \mathcal{F}^-$, between the two situations. It follows [4,5] that the gradients are given by

$$-\frac{\partial F}{\partial W_{ij}} = \langle S_i S_j \rangle^+ - \langle S_i S_j \rangle^-, \tag{8}$$

where $\langle ... \rangle^\pm$ denotes the average with respect to the respective equilibrium distribution. At zero temperature ($T = 0$), F becomes equal to the energy difference between the two groundstates, $F = E^+ - E^-$, and we simply have

$$-\frac{\partial F}{\partial W_{ij}} = S_i^+ S_j^+ - S_i^- S_j^-, \tag{9}$$

where $\{S_i^+\}$ and $\{S_i^-\}$ denote the respective groundstate configurations. This $T = 0$, i.e., deterministic version of the Boltzmann machine learning rule has been proposed by Le Cun [7].

3. Implementation of Learning Procedures

The learning procedures introduced in the previous section are gradient descent schemes, i.e., they suffer from all the problems associated with gradient descent on a complicated landscape. If, for example, the learning parameter η in Eq.(3) is chosen small, the learning process becomes very slow, while large values of η lead to oscillations which prevent the algorithm from converging to a good solution. If the error surface, moreover, contains many local minima, it may be very difficult to find a good minimum with a gradient descent technique.

For these reasons, the above learning procedures are usually not implemented in the simple form of Eq.(3). One way to improve a stochastic gradient descent method is to smooth the weight changes by overrelaxation, i.e., by introducing a so-called momentum term [5,6],

$$\Delta W_{ij}(k+1) = -\eta \frac{\partial F}{\partial W_{ij}} + \alpha \Delta W_{ij}(k), \tag{10}$$

where $\Delta W_{ij}(k)$ refers to the weight change after the presentation of the k-th learning example. An additional smoothing of the procedure is obtained by accumulating the weight changes over some number n of learning examples before the weights are actually changed. The inclusion of a weight decay term,

$$\Delta W_{ij} = -\eta \frac{\partial F}{\partial W_{ij}} - \beta W_{ij}, \tag{11}$$

prevents the algorithm from generating very large weights which may create such high barriers in the error surface that a solution cannot be found
within reasonable time. To avoid that the algorithm becomes trapped in
a local minimum, one can further change the weights from time to time
(i.e., with a small probability p) by some random amount.

All these tricks can, of course, be combined, and they help to suppress
the occurrence of oscillations and to escape from bad local minima, but
only to a certain extent. In complex situations, one may be forced to use
more sophisticated optimization strategies than gradient descent.

4. Performance Analysis

In general, learning in a neural network is a stochastic process. One
starts, for example, from a random distribution of weights, and the learning examples are usually presented in random order. The performance of
a learning procedure is thus conveniently characterized by averaged quantities, and the most obvious measure is the mean learning time, i.e, the
mean number of learning example presentations needed until the network
performs a given task with a desired accuracy. More detailed information, however, can be obtained from an analysis of learning curves which
describe, e.g., how the magnitude of the error measure F decreases, or
how the fraction of correct output answers increases during the learning
process.

The performance of a learning procedure not only depends on the
difficulty of the problem and on the structure and size of the network,
but also on a large number of implementation details. These include the
type of units used, the definition of the error measure, the form of the
initial weight distribution, and the choice of parameters such as $\eta, \alpha, \beta, ...$
(see section 3).

The analysis of the learning behavior is thus a very complex problem,
and analytical results have so far only been obtained for some special
classes of neural network structures, e.g., for Hopfield-type networks [11]
or for perceptron-like architectures [12].

For networks which contain hidden units, the existing results all refer
to empirical investigations. Tesauro et al [13,14] have considered the
n-bit parity problem whose complexity (as measured by the predicate
order k [15]) is equal to the number of inputs, $k = n$. A feedforward
network containing one layer of hidden units is used, and the network is
trained with the backpropagation algorithm. The results indicate that

the learning time increases exponentially with the order of the problem, $\tau \sim 4^k$, and that the dependence on the learning set size N follows a power law, $\tau \sim N^\gamma, \gamma \approx 4/3$, if $N \ll N_c$. At $N_c \sim 2nh$, where h denotes the number of hidden units, the network approaches its capacity limit, and τ diverges. The investigations carried out by Benedict [16] refer to networks of a similar structure, but are concerned with first order content addressable memory tasks. The number of hidden units, h, is chosen to be equal to the learning set size, N, and for $h = N = \frac{1}{2}n$ ($n =$ number of input units) the learning time is found to increase as $\tau \sim n^\alpha, \alpha \approx 1.86$. As a function of $N(= h), \tau$ saturates above $N \approx \frac{1}{2}n$.

The dependence of the learning behavior on the type of units used has been studied by Stornetta and Huberman [17] and by Bernasconi [18]. At least for the Boolean problems considered in these investigations, networks with $(-1, +1)$ units are shown to learn significantly faster than the corresponding $(0, 1)$ versions. Bernasconi [18] has further demonstrated that an adaptation of the weight-structure to the symmetries of a problem can lead to a drastically improved learning behavior.

This short review of some recent results is certainly not exhaustive, but it gives an idea of the variety of aspects that have to be included in an analysis of learning in nonlinear neural networks.

5. An Example

To illustrate how the different factors mentioned in the previous section can affect the learning behavior, we consider a problem that has recently been analyzed by Lapedes and Farber [19]. The task our network is supposed to learn is to predict a chaotic time sequence that is generated recursively by the well-studied logistic map [20],

$$X_{t+1} = 4\lambda X_t(1 - X_t), \tag{12}$$

where $0 \leq X_t \leq 1$ and $0 \leq \lambda \leq 1$. The sequence exhibits a chaotic behavior if $0.8925 \leq \lambda \leq 1$.

We choose the same network structure as Lapedes and Farber [19], i.e., a feedforward network containing a single layer of five hidden units, see Figure 1. The current value of X_t is fed into the input unit, and the other units are updated according to Eq.(1), where the nonlinear activation function f is chosen as $f(x) = 1/(1 + exp(-x))$. Using error backpropagation, the 11 weights and 6 thresholds are then adjusted until the value of the output unit predicts X_{t+1} as accurately as possible.

The learning set consists of 500 successive X_t-values, say $\{X_1, .., X_{500}\}$, generated by Eq.(12), and the individual learning examples are given by the respective pairs $\{X_t, X_{t+1}\}$. Typical results for the prediction capability of the network after learning are shown in Figs. 2 and 3. In the corresponding simulations, λ has been chosen equal to 0.913. It follows

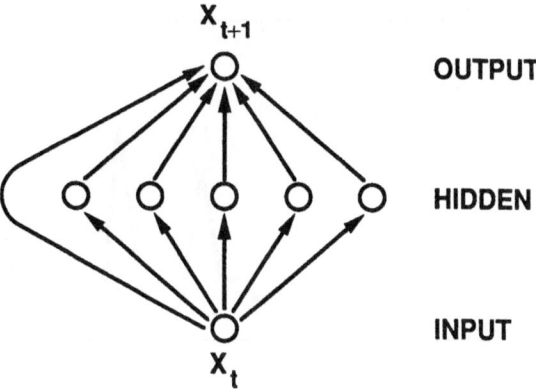

Fig. 1. The neural network used to predict a chaotic time sequence generated by Eq. (12).

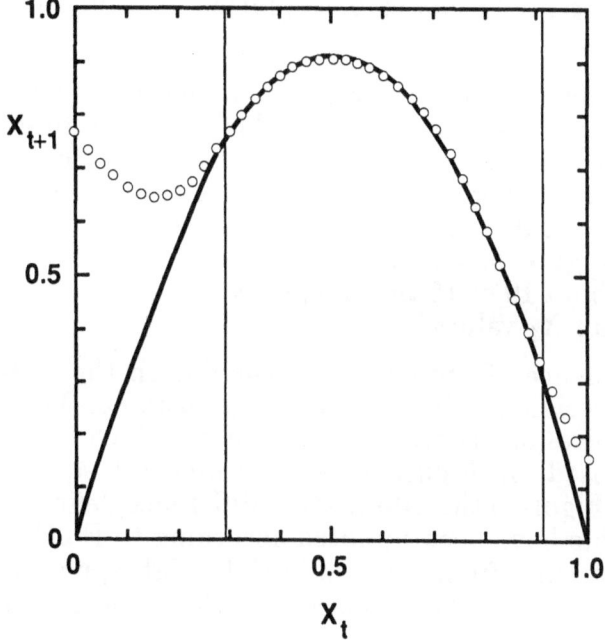

Fig. 2. Prediction of X_{t+1} given X_t. The continuous curve represents Eq. (12), with $\lambda = 0.913$, and the circles refer to the prediction of the network after learning.

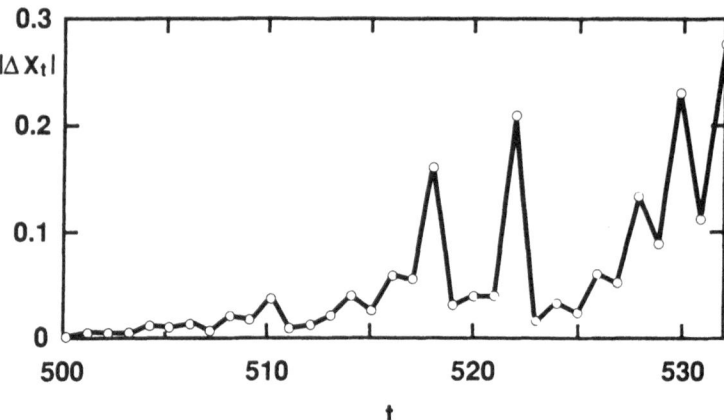

Fig. 3. Long term prediction capability of the neural network. Starting from X_{500}, the network recursively generates $X_{501}, X_{502}, ...$ without readjustment of the current X_t-value, and the resulting prediction error is plotted as a function of t.

that the learning set only contains X_t-values between 0.290 and 0.913 if the starting value is chosen within this range. Figure 2 demonstrates how accurately the network has implemented the recursion of Eq. (12), of course only in the region where learning examples have been provided. For $0.3 < X_t < 0.9$, the mean prediction error is $\langle |\Delta X_{t+1}| \rangle = 0.006$. As shown in Figure 3, this accuracy is sufficient to lead to a reliable prediction of about 10 or 15 time steps into the future without readjusting the intermediate X_t-values.

The dependence of the learning behavior on the width of the initial distribution of weights is illustrated in Figs. 4 and 5. The two figures refer to different choices of the error measure F on which the backpropagation algorithm is based. In Figure 4 we have used the squared output error, Eq.(4), and in Figure 5 the information difference defined in Eq.(6). In all runs, the weights have been updated according to Eq.(10), with $\eta = 0.05$ and $\alpha = 0.9$, starting from random initial weights, uniformly distributed in an interval $(-W_0, +W_0)$. A learning cycle consists of one pass through the entire learning set.

The results of Figs. 4 and 5 demonstrate that the learning time can be shortened considerably by increasing the width of the initial weight-distribution. (Above $W_0 = 2$ to 3, the decrease in learning time starts to saturate and then learning becomes slower again). The very slow learning

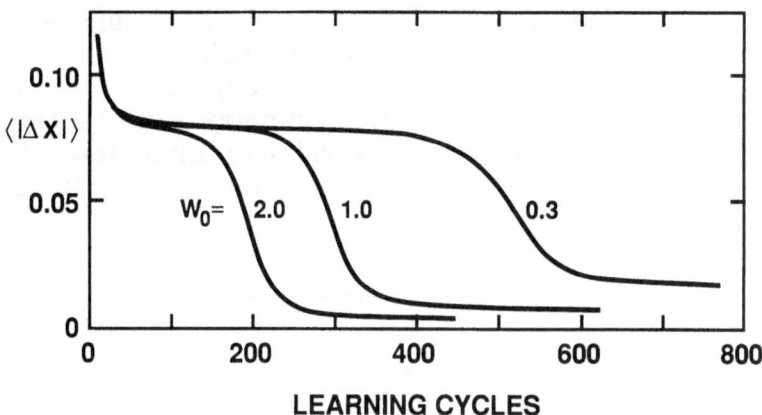

Fig. 4. Mean one-step prediction error, $\langle|\Delta X|\rangle$, vs. number of learning cycles for different scalings of the initial weight-distribution. The squared output error, Eq. (4), is used as error measure in the backpropagation algorithm.

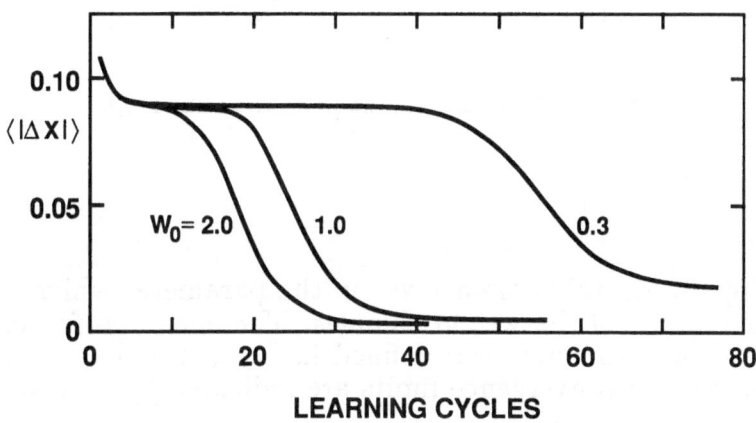

Fig. 5. Same as Figure 4, but using the information difference, Eq. (6), as error measure.

for small values of W_0 can be understood as follows. If $W_0 = 0$, it can easily be shown that the five weights from the input unit to the hidden units are always changed by equal amounts, i.e., they remain equal during the entire learning process. The same is true for the five weights from the hidden units to the output unit and for the five thresholds

52

associated with the hidden units. The network thus explores only a low-dimensional submanifold of the weight space and is unable to approximate the recursion of Eq.(12) to a reasonable accuracy. If now W_0 is not equal to zero, but still quite small, the above symmetry is only slightly broken, and it takes a very long time until the learning examples force the network to leave the neighborhood of the symmetry-induced submanifold in weight space.

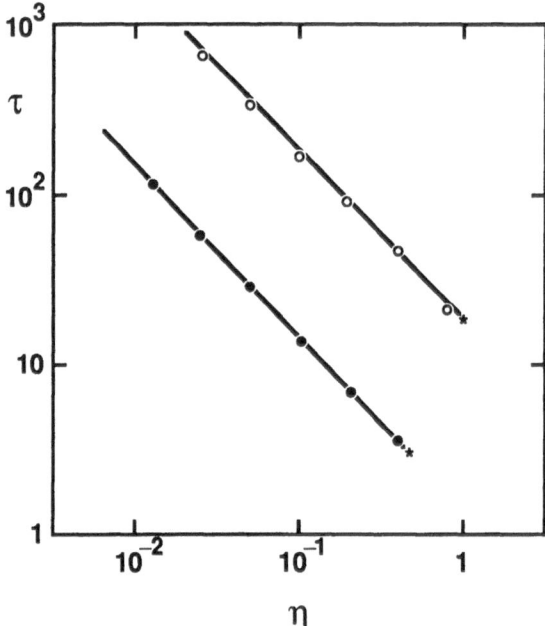

Fig. 6. Learning time τ vs. η, the parameter which scales the magnitude of the weight changes. The open and full circles refer to the error measures defined in Eqs.(4) and (6), respectively, and the convergence limits are indicated by an asterisk.

A comparison of Figs. 4 and 5 further shows that the choice of the error measure F has a big effect on the learning speed. In our example, the learning time decreases by about an order of magnitude if we use the information difference, Eq.(6), instead of the squared output error, Eq.(4), to determine the weight changes in the backpropagation algorithm. As the other parameters of the algorithm (η and α) are held fixed in this comparison, one may argue that the difference is merely due to the steeper gradients associated with the information difference, and that this can be compensated for by choosing larger η-values when using the squared output error. The argument is certainly correct, but only to a certain extent, as is shown in Figure 6. Here we plot the learning time

τ as a function of η, where τ is defined as the number of learning cycles needed until the mean prediction error, $\langle |\Delta X| \rangle$, becomes smaller than 0.006. In all runs, the values of α and W_0 have been set to $\alpha = 0.9$ and $W_0 = 2.0$, respectively. The results show that with Eq.(4) as error function for the learning procedure, η has to be increased by about a factor of 10 if we want to achieve the same learning time as with Eq.(6). Taking into account the convergence limits, however, it is also seen that the information difference tolerates much higher learning speeds than the mean squared output error. At least for this example, the information difference thus appears to represent the more robust error measure.

6. Conclusion

Learning in neural networks can be viewed as a nonlinear stochastic optimization process. With our considerations and illustrations we have attempted to provide an introduction to the concept of supervised learning from examples and to discuss some of the problems associated with corresponding learning procedures. The performance of such procedures depends on a large number of factors and implementation details, so that their analysis represents a very complex matter. Existing results all refer to empirical studies and are restricted to a few specific problems. While these investigations give valuable information about the learning behavior of neural networks with hidden units, the derivation of general analytic results is still an unsolved problem.

REFERENCES

[1] D. J. Amit, H. Gutfreund, and H. Sompolinsky, Phys. Rev. Lett. 55, 1530 (1985); Phys. Rev. A 35, 2293 (1987).

[2] E. Gardner, Europhys. Lett. 4, 481 (1987); J. Phys. A 21, 257 (1988).

[3] E. Gardner and B. Derrida, J. Phys. A 21, 271 (1988).

[4] D. H. Ackley, G. E. Hinton, and T. J. Sejnowski, Cognitive Science 9, 147 (1985).

[5] "Parallel Distributed Processing: Explorations in the Microstructure of Cognition", vol. 1: "Foundations", ed. by D. E. Rumelhart and J. L. Mc Clelland (MIT Press, Cambridge), 1986.

[6] D. E. Rumelhart, G. E. Hinton, and R. J. Williams, Nature 323, 533 (1986).

[7] Y. Le Cun, in "Disordered Systems and Biological Organization", ed. by E. Bienenstock, F. Fogelman Soulié, and G. Weisbuch (Springer, Berlin), 1986, pp. 223-240.

[8] J. J. Hopfield, Proc. Nat. Acad. Sci. USA $\underline{84}$, 8429 (1987).

[9] E. B. Baum and F. Wilczek, in "Neural Information Processing Systems", ed. by Dana Z. Anderson (American Institute of Physics, New York), 1988, pp. 52-61.

[10] J. Bernasconi, unpublished, 1989.

[11] M. Opper, Phys. Rev. A $\underline{38}$, 3824 (1988); Europhys. Lett. $\underline{8}$, 389 (1989).

[12] J. A. Hertz, A. Krogh, and G. I. Thorbergsson, Preprint (1989).

[13] G. Tesauro, Complex Systems $\underline{1}$, 367 (1987).

[14] G. Tesauro and B. Janssens, Complex Systems $\underline{2}$, 39 (1988).

[15] M. Minsky and S. Papert, "Perceptrons", (MIT Press, Cambridge), 1969.

[16] K. A. Benedict, J. Phys. A $\underline{21}$, 2643 (1988).

[17] W. S. Stornetta and B. A. Huberman, in "Proc. IEEE First International Conference on Neural Networks", 1987, pp. II-637 to II-643.

[18] J. Bernasconi, in "Neural Information Processing Systems", ed. by Dana Z. Anderson (American Institute of Physics, New York), 1988, pp. 72-81.

[19] A. Lapedes and R. Farber, Preprint LA-UR-87-2662, 1987.

[20] M. Feigenbaum, J. Stat. Phys. $\underline{19}$, 25 (1978).

MATHEMATICAL MODELLING ON RANDOM GRAPHS OF THE SPREAD OF SEXUALLY TRANSMITTED DISEASES WITH EMPHASIS ON THE HIV INFECTION[1]

Philippe Blanchard, Georg F. Bolz, Tyll Krüger

Theoretische Physik and BiBoS

Universität Bielefeld

D-4800 Bielefeld 1, FRG

I. Introduction

In this paper we present general results dealing with the description of a new mathematical model based on the concept of random graphs intended to describe the dynamics of sexually transmitted diseases. The model we develop is especially well adapted for the study of the Human Immuno-deficiency Virus (HIV) infection but can also be applied to quite different communicable diseases, not necessarily sexually transmitted. Moreover the same mathematical framework can be used to describe a large class of propagation phenomena (for instance the edges of graphs can be viewed as communication lines in a queuing network (see e.g. [1]) or as relationships in a social network (see e.g. [2]). The epidemic models we consider can be linked to site percolation problems [3] on the relevant graphs where edges are loaded with the corresponding transmission probabilities.

For shortness and simplicity we will mainly concentrate on the case of AIDS. We now summarize the paper.

In Section I we give a short description of the characteristic properties of our model and explain its usefulness for epidemiological research. We discuss briefly the main differences between our model and the standard epidemiological models and list the problems and questions to which the model is able to give quite accurate answers.

Section II is devoted to the general underlying mathematical structure. We introduce the relevant spaces of random graphs and explain the construction of the discrete stochastic process describing the spread of the infection.

In Section III we show first briefly how classical models arise as a special case of our model and introduce some simplified models. Furthermore we discuss

---

[1]This work has been supported by Bundesgesundheitsamt (Berlin) and by Bundesminister für Forschung und Technologie (Bonn) within the context of the project "Forschungsförderung des BMFT zum Krankheitsbild AIDS"

some purely mathematical aspects of the model and their connection with relevant aspects of the epidemic dynamics.

Section IV is devoted to a short discussion of computational aspects of our framework.

Acknowledgements

The authors are grateful to S. Albeverio, A. Dress, C. Römer, Y. Suhov, for their interest in our work. We want to thank K. Dietz and B. Voigt for helpful advices and very constructive discussions.

I. General discription of the model and its main properties

The study of mathematical models for epidemics has a long story starting with En'ko [4] and Ross [5]. For a recent survey of classical models we refer to [6], [7], [8].

Most of the standard models are coupled systems of deterministic nonlinear ordinary differential equations relying on the "mass–action" principle and similar to the Lotka-Volterra equations. In its continuous time form the mass action principle states that if S(t) and I(t) are the respective numbers of susceptibles and infectives in a given population then the rate of change of the infectives is proportional to the number of possible encounters S(t) I(t) between susceptibles and infectives, i.e.

$$\frac{dI}{dt} = \gamma\ S(t)\ I(t)\ ,$$

γ being the infection parameter.

For understanding the epidemiology of malaria Ross [5] proposed in the early 1900 a model of this type taking into account the coupling between the human and mosquito populations of a region. Since then deterministic and probabilistic epidemic models have relied on this mass–action effect in one form or another.

AIDS, or Acquired Immunodeficiency Syndrome, is the final stage of the disease caused by infection with HIV, a retrovirus with a very long asymptomatic period [9]. AIDS is spread sexually, through blood and from infected mother to child. Sexual transmission is the most common mode of transmission and will remain so in the future. Due to the chromosomal integration of the proviral DNA into the host cell an HIV–infected person remains infected and infectious for life. With more than 140000 cases reported from all over the world by January 1989 AIDS has become a worldwide epidemic affecting industrialized and developing countries. WHO estimates the real number of AIDS cases to date is actually closer to 280000.

For sexually transmitted diseases standard epidemiological models of the mass–action type have been used. See e.g. May and Anderson [31], Weyer [10],

Hethcote and Yorke [11], Dietz [12], Hadeler and Dietz [13], Hyman and Stanley [14], Castillo-Chavez [36], J.A. Jacques and C. Simon [37], K.P. Hadeler [38]. Dietz's model takes also into account the effects of the duration of partnerships. However for HIV epidemiology a mathematical model is needed which reflects the complex structure of sexual contacts between individuals and is able to take into account large variations in behaviour as well as in the disease progression reflecting the population risk structure (age, sexual activity, drug use, ...). See also [15], [16], [17], [18], [19], [20], [21], [22], [23], [24] and [31].

Therefore the model we proposed in [25] (see also [26], [27] and [34]) is of completely different spirit. We consider the individuals of a given population as vertices of a graph, whose edges are supposed to represent their sexual contacts in the time interval [0,T] (T can in some cases be taken infinite) during which the model is applied. While the consideration of random graphs in the context of epidemiology is not entirely new (see e.g. [28] and [29]), our approach is novel in linking the characteristics of the random graph and of the dynamics on it to social and medical data as well as in exploiting the richness of the structure of stochastic processes over random graphs to model a complicate epidemiological situation. Let us describe more precisely the situation we have to modelize. In real life the mesh of sexual contacts is generated by a very complex "local" process of pair formations (the so called matchings for graph theorists) and of pair separations. About this phenomenon of pair formation and separation no empirical sociological data are available. All we can estimate at best are "integrated" data for the graphs generated by sequences of such matchings. From this point of view it is therefore natural and reasonable to start with a mathematical framework incorporating directly these integrated data.

For purpose of illustration let us look at a very simple example in which $x_1, ..., x_6$ are the individuals and as time interval we take $T = 7$ months (see figure 1). The numbers along the edges (i,j) indicate the month during which a sexual contact between the individuals (vertices) x_i and x_j is realized. The graph we consider is generated by a sequence of 7 matchings. Let us remark that our model allows without problem the possibility to take into account multiple partnerships at a given time. However from a mathematical point of view it suffices to consider the situation where at any given time at most one contact is realized for each individual.

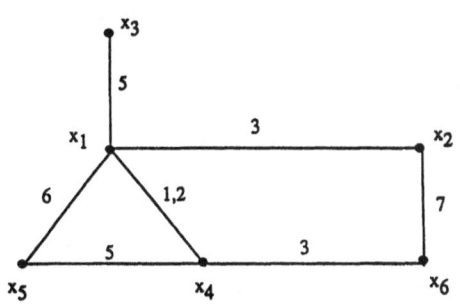

Figure 1

What kind of data can be used to generate a graph which is similar to the sexual contact graph of real life? Most important is the distribution of the contact rate during the considered time interval of length T. In other words we need a function which tells us for all n (n = 0, 1, 2, ...) how many people of a certain group have had n different sexual contacts during the time T. If we look at all graphs which satisfy the above contact rate distribution (on the graph this means now by definition the number of vertices (individuals) having n edges (sexual contacts)), we will find that most of them look very similar; that is to say that they are typical (generic). Only a few ones are very different and can be regarded as being exotic or atypical. The set of all graphs satisfying a given contact distribution together with a probability measure will be called a "random graph space". For more details on concepts in random graph theory see [30]. In Section II the precise mathematical description of various relevant random graph spaces we consider will be given.

Furthermore for the investigation of the epidemics of HIV infection we have to introduce more structure on the set of relevant graphs. Indeed we must take into account the existence of several groups of individuals (homosexual males, prostitutes, bisexual males, intravenous drug users, ...) having a sexual contact structure which is highly different from the usual heterosexual one. These highly active groups play – at least in the first stages of the epidemic – a very important role. If mixing is small then the individuals belonging to high-risk group are nearly all infected before the HIV–infection moves into lower-risk groups. If there is large mixing many more lower activity individuals will be infected. Therefore the initial data we need are the contact rate functions for each subgroup as well as the coupling rates between two different groups. Of course the age of the individuals in each subgroup should be also taken into account by the very nature of human finding sexual partners. Since the dynamics of the AIDS-epidemic involves long time scales individuals will leave a given behavioral risk group and this flow between the different risk groups is an additional source of mixing.

But given group dependent distributions of the contact rate and informations about the coupling between the different groups we are now in a position to generate a set of random graphs conforming with these given data. Up to now we have only set up the underlying structure for the epidemics. Let us now explain how the dynamics is generated in our model.

All we need is some knowledge of the times at which edges are activated (realization of a sexual contact) like in Fig. 1. Again we introduce this information by using a distribution which describes the duration of partnership of an individual of a certain group at a given age. In this way we can get time valued edges for the whole time interval T. Of course this distribution of the partnership duration is much less known than the contact rate distribution. Nevertheless the available data are better than nothing. For instance, young people between 17 and 25 years have typically several numbers of short contacts and heterosexual males between 40 and 60 are more frequently contacting prostitutes.

Finally to generate the dynamics of the epidemic of HIV infection we have to take into account some epidemiological and medical parameters. The most important are:

- probability that infection is transmitted per contact from an infected individual to a susceptible partner for the different groups
- distribution of the unusually long incubation period (of the order of 8–9 years with a wide standard deviation)
- duration of infectiousness
- influence of the different types of HIV-virus and genetic variability of HIV RNA sequences
- variable infectiousness (the infectivity seems to be much lower in the middle stage of the infection)
- proportion of virus carriers that develops AIDS
- mortality
- influence of other venereal diseases.

Observe that probability enters into our model as it enters into real life. First the underlying contact structure is varied randomly according to sociological data expressing the sexual behaviour of the individuals of the different groups. Secondly the dynamics is also generated at random using probability distributions of the sexual and epidemiological parameters which must be taken into account to simulate the important features of AIDS epidemic. But since we are working with an ensemble of graphs, the probabilities we need are assigned in a natural way to the elements of the underlying random graphs.

The dynamics we consider in actual computer calculation is a discrete one but with a short time step (one day) and the simulation on the computer of the spread of the infection is fast. Let us emphasize that the standard modelling of epidemics using ordinary differential equations appears as a special limiting case of our model. Indeed if we take only complete graphs, respectively complete n-partite graphs, sexual contacts are now realized uniformly in the sense that everybody has contact to everybody. In Section III it is shown how to obtain in this way from the bipartite complete graphs (modelling heterosexual transmission) the discrete version of the Lotka-Volterra system for the expectation of the infection density. Note that the classical models can only work with the expectation values of the relevant data distribution.

The very fine structure of the model we propose allows the accurate study of many new questions, as well as the standard ones. Let us list a few of them. Of course, we want first to know the growth rate of the total number of infected people as well as the number of infected persons inside the different groups. We are also interested in the relevance and the sensitivity of the different sociological and medical input data on the dynamics of the epidemic. As emphasized a mathematical model suggests what kinds of epidemiological data are needed to make predictions of future trends. Furthermore we want to study clustering properties and the existence of almost isolated islands on the underlying graphs, which reflect the very large inhomogeneity in the sexual contact structure of the population.

Cycles on the random graphs which correspond in our model to multiple infections (a fact which is particularly relevant inside the high risk groups) should be studied because they may be important for the average value of the incubation time. The absence of cycles at the very beginning of an epidemic is reflected in the exponential – like growth of the number of HIV-carriers.

The incidence of the existence of different types of HIV virus can also be handled in our model. Furthermore the model can be easily adapted to take into account inhomogeneous local structure (hospitals, prisons, ...). By a proper (geographical) embedding of the graph in the plane we can use our model to study spatial distributions of the epidemic respectively the propagation front of HIV-infection.

Using our model the influence of the distribution of the incubation time on the growth rate of HIV infection and AIDS epidemic can be studied. The model can help to assess the efficacy of different preventive measures since it allows to estimate the influence of global changes in sexual behaviour, like reduction in number of partners and practicing of "safer sex"; moreover it allows, e.g., to trace infected individuals as well as their past and future contact partners in order to study the effects of individual recommendations in prevention strategies.

Finally, the model should serve as a gauge for other, mathematically much simpler models, and can be used to test the validity of certain approximations made in other models. Of course, every kind of numerical prediction depends on the quality and the accuracy of the needed epidemiological (social and medical) parameters. Accurate informations are up to now often missing. But nevertheless our model is also able to give lower and upper bounds for all calculated quantities and to help to identify the most sensitive parameters.

II. Random Graph Epidemics
II.a The Underlying Graph Structure

As already mentioned, the basic probability space we consider in our model is a space of random graphs (for the definition of the simplest relevant graph theoretical concepts we refer to [32]). All graphs we consider are labelled graphs. Let $V = \{x_1, x_2, ..., x_N\}$ be the set of vertices (individuals) which splits into several relevant groups V_i

$$V = \bigcup_{i \in I} V_i .$$

Since edges between vertices describe a sexual contact, the basic given data generating the underlying graphs are the degree distributions inside the V_i as well as the couplings between pairs (V_i, V_j) of subgroups of the vertices. Let us therefore introduce the following functions

$$n^0 : \mathbb{N} \rightarrow \mathbb{N} , \quad n^0(z) = \# \{x_i \in V | d(x_i) = z\}$$

$$n_i^0 : \mathbb{N} \rightarrow \mathbb{N} , \quad n_i^0(z) = \# \{x_j \in V_i | d(x_j) = z\}$$

$$n_{ij}^0 : \mathbb{N} \rightarrow \mathbb{N} \times \mathbb{N}, \quad n_{ij}^0(z) = (\#\{x_k \in V_i , d_j(x_k) = z\} ; \#\{x_\ell \in V_j , d_i(x_\ell) = z\}).$$

We denote by $\#\{ \ldots \}$ the number of elements and by $d(x)$ the vertex degree of x. Introducing the adjacency matrix

$$\phi(x_n , x_m) = a_{nm}$$

of the graph with $a_{nm} = 1$ or 0, according to whether (x_i , x_j) is or is not an edge of G, we define

$$d_j(x_n) = \sum_{x_m \in V_j} \phi(x_n , x_m) .$$

All graphs G which satisfy the above distribution of contact rates n_i^0, n_{ij}^0, i,j $\in$ I, are the elements of the random graph space $G_{n_{ij}^0}^{n_i^0}$ which becomes a probability space by giving each element $G \in G_{n_{ij}^0}^{n_i^0}$ the same probability

$$P(G) = \frac{1}{|G_{n_{ij}^0}^{n_i^0}|} ,$$

$| \ldots |$ denoting the total number of elements.

Although for sexually transmitted diseases the degree distribution n_i^0 can often be approximately estimated (at least n^0 for the whole ensemble V of individuals; the coupling distribution functions n_{ij}^0 between the different subgroups are not so well known). For this reason it is useful to look at a larger class of random graphs, for which only the number of edges connecting different groups and the number of vertices having edges to different groups are known.

Let

$$m_{ij} = \tfrac{1}{2} \sum_{x_{i'} \in V_i, x_{j'} \in V_j} \phi(x_{i'}, x_{j'})$$

and

$$V_i \wedge V_j = (V_i \cap V_j) \cup \{x \in V_i; K(x,1) \cap V_j \backslash V_i \neq \varnothing\} \cup$$

$$\cup \{x \in V_j; K(x,1) \cap V_i \backslash V_j \neq \varnothing\}$$

$$V_{ij} = |V_i \wedge V_j|$$

with $K(x,1) = \{y \in V; \phi(x,y) = 1\} \cup \{x\}$.

For most cases the intersection $V_i \cap V_j$ will be empty.

The graph spaces we consider now are

$$G^{n^0}_{m_{ij}, v_{ij}}, \quad G^{n^0_i}_{m_{ij}, v_{ij}}, \quad G^{n^0_i}_{m_{ij}}, \quad G^{n^0_i}_{v_{ij}}$$

with the probability space structure as above.

Our first aim is now to give some conditions on n^0_i, n^0_{ij} ensuring that $G^{n^0_i}_{n^0_{ij}}$ is not empty. For simplicity we will state the condition for one group of graphs G^{n^0} but the result we find can easily be generalized. Observe that for $n^0(z)$ even with

$$n^0(z) \, z \leq n^0(z) \, [n^0(z) - 1] \qquad (*)$$

a graph $G^{n^0} \in G^{n^0}$ with the following structure exists:

G^{n^0} splits into disconnected components $G_i^{n^0}$ such that for all $x \in G_i^{n^0}$ $d(x) = i$. Of course from this graph G^{n^0} the whole space G^{n^0} can be generated by transforming the edge pairs $(x_i, x_j)(x_m, x_n)$ into the edge pairs $(x_i, x_m)(x_j, x_n)$ or $(x_i, x_n)(x_j, x_m)$. Assume now that the condition $(*)$ is not satisfied for some G. It follows that in $G_{i_0}^{n^0}$ there are $q^-_{i_0}$ multiple edges which must be removed by the above pair transformation.

Let $q_{i_n} = \tfrac{1}{2} n(i_n) = q^+_{i_n} + q^-_{i_n}$

$$q^+_{i_n} = \begin{cases} \tfrac{1}{2} n(i_n) \, i_n & \text{for } q^-_{i_n} = 0 \\ \tfrac{1}{2} n(i_n)(n(i_n) - 1) & \text{for } q_{i_n} > 0 . \end{cases}$$

Therefore the following condition is sufficient to ensure that G^{n^0} is not empty:

$$\sum_{n=0}^{m} q_{i}^{-} \leq \sum_{n \supset 0}^{m} q_{i_n} \Theta (1 - q_{i_n}^{-})$$

with $\Theta(x) = \begin{cases} 0 & x \leq 0 \\ 1 & x > 0 \end{cases}$.

In principle the $q_{i_n}^{+}$ edges in $G_{i_n}^{n^0}$ can be also used to remove $q_{i_m}^{-}(i_m \neq i_n)$ edges obtaining in this way the following necessary and sufficient condition

$$\sum_{n=0}^{m} q_{i_n}^{-} \leq \sum_{n \supset 0}^{m} q_{i_n} \theta(1 - q_{i_n}^{-})$$

$$- \min_{\pi \in S_n} [\sum_{n=0}^{m-1} \mid q_{i_{\pi(n+1)}}^{+} - q_{i_{\pi(n)}}^{-} \mid + \mid q_{i_{\pi(0)}}^{+} - q_{i_{\pi(m)}}^{-} \mid] .$$

A more difficult task is the estimation of the number of elements $|G_{n_{i_j}^0}^{n^0}|$ of $G_{n_{i_j}^0}^{n^0}$ since already

for regular graphs the size of the automorphism group is in most of the cases not explicitly known. For asymptotic estimates $|V| \rightarrow + \infty$ in a similar random graph space we refer to [33].

In view of the applications we shall now consider a further class of random graphs, which take into account the fact that in practice the distribution functions n_i^0, n_{ij}^0 are only approximately known and that furthermore in a computer realization of an element of a random graph space it is much easier to get a graph with realized degree distribution $n(G) \sim n_0$ but not exactly $n(G) = n_0$ for n_0 given.

Let us therefore introduce a distance between two degree distributions $n^0(z)$, $n^1(z)$ by

$$\rho_1(n^0, n^1) = \max_{z \in \mathbb{N}} |n^0(z) - n^1(z)|$$

and an edge distance ρ_2 by

$$\rho_2(n^0, n^1) = | \sum_{z \in \mathbb{N}} z(n^0(z) - n^1(z)) | .$$

Let

$$K(n^0, \varepsilon, \mu) = \{n(z) \mid \rho_1(n^0, n) \leq \varepsilon, \rho_2(n^0, n) \leq \mu\}$$

and let

$$\sum_{z} n(z) = N, \text{ N being fixed.}$$

In this way we get a space $G_{\varepsilon\mu}^{n_0}$ consisting of all graphs G such that $n(G) \in K(n^0, \varepsilon, \mu)$.

Analogous definitions hold mutatis mutandis for all other spaces of random graphs we introduced up to now. The analytical difficulties by working with the above defined spaces arise from the fact that edges cannot be chosen independently. For this reason it would be necessary to approximate G^{n_0} by similar spaces $G(p_{ij})$ having independent edge probabilities. (See Section III.d.)

II.b Epidemic Dynamics on Random Graphs

We are now in position to describe the dynamics on G. On each graph with vertex set $V(G) = \{x_i\}_{i \in I}$ we introduce a sequence $\{\varphi_n\}_{n \in \mathbb{N}}$, $\varphi_n : V(G) \rightarrow V(G)$ satisfying the following properties:

    i) $\varphi_n \circ \varphi_n = $ Identity

    ii) For each $x_i \in V(G)$ there exists $n_1(x_i)$ and $n_2(x_i)$ with $n_i \in \mathbb{N}$, $i = 1$, 2, such that $\varphi_n(x_i) = x_i$ for all $n \notin [n_1,n_2]$ and $|n_1 - n_2| < C$.

It follows from ii) that for each fixed $x_i \in V(G)$ that

$$|\text{supp } (n \rightarrow \# \{\{x_i\} \cup \{\varphi_n(x_i)\}\}\backslash\{x_i\})| \leq C$$

where $\#(A)$ denotes the number of elements in the set A and $\text{supp}(n \rightarrow f_n)$ denotes the support of the function $n \rightarrow f_n$. The pair $(G, \{\varphi_n\}_{n \in \mathbb{N}})$ is called a graph with time ordered edges.

Let us introduce the notation:

$$\text{Mat}_n : V(G) \times V(G) \rightarrow \{0,1\} \text{ with}$$

$$\text{Mat}_n(x,y) = \begin{cases} 0 & \text{if } \varphi_n(x) \neq y \\ 1 & \text{if } \varphi_n(x) = y \end{cases} \quad \text{for } x \neq y$$

and $\text{Mat}_n(x,x) = 0$ for all $x \in V(G)$.

Let us remark that, roughly spoken, the φ_n tell us at time n the elements of G which are potentially able to propagate an "infection" by having a sexual contact. Indeed $\varphi_n(x)$ represents if $\varphi_n(x) \neq x$ the sexual partner of x at time n.

To describe the possible stages of the evolution of a disease for each individual $x \in V(G)$ let us introduce at each time $n \in \mathbb{N}$ a mapping

$$X_n : V(G) \rightarrow \{0, 1, 2, ..., N\}.$$

To account for the spread of the HIV-infection and of the AIDS-epidemic we consider the following progression

$$X_n(x) = 0 \Leftrightarrow x \quad \text{is healthy at time n}$$
$$X_n(x) = 1 \Leftrightarrow x \quad \text{is HIV-infected at time n and not infectious}$$
$$X_n(x) = 2 \Leftrightarrow x \quad \text{is HIV-infected at time n and infectious}$$
$$X_n(x) = 3 \Leftrightarrow x \quad \text{knows about the infection at time n}$$
$$X_n(x) = 4 \Leftrightarrow x \quad \text{reachs stage ARC ata time n}$$
$$X_n(x) = 5 \Leftrightarrow x \quad \text{develops AIDS at time n}$$
$$X_n(x) = 6 \Leftrightarrow x \quad \text{is dead at time n.}$$

Using the map $X_n : V(G) \rightarrow \{0, ..., 6\}$ it is now easy to describe the process modeling the spread of the disease on a given graph. Let $Z^n \subset V(G)$ be the set of HIV-infectious individuals at time n

$$Z^n = \{x \in V(G) \mid X_n(x) \neq 0, 1, 6\}.$$

For each $x \in V(G)$ we define the conditional probabilities $P_x^n(k \rightarrow l)$, $k, l = 0, ..., 5$, expressing the progression of the disease. For example we have

$$P_x^n(0 \rightarrow 1) = P[X_{n+1}(x) = 1 \mid X_n(x) = 0]$$

$$= \gamma \, \text{Mat}(x, Z^n)$$

with $\quad \text{Mat}(x, Z^n) = \sum_{y \in Z^n} \text{Mat}(x, y) \, .$

In the above formula $\gamma \in (0,1]$ is the transmission probability per time step of the iteration. In the general case γ can be time dependent, reflecting for instance the conjectured fact that the infectivity is much lower in the middle stage of the HIV-infection. Moreover γ can depend on x and $\varphi_n(x)$.

By introducing the distributions of the latency period for seroconversion, of the incubation period and of the life expectancy after an individual x develops AIDS it is easy to define in the same way the $P_x^n(k \rightarrow l)$.

Remark 1 Since in real life the exact form of the sequences $\{\varphi_n\}_{n \in \mathbb{N}}$ are not known we consider whole spaces of time ordered random graphs specified by some conditions on the $\{\varphi_n\}_{n \in \mathbb{N}}$ and G. Usually we use conditions which are expressed in terms of the distribution of $\#K(x)$, $x \in G$, and the distribution of the sequence $l(x) = \{n_j(x)\}_{j \in \mathbb{N}}$ which is defined by $n_j(x) \in l$ if and only if $\varphi_{n_j(x)}(x) \neq \varphi_{n_{j(x)-1}}(x)$ as explained in Section II.a. For more details see [34].

Remark 2 We refer to [25] for another possible description of the epidemic dynamics on random graphs.

III. Some limiting cases and how to get a general theory

In this section we will study some simplified cases of random graphs to show the basic differences between the random graph description and the classical approach using differential equations, which arises as a special case of our model if we restrict ourself to complete graphs.

III.a Complete Bipartite Graph

To be more explicit let us show how the well-known Lotka-Volterra equation in its discretized version comes up in our model. Let $K_{N_M N_F}$ be the complete bipartite graph with N_M males and N_F females. Let n count the discrete time steps after which an edge contact between two individuals is destroyed and a new one is created (in mean). Let I_n^M and I_n^F be the number of infected males and females at time n. Let $N_M \leq N_F$. At each time we assume a unique and complete pairing for N_M into a subset of N_F of size N_M. We conclude that

$$E[I_{n+1}^M] = E[I_n^M] + p_1 \, E[^*I_n^F]$$

$$E[I_{n+1}^F] = E[I_n^F] + p_2 \, E[^*I_n^M]$$

p_1 and p_2 being the transmission probabilities per contact (female $\rightarrow$ male and male $\rightarrow$ female respectively) and $E[^*I_n^{F,M}]$ denoting the expected number of infected females or males having contact with non-infected males or females respectively. Since $E[I_n^M \, I_n^F] = E[I_n^F] \, E[I_n^F] - \text{Cov}(I_n^M, I_n^F)$ we have (remember the asymmetry $N_F \geq N_M$)

$$E[^*I_n^{F,M}] = N_M \, \frac{E[I_n^{F,M}]}{N_{F,M}} \, \frac{N_{M,F} - E[I_n^{M,F}]}{N_{M,F}} + \alpha \quad \text{with} \quad |\alpha| \leq \frac{1}{4} \; .$$

Setting $X_n = \dfrac{E[I_n^M]}{N_M}$, $Y_n = \dfrac{E[I_n^F]}{N_F}$ we get by neglecting now α

$$X_{n+1} = X_n + p_1 \quad Y_n(1 - X_n)$$

$$Y_{n+1} = Y_n + p_2 \, \frac{N_M}{N_F} \, X_n(1 - Y_n) \; .$$

In the same spirit we obtain arbitrary generalized Lotka-Volterra systems by considering n-partite complete graphs for which in each group the sexual behaviour is uniform, taking into account also birth and death rates. The n-partite complete graph expresses in our model a division of the population into n homogeneous such groups.

III.b A basic combinatorial lemma

Let us consider a partition of V of the form

$$V = X \cup Z$$

with $\#(V) \equiv |V|$ even. In this section we want to characterize the set of all possible complete matchings on V. For each given realization of such a matching we denote by K(X,Z) the number of edges between elements of X and Z and by K(X), resp. K(Z), the number of edges between elements of X, resp. Z. We have obviously

$$Z = 2\,K(Z) + K(X,Z)$$
$$X = 2\,K(X) + K(X,Z)$$

with $x = \#(X) = |X|$, $z = \#(Z) = |Z|$ and $v = \#(V) = |V|$. We want to know the probabilities $P[K(X,Z) = m]$, $m \in \mathbb{N}$, but to achieve this goal it is sufficient to calculate the following quantities

$$P_n = P[K(Z) = n]\,, \quad n \le \min\,[\tfrac{z}{2}].$$

The total number of matchings in V is given by

$$A(V) = \frac{v\,!}{2^{\frac{v}{2}}(\frac{v}{2})!}\,.$$

Let Q_n denote the possible number of matchings with exactly n edges inside Z.

A careful combinatorial counting leads to

$$Q_n = \binom{z}{2n}\,2n!\,\binom{x}{z-2n}\,(z-2n)!\,\frac{(x-z+2n)!}{2^{2n}n!\,2^{\frac{x-z}{2}}\left(\frac{x-z}{2}+n\right)!}\,.$$

We have of course $\sum\limits_{n} Q_n = A(V)$.

Using this we obtain

$$P_n = \frac{Q_n}{A(Z)}$$

and for the expectation of K(Z)

$$E[K(Z)] = \frac{\sum\limits_{n} n\,Q_n}{A(Z)}\,.$$

A short calculation shows that

$$\sum n \, Q_n = \tfrac{1}{2} \, z(z-1) \frac{(x + z - 2)!}{2^{\frac{x+z-2}{2}} \left(\frac{x+z-2}{2}\right)!}$$

which implies the following

Lemma Let $V = X \cup Z$. $|V|$ then the expected number of edges between elements of Z after a complete matching of V is given by

$$E[K(Z)] = \tfrac{1}{2} \frac{|Z| \, (|Z| - 1)}{|V| - 1}$$

Remark 1 Let us remark that $E[K(Z)]$ can also be written in the following form

$$E[K(Z)] = \frac{|V|}{2} \frac{|Z| \, (|Z| - 1)}{|V| \, (|V| - 1)}$$

where $\frac{|V|}{2}$ is exactly the number of edges in V after a complete matching
and $\frac{|Z| \, (|Z| - 1)}{|V| \, (|Z| - 1)}$ is nothing else than the probability that a randomly chosen edge links up elements of Z.

Remark 2 If $\#(V) = |V|$ is odd it can be shown that in this case $E[K(Z)]$ is given by

$$E[K(Z)] = \tfrac{1}{2} \frac{|Z| \, (|Z| - 1)}{|V|} \ .$$

In the applications we have in mind, Z can be viewed as the subset of infectious individuals. Let us denote by I_n the set of individuals which are first infected at time n. If the graph has the structure of a tree and this is almost the case for small time intervals we have

$$|I_n| = |I_{n-1}| + |I_{n-2}| + \dots \, .$$

Thus the prevalence of the infection increases exponentially in the early stage of the epidemic.
But cycles can appear. In other words multiple infections can happen along the epidemic cascade (edges between $I_{n-1} \cup I_{n-2} \cup \dots$) and this effect reduces the spread of the HIV-infection.

III.c The Fibonacci Approximation

Let us now describe in a very informal way the spread of the epidemic for small times (i.e. at initial stages of the disease). In other words we neglect the effects due to selfintersections of walks on the graphs (multiple infections are therefore not considered). Moreover we will make simplifying assumptions. The transmission probability per contact will be taken to be one. For the contact rate instead of $d(X_i)$ we will work with $d^+(X_i)$, the expected number of sexual partners in the future. Let D be the persistency of a contact, i.e. the mean time between changes of sexual partners. We take D to be the characteristic iteration time step. Let $\tau^+(X_i)$ be the expected time spent by X_i with its $d^+(X_i)$ sexual partners and Δ_I be the incubation time. By I_n we denote the number of newly infected individuals during the time interval $[(n-1)D,nD]$. If we start with only one infected individual X_1 at time $t=0$ the I_n are given by a generalized Fibonacci sequence, namely

$$I_{n+1} = I_n + I_{n-1} + \cdots + I_{n-\bar{n}+1}$$

where $\bar{n} = [D \frac{\Delta_I}{\tau^+}]$, [...] denotes the integer part, and

$$I_0 = 1, \; I_{-1} = 0 , \quad \cdots \; I_{-\bar{n}+1} = 0 .$$

The general expression for I_n is known [35] and given by

$$I_n = \sum_{j=0}^{\bar{n}-2} \frac{\alpha_j(\alpha_j)^n\left(1-\frac{1}{\alpha_j}\right)}{2 - (\bar{n}-1)\left(\frac{1}{\alpha_j}\right)^{\bar{n}-1}}$$

the $\frac{1}{\alpha_j}$ being the zeros of $1 - \sum_{j=1}^{\bar{n}-1} x^j = 0$.

This formula is called Binet's formula for k-Fibonacci sequences. By introducing selfintersections (i.e. multiple infections) on the random graph we get a nonlinear correction to the Fibonacci sequences, which makes the dynamics of the epidemic already in this simplified approach quite interesting.

III.d Models generated by independent matchings

We will now discuss briefly the simplest model of a class of models generated by independent matchings, which is appropriate if one is thinking about the case of a homosexual community with p age groups. In the notations we introduce in Section II.b to discuss the general epidemic process we allow all graphs with time ordered edges for which the following conditions hold for each $x \in G$.

$$\# \{supp(n \to \# \Pi_n(x))\} = p$$

with

$$\Pi_n(x) = \begin{cases} \bigcup_{m>n} \varphi_m(x) & \text{if} \quad \varphi_n(x) \neq x \\ \varnothing & \text{if} \quad \varphi_n(x) = x \end{cases}$$

and

$$|\text{supp } (n \rightarrow \# \; \Pi_n(x))| \equiv \max \; \{\text{supp } (n \rightarrow \# \; \Pi_n(x))\} - \min \; \{\text{supp } (n \rightarrow \# \; \Pi_n(x))\}$$
$$= p - 1 \; .$$

Note that $\text{supp}(n \rightarrow \# \; \Pi_n(x)) = \{n \in \mathbb{N} \mid \Pi_n(x) \neq \varnothing\}$.

The model characterized in this way has the following interpretation: The population G is divided in p age-groups G_i and at each iteration step n = 1, 2, ... we allow perfect independent and uniform matchings between the individuals. Furthermore each iteration step is connected with a shift of the elements of G_i into G_{i+1} and with the choice of a sexual partner. The number $\#(G_i)$ of individuals in each age subgroup G_i is taken constant and equal to N^*, from which it follows that the total size $\#(G)$ of the sexually active population is p N^*. We suppose that N^* is even and a perfect independent uniform matching is a set of $\frac{pN^*}{2}$ independent edges.

For this model we can consider an approximation relating the expected values of the number of infected individuals at time m and m+1. This yields to a discrete dynamical system. Since N is big it can be shown that this deterministic system for the iterated expectations leads to a good approximation. In [34] we look for stationary solutions and prove the existence of globally attractive stationary solutions depending on the value of a critical parameter R which is nearly equal to $\gamma \; \frac{p-1}{2}$, γ being the transmission probability per iteration step. It can be shown that in this case R can be interpreted as being the reproductive number considered in classical epidemiological models for sexually transmitted diseases. A more detailed exposition of the topics sketched in this section may be found in [34], where also additional material is presented. In particular the extension to bipartite graphes (heterosexual population) is exhibited.

III.e Clustering of Random Graphs and Suggestions for Further Work

As already mentioned in Section I one of the most relevant properties of random graphs is the splitting into disconnected components (resp. isolated islands). But to study the spread of epidemics the standard notion of disconnectedness of a graph is much stronger than needed. It would be enough to get connected clusters of large size which are interconnected only by a few edges. For this reason we set now up several cluster definitions which are the object of study in a subsequent paper.

We pick up an arbitrary $x_i \in V(G)$ and look for subsets $Cl(x_i,k,n) \subset V(G)$ such that

$$x_i \in Cl(x_i,k,n) \, , \quad |Cl(x_i,k,n)| \, = \, n \, , \quad \frac{1}{2} \sum_{\substack{x \in Cl(x_i,k,n) \\ y \in V(G) \backslash Cl(x_i,k,n)}} \Phi(x,y) \, = \, k,$$

and the corresponding subgraph $G(Cl(x_i,k,n)) \subset G(V)$ with vertices from $Cl(x_i,k,n)$ and edges connecting only these vertices has to be connected.

The set $Cl_{x_i}^{k,n} \, = \, \{Cl(x_i,k,n)\}$ is called the space of x_i-pointed k-n-Clusters. For n given an important subclass is the class of minimal clusters such that there does not exist a $Cl(x_i,n,k)$ with $k < k_{min}$. In this way we get the space $Cl_{x_i}^{k_{min},n} \, = \, \{Cl(x_i,k_{min},n)\}$. The distribution functions $g_1(x_i,n) \, = \, k_{min}$ for x_i, n given and $g_2(x_i,n) \, = \, \#Cl_{x_i}^{k_{min},n}$ describe now the inhomogeneous structure of the random graph with respect to the individual x_i.

To get now unpointed clusters we call individuals x_i and x_j k-n-cluster-equivalent if $Cl(x_i,k,n) \, = \, Cl(x_j,k,n)$ and obtain the space $Cl^{k,n}$. Now for n given let $\tilde{k}_{min} \, = \, \min_{x_i \in V(G)} \, g_1(x_i,n) \, = \, \tilde{g}_1(n)$ and $\tilde{g}_2(n) \, = \, \#Cl^{k_{min},n}$.

The concept of k-n-clusters is a new one and has not been studied at all in standard graph theory. The similar concept of k-edge-connectivity for graphs is not very useful for our purpose because it tells nothing about the size of the corresponding clusters. Although the analytical study of the k-n-cluster distribution in some space of random graphs seems to be difficult, numerical algorithms can be developed to give good approximations to the cluster distribution. Finally, having this, the computation of an epidemics can be almost restricted to clusters with large enough size n and small k.

Let us now briefly give some outlook for further research which shall be the content of subsequent papers devoted to the theoretical study of our model. From the computational point of view there arise two main questions. First the development of still more efficient algorithms for the creation of a typical element of the relevant random graph space.

Analytically, the approximation of our graph spaces $G(p_{ij})$ with independent edge probabilities $p_{ij} \, = \, P(\Phi(X_i,X_j) \, = \, 1)$ would be useful. For instance at a first level the p_{ij} can be chosen in the following way. Assume we would know $d(X_i) \, = \, d_1$, $d(x_j) \, = \, dz$, the probability p_{ij} is now

$$p_{ij} \, = \, \binom{N-2}{d_1-1}\binom{N-2}{d_2-2} \Big/ \left\{ \binom{N-2}{d_1}\binom{N-2}{d_2} + \binom{N-2}{d_1-1}\binom{N-2}{d_2-2} \right\} \, = \, \frac{d_1 \, d_2}{d_1 \, d_2 + (N-1-d_1)(N-1-d_2)}$$

provided $d(X_m) > 1$ for all $X_m \in V(G)$.

Given now a graph space G^{n_0} we can generate a space $G(p_{mn})$ such that

$$\# \{ p_{mn}, \; p_{mn} = \frac{i \cdot j}{i \cdot j + (N-1-i)(N-1-j)} \} = \begin{cases} n(i) \; n(j) & \text{for } i \neq j \\ \binom{n(i)}{2} & \text{for } i = j \end{cases}$$

which approximates G^{n_0} in the sense that the degree distribution for $G(p_{mn})$ is almost the same as for G^{n_0}.

The aim of further investigation is to derive several theorems on how well properties of the first graph space can be described by properties of the second one. At least for convex properties this approximation should be quite good. (For the case $p_{ij} = p \, \forall i,j$ and regular graphs G^r see [30]).

Furthermore it would be useful to take into account the different age groups by introducing metric dependent $\tilde{p}_{mn}$, where the age distance between males x_i^M and females x_j^F inside a partnership is reflected in a metric

$$\rho(x_i^M, x_j^F) = |\tilde{\tau}(x_i^M) - \tau(x_j^F)| \; ;$$

here $\tilde{\tau}(x_i^M) = \tau(x_i^M) + C$ is a shifted male-age according to the fact, that men contact most probably younger woman. Then $\tilde{p}_{mn} \sim p_{mn} \, e^{-\rho^2(x_m^M, x_n^F)}$ with p_{mn} representing the degree distribution as above. The resulting graph spaces are a mixture between spaces $G(p_{ij})$ and graphs frequently studied in multidimensional percolation models, for which analytical tools similar to Stepanov (see [39]) as well as to percolation methods (see e.g. [40]) shall be developed.

Finally we mention the problem of coloring graphs which may be one way to describe analytically the influence of partnership lenght distribution on the epidemic dynamics.

IV. Computer Simulation

In this last section let us take a look at our modelling of HIV-infection using random graphs from a simulational point of view. First of all, let us again emphasize the flexibility of our model, which can be suited to a variety of qualitatively different epidemiological assumptions. For more details see [25] and [27]. The underlying configuration which represents the individuals of the model population, divided into subgroups according to the population risk structure, is a realization of the random graphs defined by the sociological parameters. The discrete dynamics is a realization of the stochastic process governed by the sociological and medical parameters (eventual realization of one of the potential sexual contacts for each individual, eventual transmission of HIV-infection during this contact). The natural time step for the dynamics is 1 day, while the time scale for measuring quantities characterizing the state of the model population is of the order of several months, and the dynamics runs for several years. Note that, here, the edges represent the different partners during the same period of time while, in each time step, the vertex eventually selects one of its edges for an actual contact.

We have started to study a simple model which is modest in its demand of memory and computation time, yet complex enough to map main features of the real development. This simple model includes the following 7 groups: homosexual men, bisexual men, heterosexual men, heterosexual women, male intravenous drug users, female intravenous drug users, heterosexual women having contact to bisexual men. It comprises 20000 individuals having 5 partners at most. Number of partners, frequency of contacts and probability of infection per contact are specific for a group but homogeneous over this group. Changes in behaviour and natural vitality dynamics are not yet be included. The results presented in [27], besides demonstrating that this type of computer simulation is practibable also for much larger model populations, show that the graph structure does play an important role in the dynamics of the epidemic. In comparison with what is expected from differential equation models the most remarkable features of simulation results obtained by our model are: transport from the highly promiscuous part of the homosexual population into the heterosexual population is slowed down by the limited transmission capacity of those groups acting as mediators; inside the majority of the heterosexual population, the epidemics grows linearly rather than exponentially due to low partner number and low rate of partner change. A more detailed exposition of the topics sketched in this section can be found in [27], where also additional material is presented.

References

[1] Y. Suhov "Statistical Physics and Queuing Network", Proceedings IAMP Conference Swansea 1988
[2] S. Wassermann "Analyzing social networks as stochastic processes" JASA 75 280 (1980)
[3] R. Durret "Crabgrass, Measles and Gypsy Moths: An introduction to modern probability" Bulletin AMS Vol. 18 117-143 (1988)
[4] K. Dietz "The first epidemic model: A historical note on P.D. En'ko" Austrl. J. Statist. 30A (1988) 56-65
[5] R. Ross "The prevention of malaria" Murray 1911
[6] N.T.J. Bailey "The mathematical theory of infectious diseases and its application", Charles Griffin 1975
[7] J.C. Frauenthal "Mathematical Modelling in Epidemiology" Springer 1980
[8] N.T.J. Bailey "Introduction to the Modelling of Veneral Disease" J. Math. Biology 8, 301-322 (1979)
[9] M.G. Koch "AIDS - Vom Molekül zur Pandemie" Spektrum der Wissenschaft Verlag, Heidelberg (1987)
[10] J. Weyer "Ein Mehrgruppenmodell zur Simulation der epidemischen Dynamik von AIDS" AIFO 3, 154-156 (1988)
[11] H.W. Hethcote, J.A. Yorke, Lect. Notes Biomathematics 56 1-105 Springer (1985)

[12] K. Dietz "On the transmission dynamics of HIV", Mathematical Biosciences 90 397-414 (1988)

[13] K.P. Hadeler, K. Dietz "Epidemiological models for sexually transmitted diseases" J. Math. Biol. (1988) 26 1-25

[14] J.M. Hyman, E.A. Stanley "Using mathematical models to understand the AIDS-epidemic" Mathematical Biosciences 90, 415-473 (1988)

[15] J.U. Niehoff, N. Sönnichsen "AIDS-Eigenschaften einer echten Epidemie" 2. Klin. Med. 42 Heft 24, 2141-2145

[16] M.G. Koch, J. L'âge-Stehr, J.J. Gonzalez, D. Dörner "Die Epidemiologie von AIDS" Spektrum der Wissenschaft, August 1987 38-51

[17] D. Dörner a) "Ein Simulationsprogramm für die Ausbreitung von AIDS" Projekt Systemdenken DFG 200/5/III/86
b) Addendum zum a) Universität Bamberg

[18] M.G. Koch, U.v. Welck "AIDS-Spread Simulations and Projections" Projektstudie März 1987

[19] S.P. Blythe, C.C. Castillo-Chavez "Like-with-like preference and sexual mixing models" Preprint 1988

[20] S.A. Colgate, E.A. Stanley, J.M. Hyman, S.P. Layne, C. Qualls "A behavior based model of the initial growth of AIDS in the United States" Los Alamos Preprint LA-UR-88-2396

[21] J.M. Hyman, E.A. Stanley "The effect of social mixing patterns on the spread of AIDS" Los Alamos Preprint 1988

[22] L. Sattenspiel "Population structure and the spread of disease" Human Biology 59 (1987)

[23] C. Castillo-Chavez, K. Cooke, W. Huang, S.A. Levin "The role of long incubation period in the dynamics of acquired immunodeficiency syndrome (AIDS) Part 1: Single population models", Preprint 1988

[24] A. Flahault, A.J. Valleron "The role of air transport in the global spread of HIV infection" Preprint INSERM Paris VII (1988)

[25] Ph. Blanchard, G.F. Bolz, T. Krüger "Simulation on random graphs of the epidemic dynamics of sexually transmitted diseases" BiBoS Preprint 291/1987

[26] Ph. Blanchard "A stochastic growth model on Random graphs to understand the Dynamics of AIDS-epidemic", in "Stochastic Methods in Mathematics and Physics", Proceedings of the XXIV Karpacz Winter School of Theoretical Physics, Ed. R. Gielerak, W. Karwowski, World Scientific (1989)

[27] G.F. Bolz "Simulation on random graphs of the epidemic dynamics of sexually transmitted diseases - A new model for the epidemiology of AIDS", in "Stochastics Algebra and Analysis in Classical and Quantum Dynamics", Eds. S. Albeverio, Ph. Blanchard, D. Testard, Mathematics and Its Applications, Reidel (1989)

[28] I. B. Gertbakh "Epidemic Process on a Random Graph: Some preliminary results" J. Appl. Prob. 14 427 (1977)

[29] P. Tautu "The evolution of epidemics as random graphs"

[30] A. Bollobás "Random Graphs" Academic Presss 1985

[31] R.M. May and R.M. Anderson "Transmission Dynamics of HIV Infection" Nature 326 137-142 (1987)

[32] E. Palmer "Graphical Evolution", Wiley, 1985

[33] A. Bender, E.R. Canfield "The asymptotic number of labelled graphs with given degree sequences" J. Combinatorial Theory (A) 24 1978 296-307

[34] Ph. Blanchard, T. Krüger "Spread of AIDS-Epidemics: A discrete model on graphs I. Stationary analysis for graphs generated by independent matchings" BiBoS Preprint 337/1988

[35] W.R. Spickermann, R.N. Joyner "Binet's formula for the recursive sequence of order k" The Fibonacci Quarterly (Nov. 1984) 327-331

[36] C. Castillo-Chavez "Effects of social mixing by propagation of the HIV.infection" Oberwolfach Meeting on "Mathematical Models for Infectious Diseases", Feb. 1989

[37] J.A. Jacques, C. Simon "The Effect of Contact Pattern in the Spread of AIDS", Oberwolfach Meeting on "Mathematicl Models for Infectious Diseases", Feb. 1989

[38] K.P. Hadeler "Heterogeneous HIV-model", Oberwolfach Meeting on "Mathematical Models for Infectious Diseases", Feb. 1989

[39] V.E. Stepanov "Combinatorial algebra and random graphs", Theory Probab. Applic. $\underline{14}$ 373-399 (1969)

[40] C.M. Newman, L.S. Schulman "One dimensional $\frac{1}{|j-i|^s}$ percolation models: The existence of a transition for $\Delta \leq 2$", Commun. Math. Phys. $\underline{104}$ 547 (1986)

SUSTAINED CHEMICAL DISSIPATIVE STRUCTURES
SOME RECENT DEVELOPMENTS

J. Boissonade
Centre de recherche Paul Pascal, Univ. Bordeaux I, 33405, Talence, France

The results reviewed in this lecture have been obtained by members of the group "Systèmes dissipatifs non linéaires" of the CRPP, Bordeaux (A.Arneodo, J. Boissonade, P. De Kepper, E. Dulos, J. Elezgaray, Q. Ouyang, J.C. Roux) and members of the "Center for nonlinear dynamics", Austin (W. Horsthemke, Z. Noszticzius, J. Pearson, H. Swinney, W. Tam, J. Vastano) as a part of a joint research project supported by the B.P. Venture Research Unit.

1   Introduction

Among the fields of physics providing illustration and support to the theoretical body of nonlinear dissipative systems, the temporal and spatial behaviour of chemical systems kept far from equilibrium by a permanent supply of fresh reactants has received special attention [1-3]. In a well mixed (homogeneous) reacting solution, the concentrations c_i of the different input or intermediate reactants follow kinetic equations of the form

$$\frac{dc_i}{dt} = f_i(\ldots, c_j, \ldots) \tag{1}$$

where the f_i are in general nonlinear functions. For instance a regular elementary bimolecular step of the form $A + B \xrightarrow{k} D$ provides the following contributions to (1)

$$\frac{dc_A}{dt} = \ldots - kc_A c_B + \ldots$$
$$\frac{dc_B}{dt} = \ldots - kc_A c_B + \ldots \tag{2}$$
$$\frac{dc_D}{dt} = \ldots + kc_A c_B + \ldots$$

where k is a kinetic constant. In principle, k depends on the temperature but in the following we shall only consider isothermal systems where the temperature is only a control parameter. Usually, even when one deals with exotic reactions, the system can be properly described by a small number of species; all the other are taken into account through appropriate kinetic constants, stoichiometric factors or specific kinetic laws. Moreover the nonlinear terms remain simple (in general polynoms or rational fractions of low degree). Thus, chemical systems are convenient experimental supports for the studies of dynamical systems with a small number of degrees of freedom. When the system is not ideally mixed, equations (1) only hold locally and transport terms must be

Figure 1: *Chemical bistability and oscillations*
a)bistability: chlorite-iodide reaction
(from ref. 6)
b)periodic oscillations: Briggs-Rauscher reaction
(from ref. 8)
c)complex periodic oscillations: chlorite-thio-
sulfate reaction (from ref. 9)
d)quasiperiodic oscillations: B.Z. reaction
(from ref. 10)
e)chaotic oscillations: B.Z. reaction
(from ref. 11)

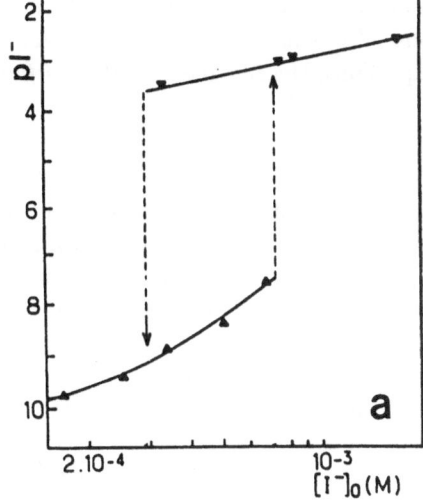

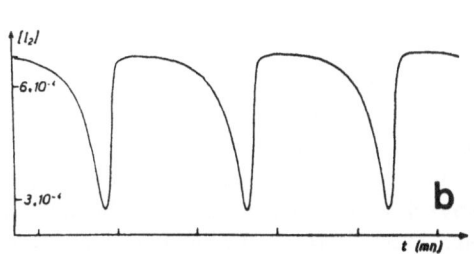

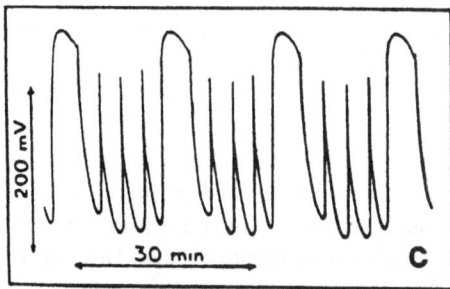

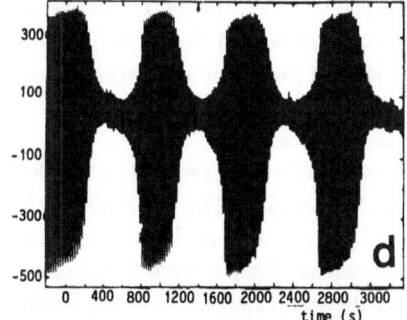

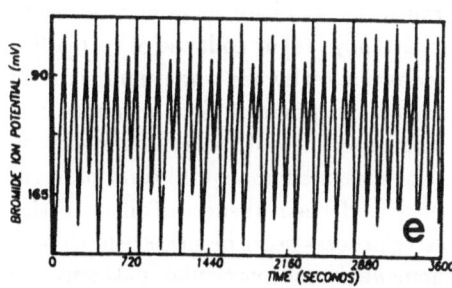

added. If there is no convective term, *e.g.* for a reacting fluid at rest, the transport limits oneself to diffusion. In this case the equations (1) change to a set of partial differential equations, the so-called *reaction-diffusion equations*:

$$\frac{\partial c_i}{\partial t} = f_i(\ldots, c_j, \ldots) + D_i \Delta c_i \tag{3}$$

where D_i is the diffusion coefficient of species i and Δ is the laplacian operator.

Most reactions only exhibit a single stable steady state for a given set of external control parameters (flows of input reactants, temperature,etc.) but when some specific type of nonlinearities , associated to autocatalytic or substrate inhibition processes are present in the kinetics, various bifurcations can occur as a function of these control parameters. Then the temporal evolution of concentrations show more complex behaviour such as multiple stationary states, oscillatory phenomena, spatial and spatiotemporal structures. These self-organization processes are predicted by dynamical systems theory and are actually observed experimentally. The most popular exotic reactions of the kind is the Belousov-Zhabotinskii reaction [4] but more than thirty other families of similar oscillating reactions are presently known [5], like the more recently discovered chlorite-iodide reaction [6]. While temporal phenomena in well mixed systems have been thoroughly studied, the present status of experimentation in spatial phenomena is much less satisfactory. Of course, the theory is less advanced but the fundamental reason of this situation has been the unsastisfied need for a manageable tool able to sustain appropriate nonequilibrium conditions.

In the case of homogeneous systems, these conditions were early satisfied with the systematic use of the so-called continuous stirred tank reactor (C.S.T.R.) [7]. The reactor is made of a reaction tank of constant volume, vigorously stirred with a paddle wheel or a Rushton turbine, and fed by pumps with constant flows of fresh reactants counterbalanced by an equal rate of output flow. In most cases the system can be considered as homogeneous; the input and output flows provide the nonequilibrium conditions necessary to obtain complex behaviour. Most dynamical phenomena expected from systems with few degrees of freedom have actually been observed and analysed in terms of general models. Let us mention bistability — there are two distinct stable stationary states in some finite range of inlet concentration (Fig. 1a) —, periodic (Fig. 1b and 1c) or quasiperiodic (Fig. 1d) oscillations and well characterized chaotic oscillations (Fig. 1e). For a short review of chaos in chemistry, see reference 12. Different scenarios of transitions to chaos have also been clearly observed such as period doubling [11,13], intermittency [14] or collapse of tori [10]. Research in this field has been so effective that one does not really see what significant advance could still be accomplished except from a quantitative point of view or maybe in the direction of hyperchaos.

The experimental spatial arrangement which would meet all the perequisites for the onset of sustained nonequilibrium patterns is much more demanding. The quest for such a device had been unsuccessfull until the works reported here. On one hand the slow diffusive process must be preserved from any perturbation caused by parasitic convective phenomena (some particular patterns observed in Petri dishes [15] have actually been proved to be of convective nature [16]), on the other hand one must feed the system with fresh reactants but this feed is quite naturally a source of convection. Because of these drawbacks most experiments were limited to transient phenomena in Petri dishes without feed. The concentrations of *major reactants, i.e.* the initial species, are supposed to be in large exces so that they can be considered as constant only for a short period of time but the system will ultimately approach equilibrium. The control parameters drift with time so that the potentaialities of these experiments are limited to the study of patterns developing in a very short time, in practice those resulting from excitability phenomena as the target or spiral patterns of traveling waves in the B.Z. reaction [17-20].

Recently the groups in Texas and Bordeaux have overcome these difficulties and built several open sustained spatial reactors. This should bring a decisive breakthrough and open new perspectives in the study of spatial and spatiotemporal patterns. In this paper we shall introduce these experimental systems with no emphasis on the technical details which can be found in the original publications. We also shall present several examples of sustained reaction-diffusion patterns obtained with these pieces of apparatus. A comprehensive interpretation of these observations in terms of the dynamical systems theory is still a challenge. In the next paragraphs we shall successively report on the implementation of three different types of reactors, respectively a *film reactor*, a *front reactor* and a *turbulent diffusion* reactor but we shall first supplement this introduction by a short survey of well documented reaction-diffusion structures. As it is difficult to establish a single classification between the various spatial and spatiotemporal behaviour obtained in numerical simulations we shall only take a glance at two radically different type of structures which have actually been extensively studied from a theoretical point of view.

The *Turing* structures first introduced by Turing as early as 1952 [21] and well analysed by the Brussel's school [1,22] result from a symmetry breaking instability. The stability of the homogeneous state of a uniformly constrained infinite system derived from equation (2) is given by the sign of the eigenvalues of the operator $[A] - [D]k^2$ where $[A]$ is the jacobian matrix $\left\{ \dfrac{\partial f_i}{\partial c_j} \right\}$, *i.e.* the stability matrix of the linearized homogeneous problem, $[D]$ the diagonal matrix of diffusion coefficients and k the wave number of the perturbation. When all the diffusion coefficients are equal $(D_i = D)$, these eigenvalues are given by $\Omega_i = \omega_i - k^2 D$ where the ω_i are the eigenvalues of $[A]$: since $Re(\Omega_i) \geq Re(\omega_i)$ the first bifurcation necessarily occurs at $(k = 0, Re(\omega_i))$. Thus the system first bifurcates to another homogeneous state, generally an unsteady oscillatory state. If the diffusion coefficients are different, it can exist a range of values $k > k_0$ $(k_0 > 0)$ where simultaneously $Re(\Omega_i) > 0$ and $Re(\omega_i) < 0$. In this case, the system first bifurcates to a nonhomogeneous stationary state with wavenumber k_0. These conditions are met in systems ruled by the competition between an activator and an inhibitor when the inhibitor diffuses faster than the activator. This is a common situation in biological systems where many processes are activated by enzymes immobilized in a matrix (zero diffusion). Gierer and Meinhart [23] have proposed an elementary model to illustrate these properties and various biological patterns ranging from the zebra strips to some early stages of morphogenesis have been modelled in this way. Unfortunately, there is no convincing example of a Turing structure produced from a simple reaction on a laboratory bench. The lack of open reactors has been a specially stringent hindrance. Moreover, most standard small size chemicals have comparable molecular diffusion coefficients $\sim 10^{-5} \mathrm{cm^2 s^{-1}}$. The necessity of equal diffusion coefficients can be discarded in finite systems with fixed concentrations at boundaries (*Dirichlet condition*) for which the mode $k = 0$ is naturally excluded (Ref. 1, p. 102) but as the pattern wavelength is imposed by geometrical constraints and not by the chemical kinetics they cannot be considered as genuine Turing structures[24]. Spatial and spatio-temporal patterns can also be obtained with equal diffusion coefficients if one starts from a non-uniform or a non-symmetric feed but in practice these systems are mostly relevant of other type of patterns discussed further. Beyond the first bifurcation, numerous patterns of increasing complexity can be obtained. Good illustrations can be found in reference 22. Among these, one finds stationary structures which can be attained directly from the homogeneous steady state by a finite amplitude perturbation [24].

Although their status is not so well-defined in the literature there is a second class of totally different structures that for sake of simplicity we shall call *front structures*. They share a common feature: they consist of a sequence of quasidiscontinuous concentration jumps, referred as *reaction-diffusion fronts*. These concentration jumps correspond to rapid switches between steady or quasisteady states. They can originate in initially homogeneous media from a local perturbation (excitable media) or from externally imposed gradients. The case of excitable media has been extensively studied in relation with the propagation of electric activity in nerve fibers and target or spiral patterns of autowaves in the B.Z. reaction. The basic features of an excitable system are appropriately described by the following *propagator-controller* [25] model:

$$\frac{\partial u}{\partial t} = \epsilon^{-1} f(u, v) + \epsilon^2 D_u \Delta u \tag{4}$$

$$\frac{\partial v}{\partial t} = g(u, v) \tag{5}$$

where the diffusion of the second variable is neglected without significant changes in the conclusions, ϵ is a small parameter and $f(u, v) = 0$ is the equation of a pleated slow manifold with two stable branches $u_-(v)$ (defined for $v > v_{min}$) and $u_+(v)$ (defined for $v < v_{max}$). The sign of the functions and the meaning of these symbols are given in Figure 2. In homogeneous conditions, the stationary states are located at the intersections of the nulclines $f(u, v) = 0$ and $g(u, v) = 0$. According to the initial conditions, the state of the system is rapidly drawn to the slow manifold and afterwards slowly follows branches u_+ and u_-, possibly jumping from u_+ to u_- (at $v = v_{max}$) or from u_- to u_+ (at $v = v_{min}$). If at a point where $v = v_0$ one creates a discontinuity such as $u = u_+(v_0)$ on one side and $u = u_-(v_0)$ on the other side, the front thus created propagates in one direction or the other according to the value v_0 with an instantaneous velocity $c_-^+(v)$ where v is the instantaneous value of the slow variable at the front. The dynamics of these fronts have been extensively studied [26-30]. Tyson and Fife [31] have shown that in a large domain of parameters, the B.Z. reaction dynamics can actually be described by the equations 4–5 where u and v are respectively proportionnal to [HBrO$_2$] and [Ce^{4+}]. The specific case of excitable systems is easily understood in this frame: there is a single stationary state $S_0(u_0, v_0)$ on $u = u_-$, with v_0 close to v_{min} (Fig. 2). Let us first ignore the space variable and diffusion. A finite but small perturbation $\delta u > 0$ beyond the unstable branch produces a rapid jump to u_+ then the system slowly moves along this branch until $v = v_{max}$ where it switches back to u_- and slowly relaxes to S_0. Figure 3 displays the typical shape of the pulse $u(t)$. Introducing now the space variable and the diffusion, we assume that a small area is excited as above. Associated with the initial switch from u_- to u_+ an outwards propagating front— the so-called *trigger wave*— develops. This can be easily understood from a physical point of view. Since a small amount δu is sufficient to induce a switch and since the gradient of u inside the front is large, only a

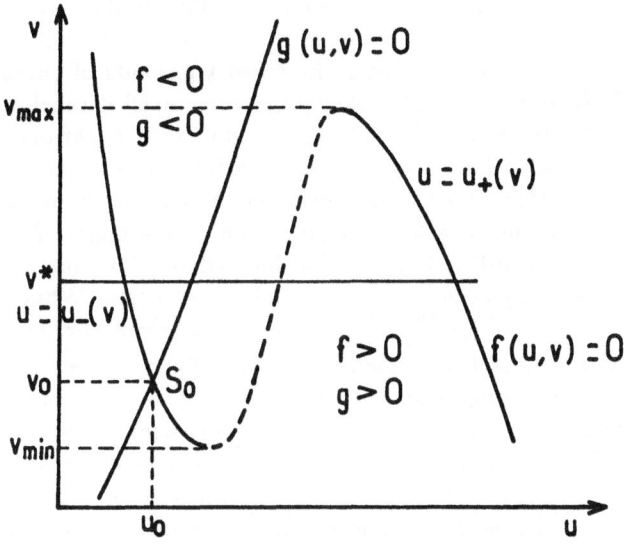

Figure 2

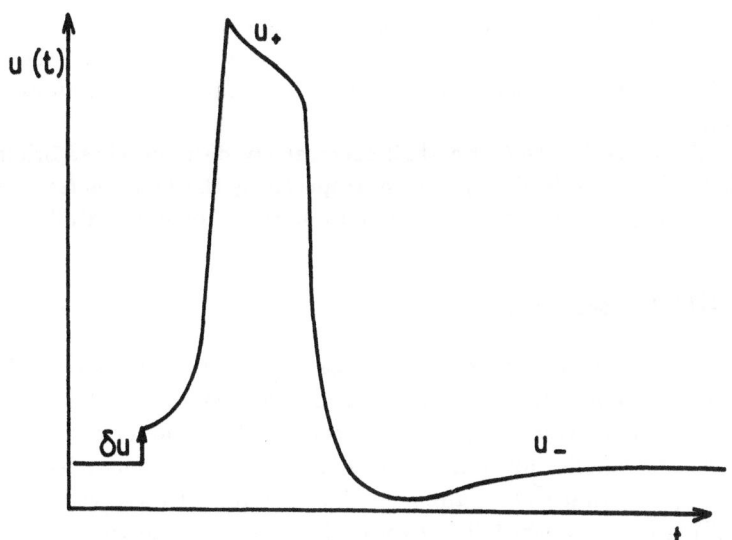

Figure 3: *Typical shape of an excitation pulse*

very short time is necessary for enough "species" to diffuse and to create a perturbation δu triggering a switch forwards. The front propagates at a constant velocity $c_+(v_{min})$ without attenuation or deformation, after the rapid transients of the initial excitation have been damped. Behind the front v slowly increases until it reaches the value v_{max} where the system goes back onto branch u_-, producing a backfront. It will not be discussed here whether, at a later time, the slow variable v remains equal to v_{max} or if it decreases to a lower value. For the B.Z. reaction, it actually happens that $v = v_{max}$ so that the backfront remains at a fixed distance behind the trigger front (*phase wave*). This distance does not depend on the diffusion but only on the kinetic law. The return to the stationary state along branch u_- is divided in two parts. While v remains large, *i.e.* $v > v^*$ where v^* is defined by $\int_{u_-}^{u_+} f(u, v^*) du = 0$ (Maxwell rule) the medium cannot be excited again; this defines the *refractory* region. When $v > v^*$ the system can be excited again with a decreasing amplitude perturbation as v decreases. In a uniform two-dimensional space, the response to a single perturbation produces a circular wave propagating outwards the excitation point. If another external perturbation breaks the ring at a particular point, a pair of spirals develop around the two free tips. These spirals have been first observed by Winfree [18]. Similarly, a periodic excitation at one point produces patterns of growing concentric rings, the so-called *target patterns* [17,19,20]. In practice this periodic excitation can be produced by the system itself since a minor change in $g(u, v)$ can move the stationary state S_0 onto the unstable branch driving the system into oscillations between u_+ and u_-. A small inhomogeneity like a dust or a catalytic particule could be sufficient to create such a wave emitting center. Although there are some competing theories which consider only homogeneous media, this interpretation of the B.Z. target patterns proposed by Tyson and Fife [31] is the most popular. Until recently target and spiral patterns were the sole spatial structures which had been extensively studied experimentally but, as mentionned earlier, the reacting film was never fed with fresh reactants so that the non-equilibrium conditions are met but for a brief while.

After this short presentation, we shall successively describe three original ways to overcome this fundamental feeding problem, respectively the *film reactor*, the *front reactor* and the *Couette flow reactor* and report on a patterns experimentally obtained.

2   The film reactor

A film reactor mimics almost exactly the usual theoretical description of a spatially extended chemical sytem. One imagines a two-dimensional thin film, in which only reaction and diffusion processes take place, uniformly fed by contact with a large reservoir of homogeneous non-equilibrium reacting medium. This apparently leads to stringent and contradictory experimental requirements but the difficulties have been recently by-passed by the Texas group in the following way [32]. The film is made of a polyacrylamide gel, a rather chemically inert transparent polymer matrix in which reaction and diffusion processes are similar to those in pure water. The gel support medium excludes convective processes. This film is used is set on top of a cylindrical CSTR which provides permanent homogeneous nonequilibrium chemical conditions. Finally, in order to avoid that the diffusive transport in the film be troubled by the turbulent mixing in the CSTR tank, a thin membrane made of parallel glass microtubes array is disposed as an interface

Figure 4: *Sustained spiral pattern in a film reactor (with permission of H. Swinney)*

between the fluid and the gel. The tubes are perpendicular to the film so that they only allow a transverse transport of reactant towards or outwards the gel without direct interference with the diffusion process in the plane of the film.

A remarkable result has been obtained when the B.Z. reaction is performed in the CSTR with appropriate reactants concentrations. After a somewhat long transient where several spirals based structures form, it happens that well defined structures made of a single spiral remain (Fig. 4). The spiral rotates indefinitely —experiments were carried on for several days— and the pattern is no more disturbed by the presence of other spirals. It becomes possible to make systematic studies of genuine asymptotic states and to follow their changes by continuous variation of control parameters. A bifurcation diagram has already been determined (Horthemke and Swinney, private communication).

3   Front reactors

The basic idea of a front reactor is to localize all the significant dynamical phenomena inside a narrow stationary reaction front in order both to decrease the effective dimensionality of the problem and to remove the "active" region away from the boundaries were the perturbations associated to the feed may disturb the dynamics. The basic configuration is a long rectangular strip of reacting medium — here again gels are specially appropriate — of length L and width l. Let us note x and y the coordinates on the axis respectively parallel to the large side ($0 < x < L$) and to the small side ($0 < y < l$). The solutions of initial reactants are divided into two essentially non reacting mixtures, say A and B. These solutions are set in contact with the strip respectively at the two opposite large edges. The slow molecular diffusion processes let A and B meet inside the strip where they react. Concentrations at the boundaries are easily maintained constant by a continuous feed of A and B. To set the ideas down let us imagine that A is an oxydizing and B a reducing medium. The redox-reactions in which we are generally interested are characterized by switching processes as soon as a threshold is attained. Thus the asymptotic concentration profile along $\overrightarrow{Oy}$ results in an oxydized and a reduced state separated by a steep front where most of the chemical transformations take place. This front can of course be considered from the point of view of singularities dynamics, like in the propagator-controller theory but it is sometimes more fruitful to think of the width a of the front as the effective width of the system. Far from the front the state of the system tend to approach the boundary conditions, does not depend on x and is extremely stable. Instabilities can only develop inside the active region. Let us for instance assume that a Turing instability could develop in a uniform environment. This is only possible when the intrinsic wavelength λ is smaller than the size of the system. In a uniform environment one should expect a structure with a wavelength $\lambda < l < L$ but the effective size along $\overrightarrow{Oy}$ is not l but $a \ll l$ so that more likely this intrinsic wavelegth is such that $a < \lambda < L$. Only a transverse mode along direction $\overrightarrow{Ox}$ can eventually develop. The effective dimensionality of the structure decreases. This mode condensation of Turing structures in *anisotropic* media has been discussed by Dewell *et al.* [33] and in the particular case of the front reactor geometry where the anistropy results from an externally imposed concentration gradient by Boissonade [34]. The last reference provides an example of how a Turing structure should look like in a front reactor. The pattern is obtained from numerical simulations performed with the popular *Brussellator* model.

$$A \xrightarrow{k_1} X$$
$$B + X \xrightarrow{k_2} Y$$
$$2X + Y \xrightarrow{k_3} 3X$$
$$X \xrightarrow{k_4} Products$$

with mass action law kinetics. At the boundaries $y = 0$ and $y = 43$ the concentrations of the major species A and B are kept constant whereas those of the intermediate species X and Y are zero. Inside the medium the stationary state can be characterized by the concentration of species X. In these particular computations periodic boundary conditions are chosen in direction $\overrightarrow{Ox}$. The diffusion of the *inhibitor* Y is faster that the diffusion of the *activator* X $(D_y/D_x = 5)$ in order to allow for emergence of a genuine Turing structure. For a proper set of the control parameters a Turing pattern of wavelength $\lambda = L/3$ develops in the sole direction $\overrightarrow{Ox}$ (Fig. 5) in agreement with the statement above. To visualize the pattern we assume that an indicator changes from clear to dark when the concentration [X] is larger than some threshold value. Figure 6 shows that the pattern appear as a spatially modulated front. When a bifurcation parameter such as [B] at one boundary is decreased, the amplitude of this modulation goes to zero at the bifurcation point. This modulation is a perfect illustration of a *symmetry breaking instability*. In the disordered phase the front is a straight line parallel to the boundaries, in accordance with the natural symmetries of the system; in the ordered phase, the front position is modulated periodically along $\overrightarrow{Ox}$ direction. The front reactor seems to be most appropriate to evidence of Turing structures.

More generally, this device appears to be a practical way to sustain quasi-one dimensionnal structures since the essential dynamics are constrained in a narrow strip along the front. One easily imagines that in this geometry, an excitation wave is forced to travel along the front. This is actually observed with the B.Z. reaction (Fig. 7). Noszticzius *et al.* [35] have proposed a specially ingenious type of front reactor: the strip of gel is

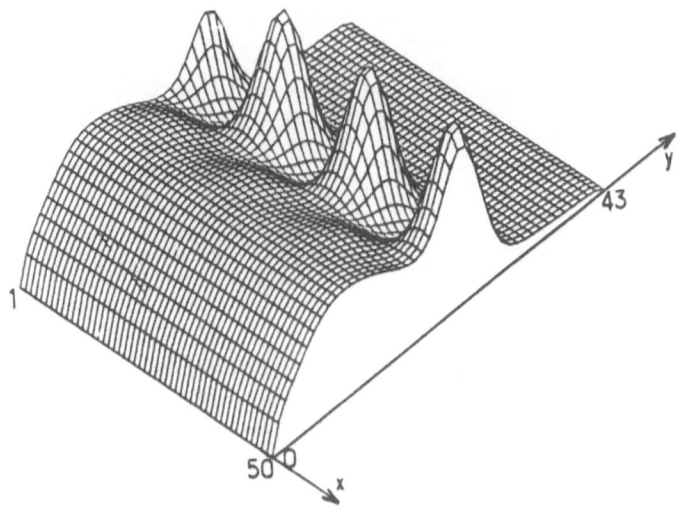

Figure 5: *Localized Turing tructure: concentration profile of species X*

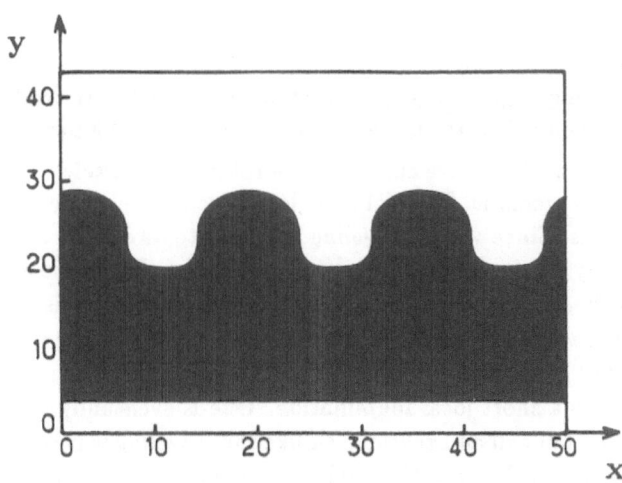

Figure 6: *Localized Turing structure: modulated front pattern*

Figure 7: *Rectangular front reactor: excitation wave at the reaction front*

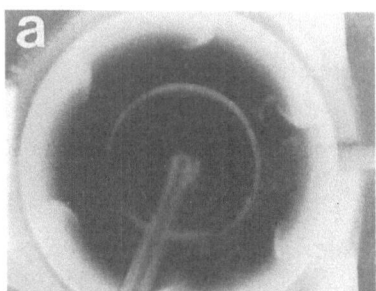 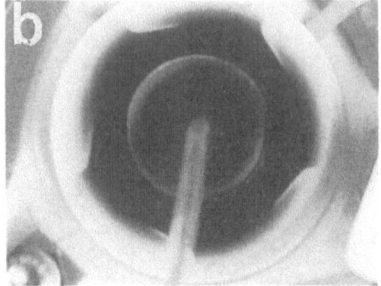

Figure 8: *Annular front reactor: rotating periodic structures of six (a) or five (b) excyclons*

ring-shaped, the feeding edges being located at the inner and outer radii. If one neglects the curvature this is equivalent to the rectangular geometry with periodic boundary conditions along $\overrightarrow{Ox}$. A traveling wave can therefore rotate indefinitely. The story is slightly more complex that it seems at first sight. A local excitation always produces a pair of counterrotating waves. Since waves traveling in opposite directions anihilate themselves when they meet, they necessarily disappear after a while . To create a permanent wave, one of them has to be eliminated. In the Nosztizcius experiment this is obtained by a local wash out, temporarily suppressing excitability just before a wave pass through. In the experiments reported here and performed in Bordeaux by E. Dulos and P. De Kepper the wave is suppressed by a short local illumination. One is eventually left with a single rotating wave that we shall call an *excyclon*. Unlike the traveling waves in Petri dishes they can be sustained indefinitely since the medium is readily kept far from equilibrium. By repeating the procedure an increasing number of excyclons rotating in the same direction can be created. Figures 8a and 8b exhibit periodic rotating structures with respectively six and five excyclons. This number cannot be increased indefinitely since the refractory zone imposes a minimum distance between two successive excyclons. Moreover the propagation velocity depends on the distance to the preceding wave. In infinite wavetrains created by periodic excitations , this dependance — when non-monotonous — can lead

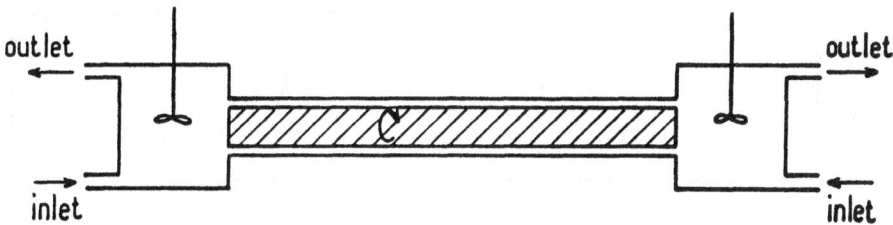

Figure 9: *Couette flow reactor (schematic)*

to complex patterns where trains of different spatial wavelength are separated by defects [36-38]. One could expect similar phenomena here: the periodic boundary condition forces a frequency in the system as would do a periodic excitation. Expulsion of super-abundant excyclons has already been observed in a preliminary work. The dynamics of these interacting waves are presenty under examination from an experimental point of view.

4   The Couette flow reactor

The basic principle of this reactor (or *CFR*), schematically represented in Figure 9, is to replace the molecular diffusion by another more controlable diffusion process, namely the *turbulent diffusion*. This is obtained by using a circular Couette flow, made of two long coaxial cylinders ($l = 30$ cm) separated by a narrow gap filled with the reacting fluid. The annular gap ($a = 0.25$ cm) is small in regard to the inner cylinder diameter (2.5 cm). The inner cylinder rotates at adjustable fixed rates, large enough for the generated turbulent flow to insure a perfect mixing both in the radial and azimutal directions, *i.e.* in a plane orthogonal to the axis. Along the axis, the turbulent transport has been shown to behave like a diffusive process with an effective coefficient D [39,40]. This Couette flow reactor thus provides a practical implementation of a quasi one-dimensional reaction-diffusion system. In the experiments reported here, the rotation rates correspond to a Reynolds number R_c ranging between $10R_c$ and $50R_c$ where R_c is the critical value at which the laminar flow loses stability; D ranges from 7.10^{-3} cm^{2}s^{-1} to 0.5 cm^{2}s^{-1}, very large values in regard to those of molecular diffusion ($\sim 10^{-5}$ cm^{2}s^{-1}). If one assume that a natural time scale is the typical period of an oscillating reaction (~ 1.min), the spatial scale of the structure is $\lambda \sim \sqrt{D\tau}$ $\sim$1–10 cm. These values are large enough in regard to the gap for a continuous diffusion approximation to be valid and very convenient for experimentation. Feeding is insured at both ends of the Couette flow by two identical CSTRs. The input and ouput flows of each CSTR are accurately balanced in order to avoid any net flow across the Couette reactor. The homogeneous states of the CSTRs define the boundary conditions of the one-dimensional reaction-diffusion system. In many experiments the internal state of the CSTRs are stationary and not very sensitive to the dynamics inside the Couette flow; this corresponds to *Dirichlet* boundary conditions.

Two different reactions have presently been studied in the CFR, namely variants of the chlorite-iodide reaction and of the B.Z. reaction. Both present a reduced and an oxydized state revealed by appropriate redox colour indicators. For instance, with the

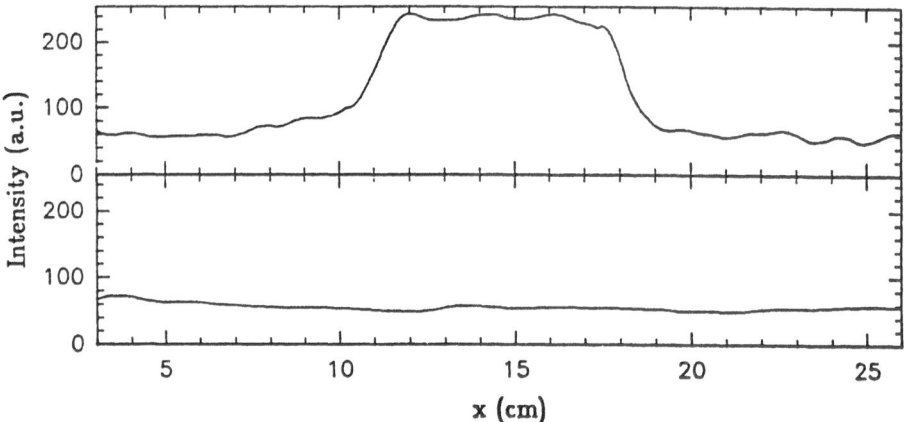

Figure 10: *Spatial bistability: two intensity profiles corresponding to a same set of control parameters*

B.Z. reaction where the ferroin is the catalyst as well as the colour indicator the former is red whereas the latter is blue. The colour intensity profiles monitored with a video camera and ultimately processed in an image analyser provide the concentration patterns evolution. Since the switch between oxydized and reduced states is much faster than the diffusion time the colour changes are generally steep so that we shall again call them *fronts* although in a 1-D system they reduce to a single point. The two CSTRs can be fed with identical or different compositions. In particular, if one CSTR is kept in an oxydized state and the other in a reduced state, there is at least one front. This is a trivial structure induced by the imposed gradient, but many non trivial patterns have also been obtained. They can all be indefinitely sustained and are genuine reaction-diffusion non-equilibrium structures. Since all diffusion coefficients are equal, these structures cannot result from the difference in transport rates like in the original Turing model. Contrary to a common belief, this type of diffusion can nevertheless destabilize a nonhomogeneous steady state. Spectacular examples have been demonstrated theoretically and experimentally in the elementary case of two coupled well-stirred reactors [41,42]. Destabilization of a stationary front is at the origin of most of the structures obtained in the CFR. It is beyond the scope of this paper to describe them all. Extensive description of the patterns , state diagrams and details on the chemistry, feed compositions and experimental procedures can be found in other publications including colour photographs [43,44]. We shall limit ourselves to a few significant examples with a reaction of the chlorite-iodide family with starch indicator. The colour intensity is directly related to the I_3^- concentration.

Spatial bistability: Both CSTRs are fed with the same compositions and kept in the reduced state. For a same set of parameters the system can stay in two different spatial stationary states: the first one is a quasiuniform reduced state, the second presents an oxydized region located in the center of the reactor (Fig. 10). This is the spatial equivalent of bistability. The transition between the two patterns occurs at different values when D respectively increases or decreases. The hysteresis defines the bistability range.

Oscillating front: The CSTRs are fed with different compositions and respectively kept in reduced and oxydized state. When a control parameter is progressively changed — here the chlorite concentration in one of the input flow — the system undergoes a bifurcation from a stationary single front pattern to a nonstationary state where the front position oscillates periodically over a finite region of the Couette flow. This is the spatial equivalent of a Hopf bifurcation. To follow such spatio-temporal phenomena we use a convenient space-time representation where the colour profile (abcissa axis) is displayed as a function of time (ordinate axis). Here, we have only retained the front position and the oxydized or reduced nature of the medium at every point of the reactor. The stationary front and the oscillating front are respectively represented on Figures 11a and 11b.

Stationary structure: This pattern, presented on Figure 11f is actually the most significant advance brought by the use of the CFR. This is the first observation of a *multipeaked* — thus non trivial — stationary dissipative chemical structure. The quest for such structures has been for a long a major objective in the field. To demonstrate its intrinsic character, the last step would be to establish an unambiguous relation between the length scale of this structure and the sole diffusion and chemical parameters

Bifurcation sequences: By continuous change of a control parameter we are able to follow the whole sequence of bifurcations leading from one state to another. For instance, in Figure 12 we report a sequence of unstationary states displaying the link between the single stationary front and the multipeaked structure. The representation highlights the continuity between the different spatio-temporal patterns. With this experimental tool, it becomes possible to explore the bifurcation diagram of spatial patterns by a continuous change of one or several parameter in the same way we use to do temporal phenomena. A two dimensional state diagram of the reaction has just been elaborated [44].

A fascinating question is the possible emergence of spatial and spatio-temporal chaos. Several prechaotic patterns have been evidenced in the CFR. Let us mention here a curious "period doubling" phenomenom in a bifurcation sequence of the B.Z. reaction obtained by continuous decrease of D, but ultimately leading to a multipeaked structure (in fact only the first three fronts are stationary)(Fig. 12)[44].

Another bifurcation sequence of the B.Z. reaction , involving frequency-locked, quasi-periodic and chaotic states has recently been evidenced and characterized by the Texas group in a shorter CFR (15 cm) directly fed at its ends, without buffering by a CSTR [45]. The bromate concentration was measured at sixteen different points with specific electrodes. The system can be understood as a continuum of coupled oscillators, the frequency of which changes continuously in the imposed gradient of constraints. Accounting for frequency locking properties, the sequence is eventually well interpreted as resulting from the coupling of two different oscillators and has been successfully simulated with a schematic model of the B.Z. reaction.

The patterns found in the Bordeaux experiments are not so well understood. Only the spatial bistability can be explained easily. In the vicinity of the feeding points the concentrations are close to those in the input flows (reduced state). In the middle of the Couette flow, far from the inlet ports, the system is in the oxydized state. The switch between the two states is much facilitated by the presence of some intermediate autocatalytic species like I_2 produced during the reaction. If the production rate of the species is smaller than its diffusion rate, the switch is inhibited. On the opposite, if the initial amount of these species is in large amount as in the oxydized state the reaction is fast and this production rate is sufficient to maintain the system in the oxydized state.

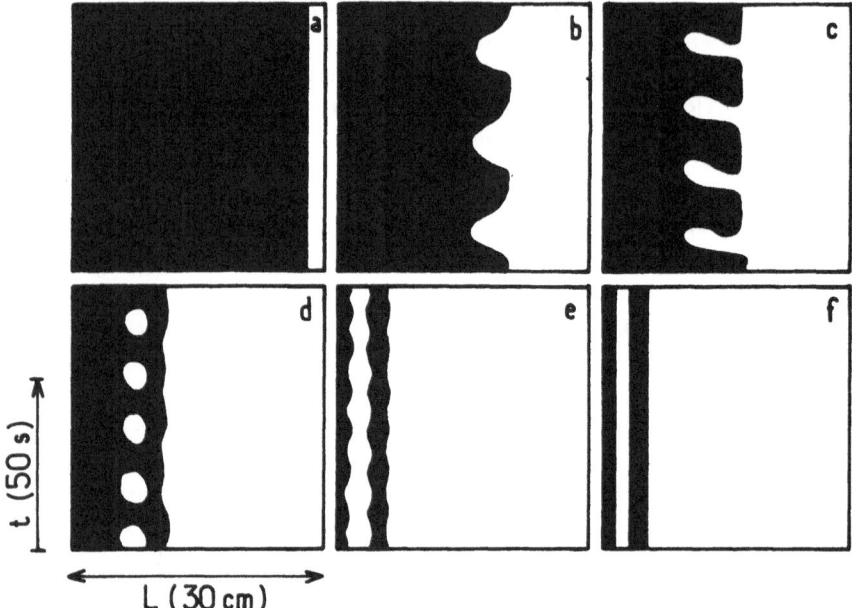

Figure 11: *Sequence of spatio-temporal patterns (chlorite-iodide reaction): Oxydized state in black; Reduced state in white*
a) Stationary front; b) Oscillating (periodic) front; c-e) Non stationary multifront patterns; f) Multifront stationary pattern

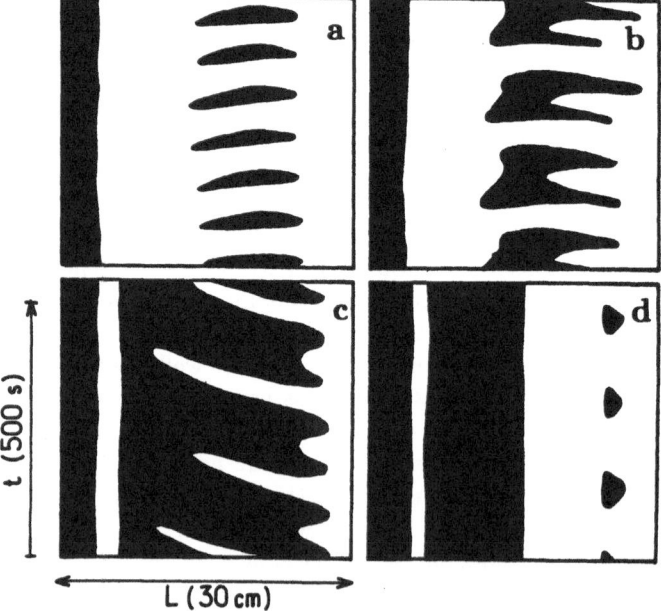

Figure 12: *Sequence of spatio-temporal patterns (B.Z. reaction) Oxydized state in black; Reduced state in white*

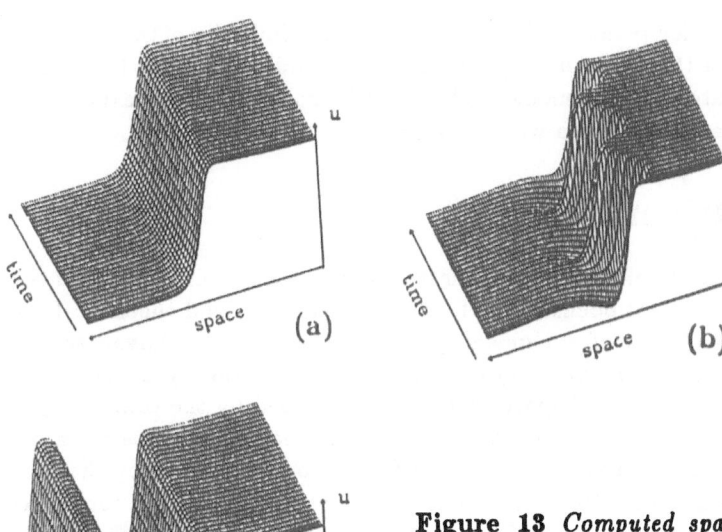

Figure 13 *Computed spatio-temporal patterns of the Arneodo-Elezgaray model (from ref. 47):*
a) Stationary front
b) Oscillating (periodic) front
c) Multifront stationary pattern

The choice between a quasiuniform or a structured state thus depends on the initial distribution of concentrations which quite naturally leads to an hysteresis phenomenom. This qualitative interpretation of bistability is supported by numerical computations on a realistic model of the reaction [44].

The theoretical body for a general interpretation of the other patterns and the bifurcation sequences is presently rather poor. With a few exceptions, the crucial role of the external gradients has not been considered and the literature is more prolific on Turing structure than on fronts dynamics. It seems legitimate to extend the propagator-controller processes, linking the front to a switch between the two sheets of a pleated slow manifold. The recent work of Elezgaray and Arneodo [46,47] is an interesting step in this direction. Although their formal model does not completely meet the experimental conditions and the specific requirements of chemical kinetic laws, it probably catches most of the physics of the patterns. The model is a two variables Van der Pol-like system with diffusive terms:

$$\frac{\partial X}{\partial t} = \epsilon^{-1}(Y - f(X)) + D\frac{\partial^2 X}{\partial r^2} \tag{6}$$

$$\frac{\partial Y}{\partial t} = -X + \alpha + D\frac{\partial^2 Y}{\partial r^2} \tag{7}$$

with $f(X) = X^2 - X^3 + X^5$. There is no gradient of constraints but, in agreement with the experiments, the two boundaries are respectively set into the two different alternative stationary states of the system. There are some striking analogies between the patterns produced with this model and those obtained experimentally. For illustration, we see that a stationary front (Fig. 13a) can be destabilized and lead to an oscillating front (Fig. 13b) beyond a bifurcation point. Still more striking is the similarity of the stationary pattern of Figure 13c with the experimental concentration profile of Figure 11f. Chaotic patterns are also generated by this model. They will be compared with experimental patterns when more quantitative data will be available.

5 Conclusion

We have made available pieces of apparatus to produce genuine sustained chemical dissipative structures. This opens a new field of experimental and theoretical investigations. From the point of view of dynamical system theory, the great advantage of working on chemical reactions is the versatility offered by a large variety of mechanisms and the extended range of time and space scales, still increased by the control of diffusion with coefficients ranging over nearly six decades (from molecular diffusion in gels to turbulent diffusion in the Couette flow). The diffusion control allowed by the use of the CFR makes possible a continuous variation of the length scale structure for a fixed geometry, changing progressively from small to extended system behaviour. From the strict point of view of chemistry, these reactors have capability to develop a new type of wet chemistry inside the front, by freezing in space kinetic steps which are normally transient. Eventually they should allow us to mimic in the laboratory some of the patterns induced by non-equilibrium processes which are common in the living world.

Aknowledgments: I thank P. De Kepper for critical reading of the manuscript and all members of the Bordeaux and Texas groups for communication of figures, photographs or other materials.

References

[1] G. Nicolis and I. Prigogine, *Self-organization in Nonequilibrium Chemical Systems* (Wiley, New York, 1977)

[2] *Oscillations and Traveling Waves in Chemical systems*, Ed. D.Field and M.Burger, (Wiley, New York, 1985)

[3] A. Babloyantz, *Molecules, Dynamics and Life*, Wiley (1986)

[4] B.P. Belousov, *Ref. Radiats, Med. 1958, Medgiz, Moscow, 145 (1959)*; M. Zhabotinskii, *Dokl. Akad. Nauk. SSSR.*, **157**, 392 (1964)

[5] A. Pacault, Q. Ouyang and P. De Kepper, *J. Stat. Phys*, **48**, 1005 (1987)

[6] P. De Kepper, I. Epstein, K. Kustin and M. Orbán, *J. Phys. Chem.* **86**,170 (1982) ;C.E. Dateo, M. Orbán, P. De Kepper and I. R. Epstein, *J. Am. Chem. Soc.*, **104**, 504 (1982)

[7] A. Pacault, P. Hanusse, P. De Kepper, C. Vidal and J. Boissonade, *Accounts Chem. Res.*, **9**,438 (1976)

[8] A. Pacault, P. De Kepper and P. Hanusse, *C.R. Acad. Sc.(Paris)*, **280B**, 157 (1975)

[9] M. Orbán and I.R. Epstein, *J. Phys. Chem.*, **86**, 3907 (1982)

[10] F. Argoul, A. Arneodo, P. Richetti and J.C. Roux, *J. Chem. Phys.*, **86**, 3325 (1987); J.C. Roux and A. Rossi, in *Non equilibrium Dynamics in Chemical Systems*, ed. A. Pacault and C. Vidal, p. 33, Springer (1984)

[11] H.L. Swinney and J.C. Roux, in *Non equilibrium Dynamics in Chemical Systems*, ed. A. Pacault and C. Vidal, p. 76, Springer (1984)

[12] F. Argoul, A. Arneodo, P. Richetti, J.C. Roux and H.L. Swinney, *Accounts Chem. Res.* **20**, 436 (1987)

[13] J.C. Roux, R.H. Simoyi, H.L. Swinney, Physica,**8D**, 257 (1983); K.G. Coffman, W.D. McCormick, Z. Noszticzius, R.H. Simoyi, H.L. Swinney, J. Chem. Phys., **86**, 119 (1987)

[14] Y. Pomeau, J.C. Roux, A. Rossi, S, Bachelart and C. Vidal, *J. Physique*, **42**, L271 (1981)

[15] D. Avnir and M. Kagan,*Nature*, **307**, 717 (1984)

[16] J.C. Micheau, M. Gimenez, P. Borckmans and G. Dewel, *Nature*, **305**, 43 (1983)

[17] A.N. Zaikin and A.M. Zhabotinskii , *Nature*, **225**, 535 (1970)

[18] A.T. Winfree, *Science*, **175**, 634 (1972)

[19] C. Vidal and A. Pacault in *Evolution of Order and Chaos*, Ed. H. Haken, p.74 (Springer-Verlag, 1982)

[20] S.C. Müller, in *From Chemical to Biological Oganization*, Ed. M. Markus, S.C. Müller and G. Nicolis, p.83 (Springer-Verlag, 1988)

[21] A.M. Turing, *Philos. Trans. R. Soc. London*, **B 327**, 37 (1952)

[22] G. Nicolis, T. Erneux and M. Herschkowitz-Kaufman, *Adv. Chem. Phys.* **88**, 263 (1978)

[23] A. Gierer and H. Meinhardt, *Kybernetik*, **12**, 30 (1972); H.Meinhardt, *Models of Biological Patterns Formation*, Ch.3, (Academic Press, New York, 1982)

[24] J.A. Vastano, J.E. Pearson, W. Horsthemke and H.L. Swinney, *Phys. Lett.* **124**, 320 (1987)

[25] P.C. Fife, in *Non equilibrium Dynamics in Chemical Systems*, ed. A. Pacault and C. Vidal, p. 76, Springer (1984)

[26] P. Ortoleva and J. Ross, *J. Chem. Phys.*, **63**, 3398 (1975)

[27] P.C. Fife, *J. Chem. Phys.*, **64**, 554 (1978)

[28] L.M. Pismen, *J. Chem. Phys.*, **71**, 462 (1979)

[29] J.P. Keener, *SIAM J. Appl. Math.*, **46**, 1039 (1986)

[30] J.J. Tyson and J.P. Keener, *Physica*, **32D**, 327 (1988)

[31] J.J. Tyson and P.C. Fife,*J. Chem. Phys.*, **73**, 2224 (1980)

[32] W.Y. Tam, W. Horsthemke, Z. Noszticzius and H.L. Swinney *J. Chem. Phys.* **88**, 3395 (1987)

[33] G. Dewel, D. Walgraef and P. Borkmans, *J. Chim. Physique (Paris)*, **84**, 1335 (1987)

[34] J. Boissonade, *J. Physique (France)*, **49**, 541 (1988);

[35] Z. Noszticzius, W. Horsthemke, W.D. McCormick, H.L. Swinney and W.Y.Tam, *Nature* **329**, 619 (1987)

[36] R.N. Miller and J. Rinzel, *Biophys. J.*, **34**, 227 (1981)

[37] K. Maginu, *SIAM J. Appl. Math.*, **45**, 750 (1985)

[38] C. Elphick, E. Meron and E.A. Spiegel, *Phys. Rev. Lett.*, **61**, 496 (1988)

[39] W.Y. Tam and H.L. Swinney *Phys. Rev.* **A36**, 1374 (1987)

[40] J.B. Grutzner, E.A. Patrick, P.J. Pellechia and M. Vera, *J. Am. Chem. Soc.*, **110**, 726 (1987)

[41] M. Boukalouch, J. Elezgaray, A. Arneodo, J. Boissonade and P. De Kepper, *J. Phys. Chem.* **91**, 5843 (1987)

[42] M. Boukalouch, J. Boissonade and P. De Kepper, *J. Chim. Physique (Paris)*, **84**, 1354 (1987); J. Boissonade, M. Boukalouch and P. De Kepper, in *"Spatial Inhomogeneities and Transient Behaviour in Chemical kinetics"*, ed. P. Gray, G. Nicolis, F. Baras, P. Borkmans and S.K. Scott, Ch.30, Manchester University Press (1989), in press

[43] Q. Ouyang, J. Boissonade, J.C. Roux and P. De Kepper, *Phys. Lett.*, **A134**, 282 (1989)

[44] Q. Ouyang, *Thesis*, Bordeaux (1989)

[45] W.Y. Tam, J.A. Vastano, H.L. Swinney and W. Horsthemke, *Phys. Rev. Lett.*, **61** (1988)

[46] A. Arneodo and J. Elezgaray, in *"Spatial Inhomogeneities and Transient Behaviour in Chemical kinetics"*, ed. P. Gray, G. Nicolis, F. Baras, P. Borkmans and S.K. Scott, Manchester University Press (1989), in press.

[47] J. Elezgaray and A. Arneodo, in *New Trends in Nonlinear Dynamics and Pattern Forming Phenomena: the Geometry of Nonequilibrium*, Ed. P.Coullet and P.Huerre (Plenum, New York, in press);

SPACE-TIME DYNAMICS IN THERMAL CONVECTION IN AN ANNULAR GEOMETRY

S.Ciliberto
Istituto Nazionale Ottica
Largo E.Fermi 6-50125 Firenze-Italy

1) Introduction

During the last years many experiments [1] have shown that the transition to chaotic time dependent regimes of several fluid instabilities may be described by low dimensional strange attractors [2]. However in these experiments the sizes of the container are of the same order of magnitude of the characteristic wavelength of the instability. In this way the fluid motion can not change its spatial pattern during the dynamics and the time dependent perturbations do not influence, on a first approximation, the spatial order . Thus on the basis of these results it is almost impossible to understand if the transition to low dimensional chaos may be seen as a precursor of the turbulent regimes, where the fluid motion presents a chaotic evolution both in space and time. To give more insight into this problem there is nowaday a great deal of interest in the theoretical [3-8] and experimental [9] study of the temporal chaotic regimes in spatially extended systems. The simplest mathematical models, in which the features of this transition may be analysed, are systems of coupled maps[4-6], one dimensional partial differential equations [3],[6],[7] and cellular automata [8]. One of the typical chaotic regimes presented by these models is spatiotemporal intermittency,that consists of a fluctuating mixture of laminar and turbulent domains with well defined boundaries. It is important to notice that such a phenomenon appears also in boundary layer flows [10] and in Rayleigh-Benard convection [11-12].

In this paper we descibe an experiment in which the space time evolution of Rayleigh-Benard convection has been studied in order to investigate how the spatial order is lost in temporal chaotic regimes and to compare the behaviours of our system with those of the above mentioned mathematical model. We have also studied the statistical properties of the onset of spatiotemporal intermittency. Our results display features typical of phase transitions similar to those obtained in numerical simulations[5-7].

The general properties of Rayleigh-Benard convection,that is thermal convection in a horizontal fluid layer heated from below, may be found in standard text books and in review papers [13], thus we briefly remind, in section 2), only the main features of this instability. The rest of the paper is organized as follows. In section 3) the experimental apparatus is described. In section 4) the different space time dynamics observed as function of the control parameter are discussed. In section 5) the results concerning spatiotemporal intermittency are reported and compared with those observed in numerical simulation. Finally conclusion are presented in section 6).

2) Rayleigh-Benard Convection

To illustrate the general features of Rayleigh-Benard convection let us consider a fluid layer confined between two horizontal solid plates and heated from below. The most relevant parameter of this instability is the Rayleigh number $Ra = \beta g \Delta T d^3/(\nu \chi)$, where β is the volumetric expantion coefficient, g the acceleration of gravity , ν the kinematic viscosity, χ the thermal diffusion coefficient, d the depth of the layer and ΔT the difference of temperture between the two horizontal plates. When Ra exceeds the threshold value Rc a steady convective flow arises, producing a pattern of parallel rolls with a well defined wavenumber q. The roll pattern is parallel to the shortest side of the cell containing the fluid, and a schematic view of the pattern is given in Fig.1. The values of Rc and q depend on the sizes of the cell and on the nature of boundary conditions, for example in the case of an infinite layer and perfect conduting plates $Rc = 1708$ and $q = 3.11/d$. Thus, from an experimental point of view, it is very important to define two other parameters, that are the aspect ratios $\Gamma_x = L_x/d$ and $\Gamma_y = L_y/d$, where L_x and L_y are the two horizontal lengths of the cell. The time dependent regimes of Rayleigh Benard convection observed at $Ra >> Rc$ are strongly influenced by the aspect ratios and also by the Prandtl number $Pr = \nu/\chi$. Indeed in the experiments in which the transition to low dimensional chaos has been studied [1] Γ was of the order of $2\pi/q$.

In the experiment that we describe in this paper the cell containing the working fluid has an annular geometry. Indeed with this geometry and a suitable choise of the radial aspect ratio, it is possible to construct a pattern that is almost a one dimensional chain of radial rolls(roll axis along radial directions, see also Fig.3,4) with periodic boundary conditions. These features of the spatial pattern are very useful in order to compare the results of our experiment, with those obtained in the above mentioned mathematical models.

3) Experimental apparatus

A schematic cross section of the cell is reported in Fig.2. The lateral walls of the cell are made of plexiglass. The outer and inner diameters of the annulus are 8 Cm and 6 Cm respectively. The depth of the layer d is 1 Cm. With these dimensions the radial aspect ratio is 1 whereas the aspect ratio along the circle, of diameter $2r_o = 7cm$, is 21.99. The bottom plate of the cell is made with a copper plate whose upper surface is finished to a mirror quality and is protected with a film of nickel to prevent oxidation. The plate is heated with an electrical resistor R_1. The upper plate is made of a sapphire window SW whose top is cooled by the water circulation Wa, that is confined on the other side by the glass window GW. This arrangement allows an optical investigation of the convective motion. The cell is inside a temperature stabilized box that reduces the thermal fluctuations of the enviroment. The temperture of the cooling water Wa is stabilized by a thermal bath. The long term stability of ΔT is $\pm 0.001°C$. The working

Figure 1: Schematic drawing of convective pattern near Rc.
The characteristic wavelength $\lambda = 2\pi/q$ is indicated.

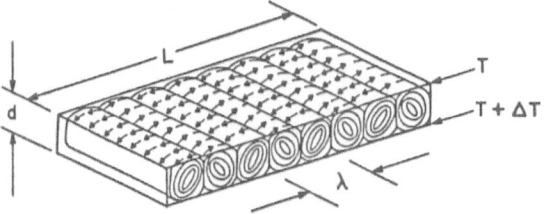

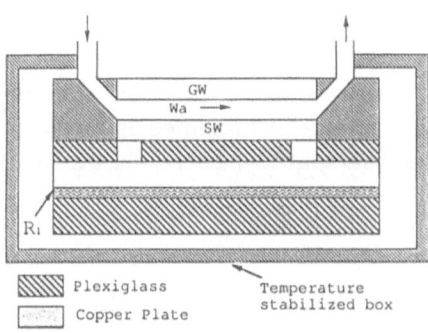

Figure 2: Schematic diagram of the cell: R_1 heating resistor, SW sapphire window, GW glass window, WA cooling water

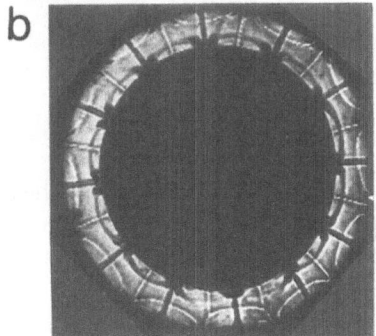

Figure 3: Shadowgraphs of typical spatial patterns. White and dark regions correspond to cold and hot currents respectively. a)Stationary spatial pattern at $\eta = 100$. b) Snapshot of the spatial pattern at $\eta = 190$ in a time dependent regime.

a

b

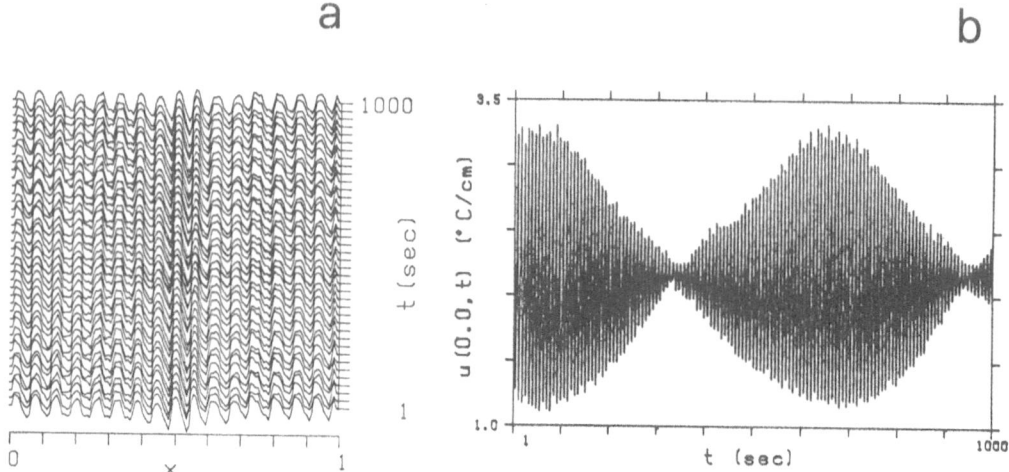

Figure 4:a) Space time evolution of u(x,t) at $\eta = 164$;b) Corresponding time evolution of the point x=0. The vertical scale has been amplified in b) because the time dependent modulation slightly perturbes the spatail pattern shown in a),where the maximum amplitude is roughly $4°C/cm$

a

b

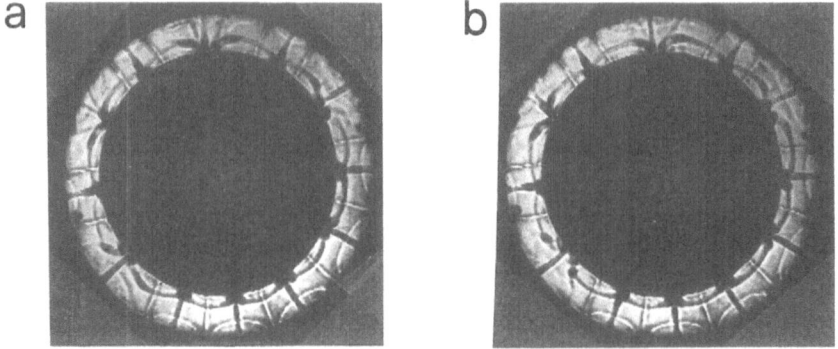

Figure 5): Snapshots of the spatial patterns at $\eta = 230$. The time interval between a),b) is 30 sec.

fluid is silicon oil with $Pr = 30$. The critical difference of temperature , computed with $Rc = 1708$,is $\Delta T_c = 0.06°C$.

The qualitative features of the patterns are determined by a digital enhanced shadowgraph technique [14]. An optical technique, based on the deflection of a laser beam that sweeps the fluid layer [15], enable us to obtain quantitative global and local characteristics of the pattern. The shadowgraph and laser beam deflection techniques are not perturbative and rely upon the changes of the index of refraction induced by the temperature field. The principle of the sweeping technique has been described elsewhere[15]. The actual set up provides the possibility of measuring on the circle of radius $r_o = 3.5cm$ (that is on the circle of mean diameter), with a twelve-bit resolution, the two components of the thermal gradient averaged along the vertical direction, in the polar coordinate reference frame r, θ. The accuracy of the measurement is about 7%, the sensitivity $0.01°C/cm$ and the spatial resolution about 1 mm. In time dependent regimes only the component of the gradient perpendicular to the roll axis has been recorded. This component will be called u(x,t), with $x = \theta/(2\pi)$.The function u(x,t) is sampled at 128 points in space. In time dependent regimes u(x,t) is recorded for at least 5000 times at interval of 1 sec. that is roughly 1/10 of the main oscillation period of the system.

3) Spatial patterns

Analysing the fluid behaviour as a function of $\eta = \Delta T/\Delta T_c$, we observe, that for η around 1, the spatial structure has about 22 rolls. This number increases with η and reaches 38 at η around 200. A detailed analysis of the wavenumber selection process has been reported elsewhere[16]. In Figs.3a we show the shadowgraph of the spatial pattern at $\eta = 100$. Dark region correspond to the hot currents rising up and white regions to the cold ones, going down. We observe that our geometry constrains the spatial structure to an almost one dimensional chains of rolls.

The spatial structure remains stationary for $\eta < 164$ where a subcritical bifurcation to the time dependent regime takes place.For $\eta > 164$ the time evolution is chaotic but, reducing η, the system presents either periodic or quasiperiodic oscillations, and at $\eta = 149$ it is again stationary.In the range $149 < \eta < 200$ the time dependence consists of rather localized fluctuations that slightly modulate the convective structure, which mantains its periodicity. This is clearly seen in Figs.3b) where a snapshot of the spatial structures at $\eta = 190$ is reported. The presence of hot and cold currents transverse to the main set of rolls merit a special comment. Such a two dimensional effect certainly influence the dynamics. However considering that the ratio between the length and the width of the annulus is roughly 22 we realise that the system can be considered almost one dimensional for what concerns the propagation time of thermal fluctuations along the circle, because the two time scales are bery well separated. Besides, we also observe that the time dependent fluid motion is still very correlated along the radius.

The space time evolution of u(x,t) and the corresponding time evolution of the point x=0 at $\eta = 164$ are shown in Fig.4a and Fig.4b. In looking at Fig.4b we clearly see that the time evolution is quasiperiodic. However this time dependent modulation is hardly seen in Fig.4a, because it sligtly perturbes the spatial pattern that mantain

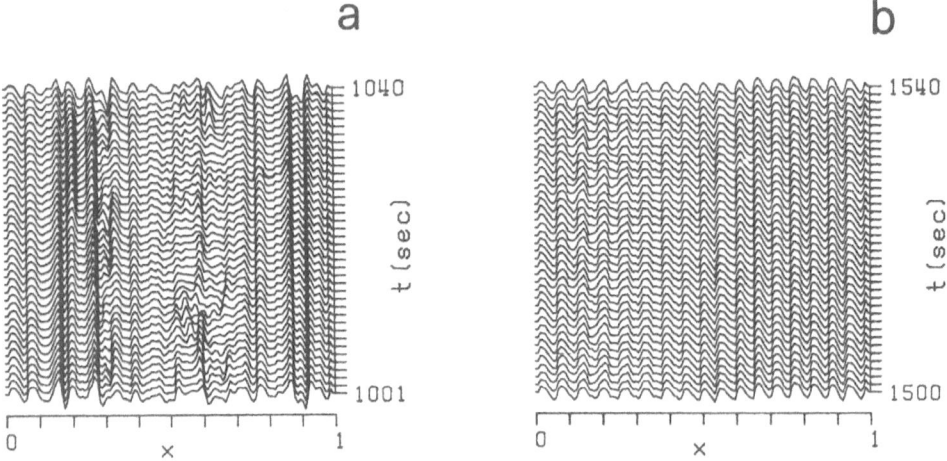

Figure 6): Space time evolutions of u(x,t) at $\eta = 216$ at two different time intervals of 40 sec each.

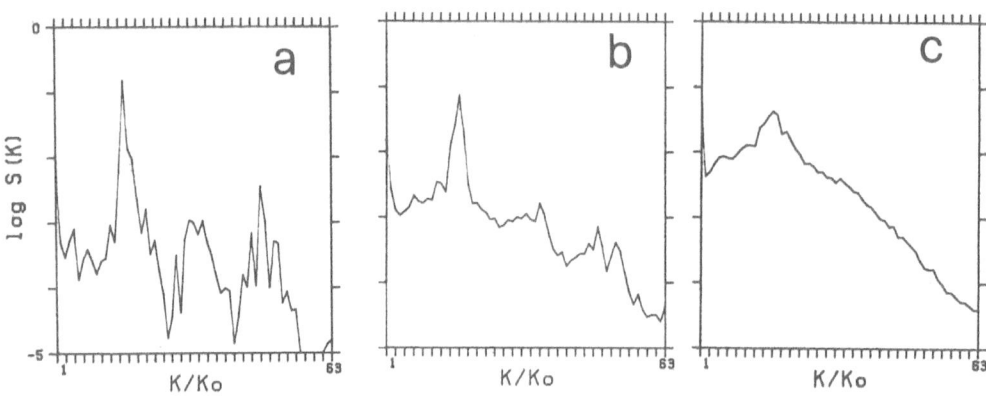

Figure 7): Spatial power spectra at different values of η:a) $\eta = 164$;b) $\eta = 216$;c)$\eta = 348$.

its original periodic structure. Increasing η the time evolution becomes chaotic but the spatial order is still mantained. The fractal dimension and the orthogonal decomposition [17] indicate that the number of degrees of fredom involved in the dynamics is around 3.

At higher η the spatial order begins to be destroyed because of the appearence of bursts, detaching from the boundary layer. This spatiotemporal intermittent regime appears at $\eta = 200$. Typical spatial patterns at $\eta = 230$ are shown in Fig.5 for two different times. They present, several domains where the spatial periodicity is completely lost (we will refer to them as turbulent) and other regions (that we call laminar) where the spatial coherence is still mantained. The space time evolution of u(x,t) at $\eta = 216$ is shown in Fig.6a),6b) at two different times. We notice that for $1000 < t < 1040$ there are strong oscillations that locally destroy the spatial order whereas for $1500 < t < 1540$ the pattern is again very regular.

The time averaged spatial Fourier spectra at $\eta = 164$, $\eta = 216$, $\eta = 347$ are reported in Figures 7a),7b), and 7c) respectively. The spectrum of Fig.7a) corresponds to a quasiperiodic regime and being the spatial structure still very ordered the spectrum presents well defined peaks. In contrast Fig.7b), corresponding to a value of η that is very close to the threshold for spatiotemporal intermittency presents a broadened third harmonic. This indicate that the most important length scales for this transition are the shortest ones. Finally in Fig.7c) the spectrum, corresponding to a value of η far above the transition point, is totally broadened because the spatial order has been destroyed.

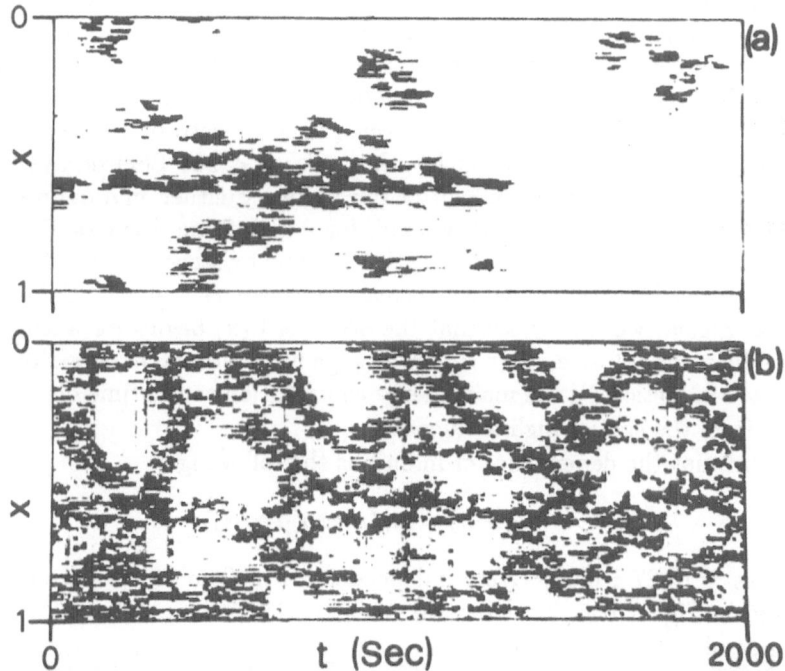

Figure 8): Binary representation,at $\alpha = 1.5°C/cm$, of the space-time evolution of u(x,t) at $\eta = 216$ a) and $\eta = 248$ b). The dark and white area correspond to turbulent and laminar domains respectively.

4) Spatiotemporal intermittency

As we discussed in the previous section, the space time evolution of u(x,t) shows that in the turbulent domains the time evolution is characterised by the appearence of large oscillatory bursts. Instead in laminar regions the oscillations remain very weak. Thus the two regions can be identified by measuring the local peak to peak amplitude,for a time interval comparable with the mean period of the oscillation. Choosing a cutoff α, and making black all the points where the oscillation amplitude is above α, we can easily represent the dynamics of turbulent and laminar regions. As an example of such a code we show the spacetime evolution of u(x,t) at $\eta = 216$, in Fig.8a,

and $\eta = 248$ in Fig.8b. We remark that the qualitative features of these pictures are rather independent of the precise value of the cutoff. We can easily verify that the code catches the main properties of the dynamics by comparing Fig.8a) with Figs. 6a) and 6b). Indeed we clearly see that at the most oscillating and disordered regions of Figs.6a),6) correspond to black points in Fig.8a whereas ordered and not oscillating regions are represented by white points.

At $\eta = 216$,Fig.8a), a wide laminar region surrounds completely the turbulent patches that remain localized in space, after their appearance. Furthermore, the nucleation of a turbulent domain has no relationship with the relaxation of another one. In contrast, at $\eta = 248$ Fig.8b),the turbulent regions migrate and slowly invade the laminar ones. This last regim that sets in for $\eta > 245$ is very similar to those obtained in theoretical models [5-7]. The change from the regime of Fig.8a) to that of Fig.8b) is reminescent of a percolation [6], that, indeed, has been proposed as one of the possible mechanisms for the transition to spatiotemporal intermittency.

Following a method also used in numerical models [5-7], we quantitatively characterize such a behaviour by computing, over a time interval of $10^4 sec$, the distibution P(x) of the the laminar domains of length x. For $\eta < 248$ P(x) decays with a power law.The exponent does not depend within our accuracy, either on α or on η . Its average value is $\mu = 1.9 \pm 0.1$. On the other hand, for $\eta > 248$,the decay of $P(x)$ for $x > 0.1$ is exponential with a characteristic length 1/m.The existence of two different regimes is clearly seen in Figs.9a),9b) which display P(x) versus x at $\eta = 241$ and $\eta = 310$. Looking at Fig.9a) we clearly see that the decay of P(x) begins for a length scale that is smaller than the roll size. This rather strange result has an explanation, because ,as we remarked in section 4), the main energy contribution to the time dependent regimes is coming from the spatial high frequencies.

We find that the dependence of m on η is the following:

$$m(\alpha, \eta) = m_o(\eta) exp(-\alpha/\alpha_o) \qquad (1)$$

with $\alpha_o = (0.87 \pm 0.06)°C/cm$ independent of η. The dependence of m_o versus η is reported in Fig.10.The linear best fit for $\eta > 246$ of the points of Fig.10) gives the following result:

$$m_o(\eta) = m_1(\frac{\eta}{\eta_s} - 1)^{\frac{1}{2}} \qquad (2)$$

with $\eta_s = 247 \pm 1$ and $m_1 = 117 \pm 2$. This equation shows the existence of a well defined threshold η_s for the appearance of an exponential decay in P(x). Besides we see that the characteristic length $1/m_o$ diverges at $\eta = \eta_s$. In the range $200 < \eta < 400$, P(x) is very well approximated by the following equation:

$$P(x) = (Ax^{-\mu} + B)exp[-m(\alpha, \eta)x] \tag{3}$$

where $m(\alpha, \eta)$ is given by 1) and μ has the previous determined value. A,B are instead free parameters that can be very easily determined. It is possible to fit our experimental P(x),in the range $0.4°C/cm < \alpha < 3°C/cm$, with $A = 10\ B \simeq 4 \cdot 10^3$ for $\eta > \eta_s$ and $B = 0$ for $\eta < \eta_s$.

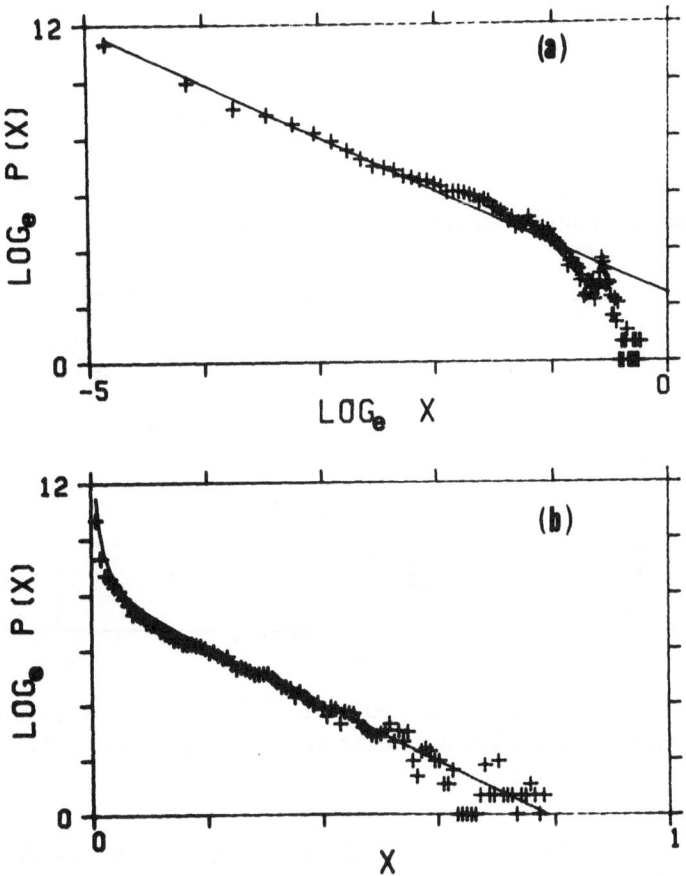

Figure 9): Distribution P(x) of the laminar regions of length x.(a) $\eta = 241$, algebraic decay with exponent 1.9; (b) $\eta = 310$ and $\alpha = 1.6°C/cm$, exponential decay with a characteristic length $1/m = 0.10$. The solid lines are obtained from Ee.3).

The features of P(x) displayed by equations 2),3)are typical of phase transitions. Therefore, being the transition point η_s very close to the point where the behaviour like that of fig.8b) sets in, we conclude that the transition to this behaviour may be a phase transition [18] The main features of P(x) for $\eta > \eta_s$ qualitatively agree with those obtained in coupled maps [5-6] and partial differential equations [6-7] in spatiotemporal intermittent regimes. Of course these models do not reproduce the values of the non-universal exponents in Eqs.2),3) [5].

The transition may also be characterized by measuring p_o that is the probability of finding a laminar point[7]. If we suppose that alaminar site is generated at a certain time with space-time independent probability p_o, the probability of finding a laminar region of length x is given by $P(x) \propto exp[xlog(p_o)/l_o]$, where l_o is a suitable characteristic length. We can verify this hypothesis by computing directly p_o on the experimental data. By following the same procedure m as a funtion of α, we find that $\log p_o$ extrapolated at $\alpha = 0$ has the following dependence on η :

$$|\log p_o| = const.(\frac{\eta}{\eta_c} - 1)^{\frac{1}{2}} \tag{4}$$

with $\eta_c = 216$. So we conclude that it has the same exponent but different critical threshold. This means that the appearence of a laminar site may be considered a statistical independent process for $\eta >> \eta_s$ and that a certain correlation exists between laminar and turbulent sites near the critical value η.

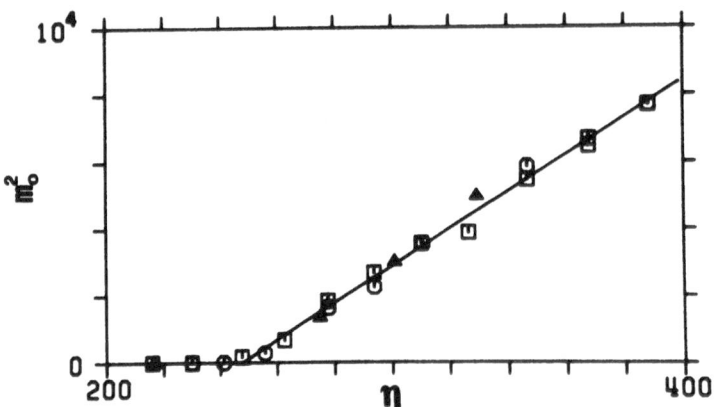

Figure 10): Dependence of $m_o{}^2$ on η ,the different symbols pertain to different sets of measurements done either increasing or decreasing η. The solid line is obtained from Eq.2).

The presence of a power law decay of P(x) for $\eta_c < \eta < \eta_s$ may be due either to finite size effects or to defects [18]. This aspect of the problem is not yet very well understood and further investigation is in progress to clarify this point.

6) Conclusion

Rayleigh-Benard convection in an annular geometry is very useful to investigate the transition from low dimensional chaos to weak turbulence because both the regimes are found as a function of the control parameter.

The onset of spatiotemporal intermittency, in our cell, displays features of a phase transition that is reminescent of a percolation. Although many aspects of this phenomenon are still to be investigated, the analogy of the behaviour of our system with that observed in coupled maps, P.D.E. and some cellular automata, suggests that these models may be very useful to understand the general features of spatiotemporal intermittency.

This work has been partially supported by G.N.S.M.

References:

[1] A.Libchaber, C. Laroche, S. Fauve, J. Physique Lett. 43, 221, (1982); M.Giglio, S.Musazzi, U.Perini, Phys. Rev. Lett. 53, 2402 (1984); M.Dubois, M.Rubio, P.Berge', Phys. Rev. Lett. 51, 1446 (1983); S. Ciliberto, J. P. Gollub, J. Fluid Mech. 158, 381 (1984).

[2] For a general review of low dimensional chaos see for example: J. P. Eckmann, D. Ruelle, Rev. Mod. Phys. 1987; P. Berge, Y. Pomeau, Ch. Vidal, L'Ordre dans le Chaos (Hermann, Paris 1984).

[3] B. Nicolaenko, in " The Physics of Chaos and System Far From Equilibrium", M. Duong-van and B. Nicolaenko, eds. (Nuclear Physics B, proceeding supplement 1988); A. R. Bishop, K. Fesser, P. S. lomdhal, W. C. Kerr, M. B. Williams, Phys. Rev. Lett. 50, 1095 (1983).

[4] G. L. Oppo, R. Kapral Phys. Rev. A 3, 4219 (1986).

[5] K. Kaneko, Prog. Theor. Phys. 74, 1033 (1985); J. Crutchfield K. Kaneko in "Direction in Chaos", B. L. Hao (World Scientific Singapore 1987); R. Lima, Bunimovich preprint.

[6] H. Chate', P. Manneville, Phys. Rev. Lett. 54, 112 (1987); Europhysics Letters 6,591(1988);Physica D 32, 409 (1988)

[7] H. Chate', B. Nicolaenko, to be published in the proceedings of the conference: "New trends in nonlinear dynamics and pattern forming phenomena", Cargese 1988.

[8] F. Bagnoli,S. Ciliberto, A. Francescato, R. Livi, S. Ruffo, in "Chaos and complexity" , M. Buiatti, S. Ciliberto, R. Livi, S. Ruffo eds., (World Scientific Singapore 1988)

[9] P. Kolodner, A. Passner, C. M. Surko, R. W. Walden, Phys. Rev. Lett. 56, 2621 (1986); S. Ciliberto, M. A. Rubio, Phys. Rev. Lett. 58, 25 (1987); A. Pocheau Jour. de Phys. 49, 1127 (1988)I; I. Rehberg, S. Rasenat , J. Finberg, L. de la Torre Juarez Phys. Rev. Lett. 61, 2449 (1988); N. B. Trufillaro, R. Ramshankar, J. P. Gollub Phys. Rev. Lett. 62, 422 (1989).

[10] M. Van Dyke, An Album of Fluid Motion (Parabolic Press, Stanford, 1982); D. J. Tritton, Physical Fluid Dynamics (Van Nostrand Reinold, New York, 1979), Chaps.19-22

[11] P.Berge',in " The Physics of Chaos and System Far From Equilibrium", M.Duongvan and B.Nicolaenko, eds. (Nuclear Physics B, proceedings supplement 1988)

[12] S.Ciliberto,P.Bigazzi,Phys.Rev.Lett. 60, 286 (1988).

[13] S.Chandrasekar, Hydrodynamic and Hydromagnetic Stability, Clarendon Press, Oxford 1961; F.H.Busse, Rep.Prog.Phys. 41 (1978) 1929; Ch.Normand, Y.Pomeau, M.Velarde Rev.Mod.Phys. 49, 581,(1977).

[14] W.Merzkirch, Flow Visualisation, Academic Press, New York 1974.

[15] S.Ciliberto,F.Francini,F.Simonelli,Opt.Commun.54,381 (1985).

[16] S. Ciliberto, M. Caponeri, F. Bagnoli, submitted to Nuovo Cimento D.

[17] S.Ciliberto,B.Nicolaenko submitted for publication.

[18] H. Muller-Krumbhaar in 'Monte Carlo Methods in Statistical Physics", edited by K. Binder (Springer- Verlag,New York 1979); D. R. Nelson ,'Phase transitions and critical phenomena' , edited by C. Domb and J.L. Lebowitz (Academic Press London 1983)

INVARIANT MEASURES IN HYDRODYNAMIC SYSTEMS WITH RANDOM PERTURBATIONS

Ana Bela Cruzeiro

Centro de Matemática e Aplicações Fundamentais, I.N.I.C.

Av. Prof. Gama Pinto 2, 1699 Lisboa Codex - PORTUGAL

1. INTRODUCTION

In several mathematical and physical situations the evolution in time of a system, given some initial conditions, is described by a set of partial (eventually non-linear) differential equations. One can of course look for individual classical solutions. Another point of view consists in considering the initial data on some probability space, which is supposed to describe the set of "physical initial conditions" and to study the evolution of the corresponding measure along the time. That is, one looks for a familly of measures indexed by time that are concentrated on the set of solutions of the given differential equations. By this procedure we obtain the so-called statistical solutions. Following the statistical approach, one of the basic questions is to look for invariant measures for the system. In fact, not only they provide suitable invariants for the motion described by the equations, but also one may expect to prove existence, uniqueness and asymptotic properties with respect to these measures.

In statistical hydrodynamics the set of equations, which descibe the motion of an incompressible fluid in a bounded domain are, as is well known, the Navier-Stokes equations. A (reasonably) complete description of the motion is given by the velocity u, the preassure p and eventually the density ρ at each time $t \geqslant 0$, when the initial and boundary conditions are fixed. In this work we shall consider that the density is conserved during the evolution and that the pression is given. The equations are then

$$
\begin{cases}
\rho \dfrac{\partial u}{\partial t} = -\rho\,(u.\nabla)u + \nu\,\Delta u - \nabla p + \rho\,f \\[2mm]
\operatorname{div} u = 0 \ \ (\text{incompressibility condition}) \\[2mm]
\dfrac{\partial \rho}{\partial t} = -(u.\nabla \rho)
\end{cases}
\tag{1.1}
$$

where $\nu \geqslant 0$ is the viscosity, f the density of the external forces. Let us consider the following situations:

1) If $\nu = 0$ and the density is considered to be constant, (1.1) reduces to the so-called Euler system. This is a conservative system and therefore one may expect to define an invariant measure of Gaussian type with the aid of an invariant quantity. In fact, once this measure is defined, it is proved that for almost all realizations with respect to it the energy turns out to be infinite, and therefore the statistical approach is quite different from the classical one. A first mathematical construction of invariant measures associated to the Euler equations in two-dimensions was first given in [2] and [4]. In [1] we proved the existence of the associated flows.

2) Supposing that $\nu > 0$ and $\rho \equiv 1$, we have the homogeneous Navier-Stokes equations. Several different types of statistical solutions have been discussed (see e.g. [8] and references therein). With respect to results on invariant measures, and because thys system is a dissipative one, it is not possible in principle to construct measures by the method that is used for the Euler case. Still, by replacing the external (deterministic) forces by random ones, and therefore by introducing some additional energy in the system, we obtain invariant measures for the corresponding stochastic flows. The random perturbation being adapted, in particular, to the dimension, it is possible to consider any dimension $d \geqslant 0$. These type of results were proved in [5]. We remark that, in the two-dimensional case, we proved in [1] that the Gaussian measure that is invariant for the Euler system is also invariant for a suitable perturbed Navier-Stokes one.

3) The more general case, namely the non-homogeneous Navier-Stokes system (1.1) will be discussed in more detail in paragraph 3, where we show how the techniques of [5] can be generalized to this situation. We mention that the regularity of the associated stochastic flows can be improved; this was discussed in a recent work ([6]).

In the Rayleigh-Bénard convection (cf. [3]), where the flow is generated by a heat flux, the density is sometimes approximated in the following way:

$$\rho = \rho_0[1 - \alpha(T - T_0)],$$

where α is the expansion coefficient and T the temperature. In this framework, the equation for the temperature is given by:

$$\frac{\partial T}{\partial t} = - (u.\nabla T) + \beta \Delta T \; ,$$

β being the thermal diffusivity. The inhomogeneous case would correspond in this situation to a constant α and to $\beta = 0$, which is certainly not a very good approximation for the problem. Nevertheless, if β is supposed to be constant, the methods described in this work can still be a good approach, if one considers, for instance, truncature functions applied to the velocity.

The author is gratefull to professor V. Mendes for having introduced her to the thermally driven flow problem, as well as for the invitation to this conference. She also acknowledges H. Fujita for usefull discussions.

2. NOTATIONS

Let us define the functional spaces that shall be used in what follows. We shall denote by V the space $V = \{u \in C_0^\infty : \text{div } u = 0\}$ and by $\mathcal{H}$ it's closure in $L^2(\Omega)$, where $\Omega \subset \mathbb{R}^d$ ($d \geq 2$) is a bounded domain with a C^∞ boundary. The operator $A = -\Delta$ is a self-adjoint, positive definite operator on $\mathcal{H}$, having a completely continuous inverse; therefore it has in $\mathcal{H}$ a complete set of orthonormal eigenfunctions $e_1, \dots , e_i, \dots$, whose eigenvalues verify $0 < \lambda_1 \leq \dots \leq \lambda_i \xrightarrow{i} + \infty$ ($\lambda_i \sim i^{2/d}$, to be more precise). We shall consider the following (Sobolev) spaces, for $s \in \mathbb{R}$:

$$\mathcal{H}^s = \left\{ u(x) = \sum_{i=1}^{+\infty} u^i e_i(x) : \|u\|_s^2 = \sum_{i=1}^{+\infty} \lambda_i^s |u^i|^2 < + \infty \right\}$$

with the corresponding (Hilbert) scalar product. $\mathcal{H}^0$ coincides with $\mathcal{H}$ and the spaces $\mathcal{H}^s$ and $\mathcal{H}^{-s}$ are dual to each other with respect to the relation $<u,v> = \Sigma u^i v^i$, $u = \Sigma u^i e_i \in \mathcal{H}^s$, $v = \Sigma v^i e_i \in \mathcal{H}^{-s}$.

Let $B: \mathcal{H} \to \mathcal{H}^{-s}$ be the non-linear operator defined by:

$$<B(u),v> = \sum_{i,j=1}^{d} \int_\Omega u^i u^j \partial_i v^j \, dx \; , \quad v \in \mathcal{H}^s$$

If we consider that $s > d/2 + 1$, by the Sobolev immersion lemma we have the following estimations:

$$|<B(u),v>| \leqslant c \, \|u\|_0^2 \, \|v\|_s \qquad \text{and}$$

$$\|B(u) - B(u')\|_{-s} \leqslant c \, \|u-u'\|_0 \, (\|u\|_0 + \|u'\|_0)$$

Also, because we suppose div $u = 0$, by an easy application of the integration by parts formula,

$$<B(u),u> = 0 \qquad \forall \, u \in \mathcal{H}^s \, .$$

Let us now take the system (1.1), in the distributions sense, with initial conditions $u(.,0) = u_0 \in \mathcal{H}$ and $\rho(.,0) = \rho_0 \in L^\infty(\Omega)$ and with zero boundary conditions. The system (1.1) or, more precisely, it's projection on the space $\mathcal{H}^{-s}$ is equivalent to the following system of equations:

$$
\begin{cases}
\rho \, du_t = [-\nu \, A \, u_t - B(u_t) + f(t)] \, dt \\
\dfrac{d\rho}{\partial t} = -(u.\nabla\rho)
\end{cases}
\qquad (1.2)
$$

The system we shall consider is a stochastic perturbation of (1.2), that is, the system we obtain by replacing the forces by some white-noise ones. For $k > d/2$, put $b_t = \sum\limits_i b_t^i \, \lambda_i^{-k} \, e_i$, where the b_t^i are independent copies of real brownian motions. We shall study the following stochastic differential system:

$$
\begin{cases}
\rho \, du_t = \rho \, db_t - [\nu \, A \, u_t + \rho \, B(u_t)] \, dt \\
\dfrac{\partial\rho}{\partial t} = -(u.\nabla\rho) \\
u_0 \in \mathcal{H} \\
\rho_0 \in L^\infty(\Omega) \quad \text{with} \ 0 < C_1 \leqslant \rho_0(x) \leqslant C_2 < +\infty \qquad \forall \, x \in \Omega
\end{cases}
\qquad (1.2)'
$$

3. THE INHOMOGENEOUS NAVIER-STOKES STOCHASTIC EQUATIONS

We obtain the solutions of (1.2)' by passing to the limit the corresponding Galerkin approximations. Define V_n to be the space generated by the first n eingenfunctions $e_1, \dots, e_n$, $u_0^n = \sum\limits_{i=1}^n u_0^i$ and $b_t^{(n)} = \sum\limits_{i=1}^n b_t^i \, \lambda_i^{-k} \, e_i$. The Galerkin approximations to (1.2)' are the solutions of the following (finite-dimensional) equations:

$$\begin{cases} \rho^n \, du_t^n = \rho^n d \, b_t^{(n)} - \left[\nu \, A \, u_t^n + \rho^n \, \pi^n \, B\!\left(u_t^n\right) \right] dt \\[2mm] \partial_t \, \rho^n = - \, (u^n . \nabla \rho^n) \end{cases} \tag{3.1}$$

starting from u_0^n and ρ_0, and where π^n denotes the orthogonal projection of $\mathcal{H}^{-s}$ on V_n.

The system (3.1) does not offer too many difficulties. The coefficients A and $\pi^n B$ are locally Lipschitz and, by writing $v_t^n = \int_0^t u_s^n \, ds$, with $u_t^n \in C([0,t] \; ; \; C^1(\Omega) \cap V)$, the equation for ρ^n is $\dfrac{d\rho^n}{dt} (t, v_t^n) = 0$. We can therefore find a solution belonging to $C^1([0,T] \times \Omega)$ that verifies:

$$C_1 \leqslant \rho^n(t,x) \leqslant C_2 \quad \forall t \quad \forall n$$

(see [6] for details). Furthermore, by Itô's formula, we have:

$$d<\rho_t^n \, u_t^n \, , u_t^n > = 2<\rho_t^n \, u_t^n \, , u_t^n \, .db_t^n > - \left[2<Au_t^n, u_t^n> - \sum_{i=1}^{n} \lambda_i^{-2k} \left[\left(\rho_t^n \right)^i \right]^2 \right.$$

$$\left. - 2<\rho_t^n \, u_t^n \, , \pi_n \, B(u_t^n)> - <(\partial_t \rho_t^n) \, u_t^n \, , u_t^n > \right] dt \tag{3.2}$$

Now let us remark that the last term is equal to $- <\left(u_t^n . \nabla \rho_t^n \right) u_t^n \, , u_t^n>$ and that, by an integration by parts argument we have:

$$<(u.\nabla \rho) \, u,u> \; = - \, 2 <\rho \, (u.\nabla) \, u,u>$$

$$= -2 <\rho \, B(u),u> \tag{3.3}$$

Therefore the two last terms in (3.2) compensate and

$$E \, \|u_t^n\|_0^2 \quad \leqslant \frac{1}{C_1} E <\rho_t^n \, u_t^n \, , u_t^n >$$

$$\leqslant \frac{C_2}{C_1} \left(\|u_0\|_0^2 + C_2 \, C' \, t \right),$$

where $C' = \sum_i \lambda_i^{-2k} < + \infty$ by the hypothesis on k. This estimation allows us to conclude that the processes u_t^n are defined, in fact, for all $t \geqslant 0$.

By the estimations on ρ_t^n we can obtain a subsequence converging towards ρ_t in $\mathcal{H}$, the limit belonging to $L^\infty (\mathbb{R}^+ \times \Omega)$. Let ν_n denote the law of u_t^n on the space of measures over $Z = L^2([0;T]; \mathcal{H}^1) \cap C([0;T]; \mathcal{H}^{-s})$. By using Itô's formula, and the (uniform) estimations on the coefficients of the equations, we obtain the inequalities

$E |||u_t^n||| \leqslant \eta \quad \forall t \in [0,T] \quad \forall n$, where η only depends on $||u_0||_0$ and T and where the norm $|||.|||$ is defined for $u \in L^2([0;T]; \mathcal{H}^1) \cap C([0;T]; \mathcal{H}^{-s})$ by:

$$|||u||| = \left(\int_0^T ||u(t)||_1^2 \, dt \right)^{1/2} + \sup_{0 < t_1 \leqslant t_2 \leqslant T} \frac{||u(t_1) - u(t_2)||}{|t_1 - t_2|^\alpha}, \quad 0 < \alpha < \frac{1}{2}.$$

(cf. [5] for a similar reasoning).

The sets $\{ |||u||| \leqslant R \}$ are precompact in Z (cf.[8]) and therefore, by Prokhovov's theorem, we obtain a limiting measure ν for the ν_n on the space of measures over Z endowed with the weak topology. There is a stochastic process associated to ν (given by Skorohod's theorem). We finally obtain, by the method exposed, the following result:

3.1 **Theorem** - The system (1.2)' has a solution (u_t, ρ_t), where $u_t \in L^2([0;T];\mathcal{H}) \cap C([0;T]; \mathcal{H}^{-s})$ and $\rho_t \in L^\infty (\mathbb{R}^+ \times \Omega)$, T > 0 being arbitrary. Futhermore, we have:

$$C_1 \leqslant \rho_t(x) \leqslant C_2 \quad \forall t \quad \forall x \in \Omega$$

4. THE INVARIANT MEASURES

To prove the existence of an invariant measure for the approximative processes u_t^n, we shall need an estimation of $E \, ||u_t^n||_0^2$ slightly better than the one obtained from (3.2). For that, and again by an application of Itô's formula,

$$e^{\beta t} E < \rho_t^n u_t^n, u_t^n > = < \rho_0^n u_0^n, u_0^n > + E \int_0^t \beta e^{\beta s} < \rho_s^n u_s^n, u_s^n > + e^{\beta s}$$

$$\left[\sum_{i=1}^n \lambda_i^{-2k} \left[(\rho_t^n)^i \right]^2 - 2 \nu < A u_s^n, u_s^n > - 2 < \rho_s^n u_s^n, \pi^n B(u_s^n) > + < \partial_s^n \rho_s^n u_s^n, u_s^n > \right] ds$$

for any constant $\beta > 0$. The two last terms compensate by (3.3) and we have:

$$E \, ||u_t^n||_0^2 \leqslant \frac{e^{-\beta t}}{C_1} \left(C_2 \, ||u_0||_0^2 + \beta C_2 \int_0^t e^{\beta s} E \, ||u_s^n||_0^2 \, ds + C_2^2 C' \int_0^t e^{\beta s} \, ds - \right.$$

$$- 2\nu\lambda_1 \int_0^t e^{\beta s} E \|u_s^n\|_0^2 \, ds \Bigg)$$

If we choose $\beta = \dfrac{2\nu\lambda_1}{C_2}$, we get

$$E \|u_t^n\|_0^2 \leqslant e^{-\beta t} \frac{C_2}{C_1} \|u_0\|_0^2 + \frac{C_2^2 \, C'}{C_1} \left(1 - e^{-\beta t}\right) \tag{4.1}$$

This estimation is the main tool to prove the existence of an invariant measure. It is in fact known that, in finite dimensions, a process whose expectation of a moment has a finite superior limit (in t) posesses an invariant measure (cf. e.g. [7]). On the other hand, the estimation (4.1) being uniform on the dimension, we can obtain a limiting measure which will be invariant with respect to the process u_t of the last section (cf. [5] for details). We can therefore prove that:

4.1 <u>Theorem</u>: There exists a probability measure μ, supported by $\mathcal{H}$, which is invariant with respect to the process u_t of theorem 3.1, in the following sense:

$$\int_{\mathcal{H}} E_u \, f(u_t) \, d\mu\,(u) = \int_{\mathcal{H}} f \, d\mu \qquad \forall t > 0 \qquad \forall f \in C_b(\mathcal{H}^{-s})$$

<u>Remark:</u> We remark that the methods used in the proof are not constructive ones. The estimations on the moments that are obtained are enough to show, for instance, that μ is not a trivial measure, namely a Dirac mass at point zero. Nevertheless, a study of the support of the measures obtained by these methods would be an interesting subject.

REFERENCES

[1] S. Albeverio and A.B. Cruzeiro, Global flows with invariant (Gibbs) measures for Euler and Navier-Stokes two dimensional fluids, preprint, Bochum (1988).
[2] S. Albeverio, M. Ribeiro de Faria and R. Høegh Krohn, Stationary measures for the periodic Euler flow in two dimensions, J. Stat. Phys., Vol. 20, No. 6 (1979), p. 585-595.
[3] R.P. Behringer, Rayleigh-Bénard convection and turbulence in liquid helium, Rev. Mod. Phys., Vol. 57, No. 3, I (1985), p. 657-687.
[4] C. Boldrighini and S. Frigio, Equilibrium states for the two-dimensional incompressible Euler fluid, Comm. Math. Phys. 72 (1980), p. 55-76.
[5] A.B. Cruzeiro, Solutions et mesures invariantes pour des équations d'évolution du type Navier-Stokes, to appear in Expo. Mathem.
[6] H. Fujita-Yashima, preprint, Pisa
[7] Y. Miyahara, Invariant measures of ultimately bounded stochastic processes, Nag. Math. J., Vol. 49 (1973), p. 149-153.
[8] M.I.Visik, A.I. Komech and A.V. Fursikov, Some mathematical problems of statistical hydrodynamics, Russ. Math. Surv. 34 (1979), p. 149-234.

ANOMALOUS TRANSPORT OF ENERGY IN TOKAMAKS AND THE BEASTS MODEL

Marc A. DUBOIS
Theory Group , DRFC , CEN Cadarache
F13108 St Paul lez Durance Cedex France

ABSTRACT :   We give an overview of the problem of anomalous
----------   confinement of thermal energy in tokamaks , and of the
failure to understand it . Then we introduce a new tool to adress this
problem : the " beasts " model , and present some results using it .
This model is potentially of interest for other problems in physics.

INTRODUCTION :   The concept of toroidal magnetic confinement of
---------------   thermonuclear plasmas is nearly forty years old , and
the goal of energy production using fusion of hydrogen isotopes is
within reach , at least at the proof of principle level , with
machines like the european JET tokamak . It is probably less well known
that the understanding of plasma confinement in such costly devices is
far from being satisfactory .

A tokamak is a toroidal vessel with a toroidal magnetic field B_T to
which is superimposed a poloidal field B_P created by a plasma carried
toroidal current I : magnetic surfaces are nested torii on which field
lines are wound with radially dependant helicity (as I is not
constant but decreases from the magnetic axis to the plasma edge) :
see Fig.1. The KAM theorem can be used to show that in the absence of
plasma magnetic instabilities , most torii are preserved : radial
plasma transport would then be due only to collisionnal diffusion .
The so-called "neo-classical" theory computes the diffusion
coefficients of energy and particles for ions and electrons , taking
into account the fact that some particles are " trapped " due to the
radial dependance of the field B_T : see Fig.2 . The neoclassical
theory distinguishes three regimes , depending upon the collision
frequency : Fig.3 .

Analysis of experimental results have consistently shown for many
years that the ions follow the neoclassical theory (within a factor 2
to 3 , which is a good agreement in plasma physics) , but that
electrons transport energy about two orders of magnitude faster than

what the theory predicts (which is NOT considered to be satisfactory)
The discrepancy is even worse when additionnal heating (by injection
of waves or of energetic particles) is used on top of the Joule
heating of the plasma .

This huge "anomalous transport" has been extensively studied ,
both from a theoretical and an experimental point of view , but it is
fair to say that it is still not understood at all : the next section
is a short summary of the various approaches to this problem .

THEORETICAL APPROACHES AND COMPARISON WITH EXPERIMENTS :

An extensive litterature exists on this topic :the interested reader
will find under [1] useful reviews of experimental tokamak physics and
of the specific problem of anomalous transport . The general guiding
idea is , for most approaches , that some turbulence is present in the
plasma and contributes to transport of particles and of energy .A
schematic classification would distinguish mainly two branches of
instabilities leading to such turbulence : electrostatic modes , which
modify the transport through the ExB term , and magnetic modes which
lead to a desorganisation of magnetic surfaces and hence to loss of
confinement (the KAM torii in real space being destroyed). Those two
approaches have been extensively followed , mainly in the linear
regime , and review papers on those theories exist [2] ; a fair
summary would be that no consensus has yet emerged , and that if
electrostatic modes are known to be unstable in the linear regime ,
they do not explain the anomalous transport satisfactorily – on the
other hand , magnetic modes , such as microtearing or rippling modes ,
are linearly stable in many of the plasma conditions , and if they are
responsible of the anomalous transport , only a non linear analysis
can be useful .

We feel that a global approach , yielding general constraints on
the theoretical models from experimental data is doubtlessly necessary
to progress in the understanding of tokamak confinement ; a good
starting point is the observation that the temperature profile is
remarkably stable against all perturbation , such as additionnal
heating , even when it is much more localised than the resistive (
Joule) heating : the " profile consistency " hypothesis , very
fashionnable in the recent times , was simply the constatation that
whatever was done to the plasma , its electronic temperature profile
could always be fitted by a gaussian . A few years ago , the so-called
" marginal stability " theory [3] attempted to explain this as due to

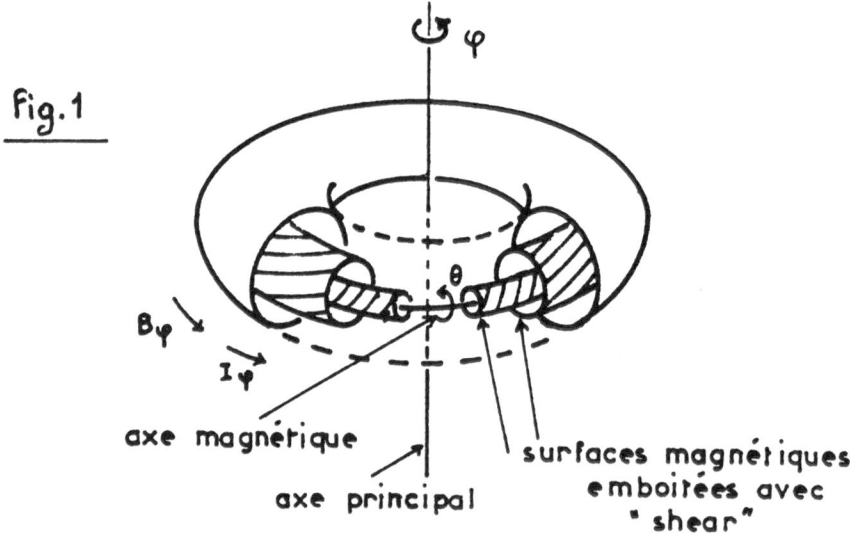

Fig.1

φ

B_φ

I_φ

θ

axe magnétique

axe principal

surfaces magnétiques
emboîtées avec
" shear "

Fig. 2

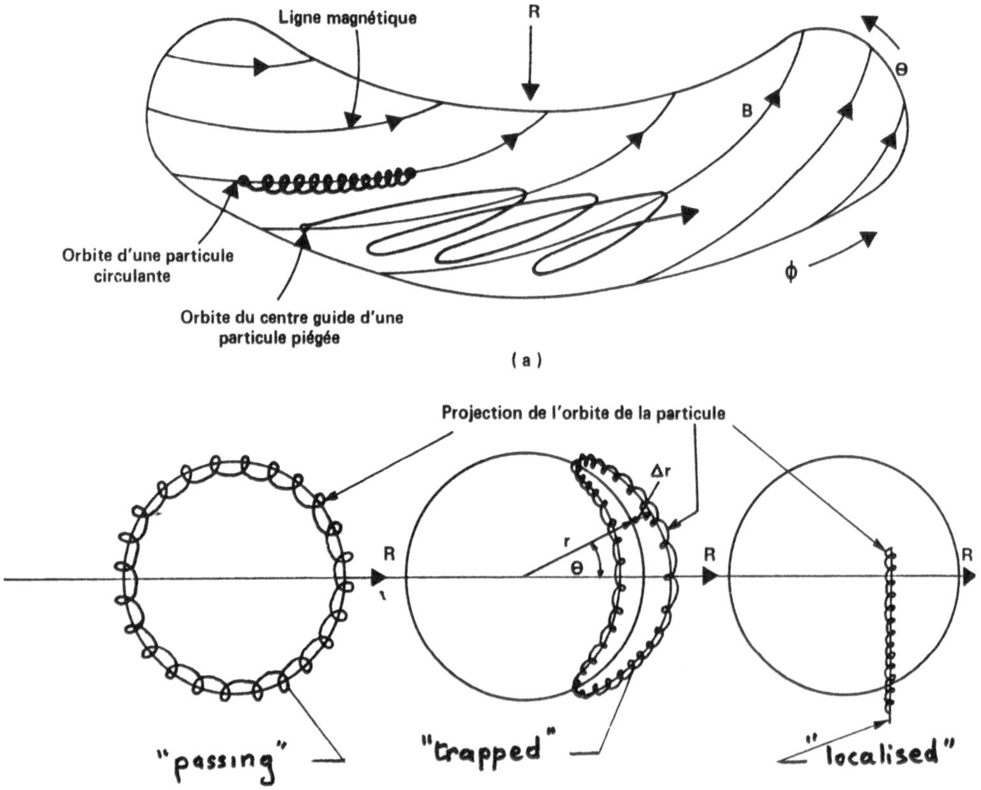

Ligne magnétique

R

B

θ

ϕ

Orbite d'une particule
circulante

Orbite du centre guide d'une
particule piégée

(a)

Projection de l'orbite de la particule

Δr

r

θ

R

R

R

"passing"

"trapped"

"localised"

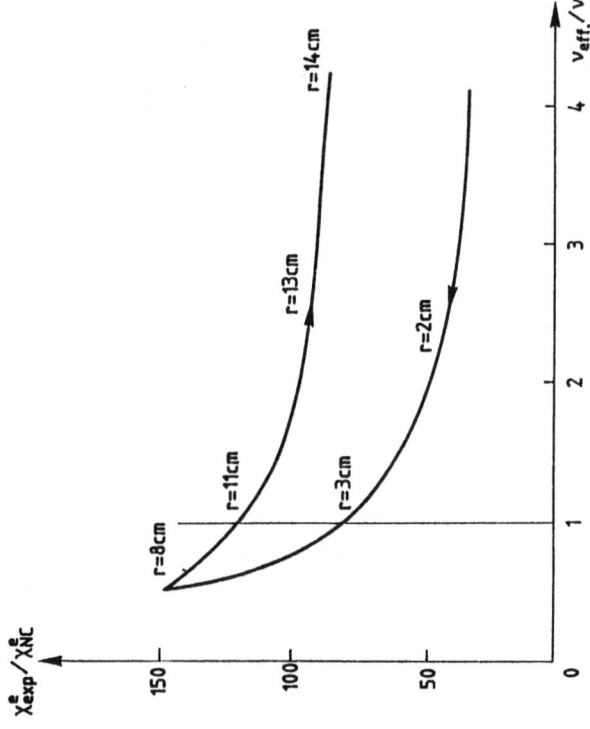

Fig. 4 - "Anomalie" χ^e_{exp}/χ^e_{NC} du coefficient de transport expérimental au coefficient néoclassique pour les électrons (la flèche indique les rayons croissants)

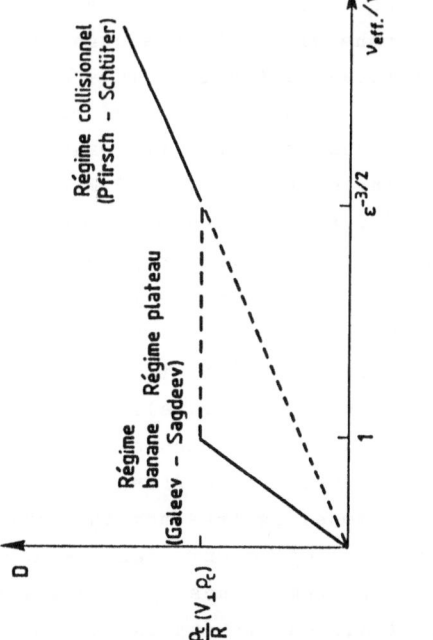

Fig. 3 - Coefficient de diffusion néoclassique dans un tokamak

the fact that the profile adapted itself so that everywhere some
instability was marginally (un-)stable : any perturbation would
therefore be self regulated . This approach has been partially
forgotten , because the chosen mode is no longer fashionnable , but
the basic idea deserves further studies .

We wish therefore to build a model of a plasma with unstabilities
dependant on the temperature (and if necessary on the density)
profile , the unstabilities leading to changes in the thermal
conductivity . Such a model can yield constraints on the functionnal
dependancies of the unstabilities growth rates as well as on the
dependancies of the transport coefficients to the modes amplitudes . A
cellular automata-like modelisation is not possible , because two
different quantities - the mode amplitudes and the profile of the
transported quantity - have to be computed at the same time , and are
dependent of each other . We take into account the fact that the
magnetic modes are localised on resonant magnetic surfaces (that is
on surfaces where the field lines close upon themselves after a finite
number of turns) to construct what we call the " beasts model " in
which a finite number of localised automata (the " beasts ") evolve
as a function of a local quantity , the flux of which is dependent
upon the amplitude of the beasts in the vicinity of the place where it
is computed .

THE " BEASTS " MODEL :   We consider a slab (cylindrical or
--------------------   otherwise) where a quantity U(r,t) is defined
as a function of r : a transverse (radial in a cylindrical case)
coordinate , and of time t . In the absence of perturbations , the
slab can be considered to consist of parallel surfaces on each of
which U is constant . A finite set (r_j) of values of the transverse
coordinate is also defined , and at each r_j a perturbation can appear
, which we call a " beast " . This perturbation is characterised by
its transverse extension $\delta_j(t)$; we define its growth rate Γ_j as $d\delta/dt$
, and we have $\Gamma_j(t) = \Gamma_j(U(r), dU/dr, ...)$. On one side (r=0.) there
is a source , i.e. an entering flux $\bar{\Phi}$ of the quantity U , while the
outer boundary (r=1.) is an absorbing well . The quantity U can
diffuse across the slab , and in the absence of perturbations (i.e.
if δ_j=0. for every j) the transverse diffusion coefficient is $D_{\perp 0}$.
The last ingredient of the model is the fact that the " beasts " can
modify locally the transport coefficient , and we have the relation :
$D_{\perp}(r,t)=D_{\perp 0}.f(\delta_j(t))$. Knowing the values of U(r) and of the δ_j at a
time t_0 , the evolution of all quantities is uniquely determined .

This simple dynamical system can be thought of as an extreme idealisation of a plasma confinement device with magnetic turbulence : the r_J are resonant magnetic surfaces , the "beasts" are magnetic islands of width δ_J, and U is the density of a population of monoenergetic electrons which have a transverse diffusion coefficient (due to collisions with stationnary ions) $D_{\perp o} = D_o$.

We give a few examples of possible behaviour of such systems : in the cases shown on fig. 5 to 8 , the δ_J at t=0 are equal to 0 , the diffusion coefficient between r_J and r_{J+1} is (see Appendix) :

$$D_{\perp J K} = \inf\{ D_{\varphi L} , D_o.1/(1-s) \} \quad \text{with } s = (\delta_J+\delta_K)/(r_K-r_J)$$

$$D_o = cste \quad \text{and } D_{\varphi L} = \alpha.\epsilon^\beta \quad \text{with } \alpha \text{ and } \beta = csts \text{ and } \epsilon = s^2$$

The upper part of each figure shows the evolution of the " beasts " widths as a function of time , and the lower part the evolution of U(r) ; the initial conditions for U(r) correspond to the profile at t=0 . The growth rate expression Γ as a function of U , dU/dr etc... is given on each figure , as well as the values of $D_{\varphi L}$ and D_o .
Several interesting features are to be noted : "cold" and "hot" fronts can be observed , propagating radially , and visible as zones where the "beasts" are in contact (e.g. Fig.6,7). An other interesting feature is that for some dependences of Γ , a quiescent , stationnary regime can appear (e.g. Fig.8) : it is possible if the Γ_J can cancel simultaneously for a continuous (non-zero) profile of U . The so called "self-consistency" of tokamaks temperature profiles could be due either to such quiescent behaviour of the perturbations , or to a dynamical averaging like in Fig.5 .
It is easy to compute an analytical expression for the propagation velocity of the cold and hot fronts if one considers that when two beasts come into contact , the diffusion coefficient is greatly increased , leading to an increased gradient at the front edge , and hence to an increased growth rate on the next site . No general expression will be given here , as it depends on the specific Γ and D expressions .

DISCUSSION AND CONCLUSIONS : The present model is the object of
------------------------------ analytical study and of numerical
experiments ; it is a simple , self consistent dynamical system to

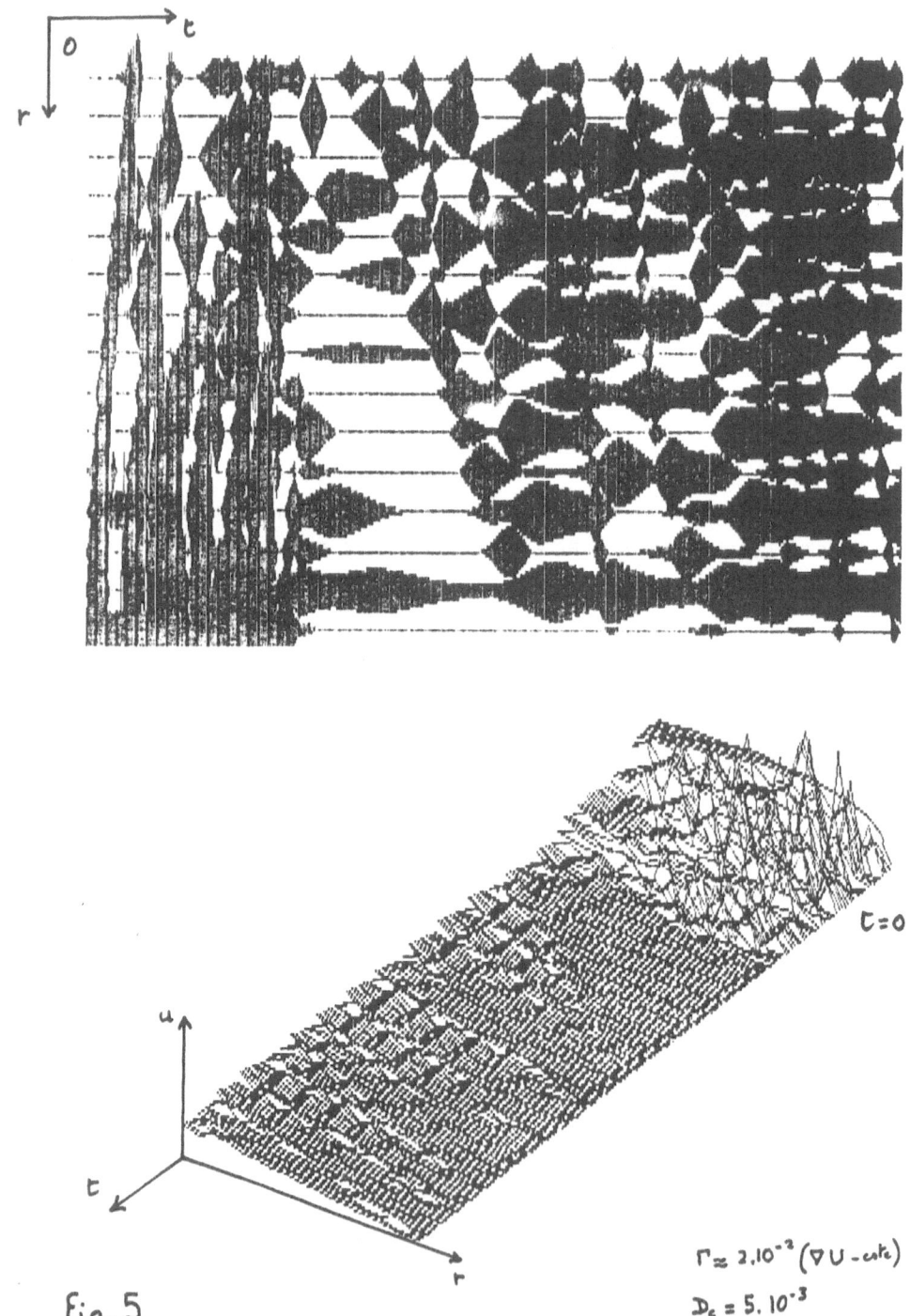

Fig. 5

$\Gamma \approx 2.10^{-3} \left(\nabla U - c \text{ste} \right)$

$D_c = 5.10^{-3}$

$\tau = 0$

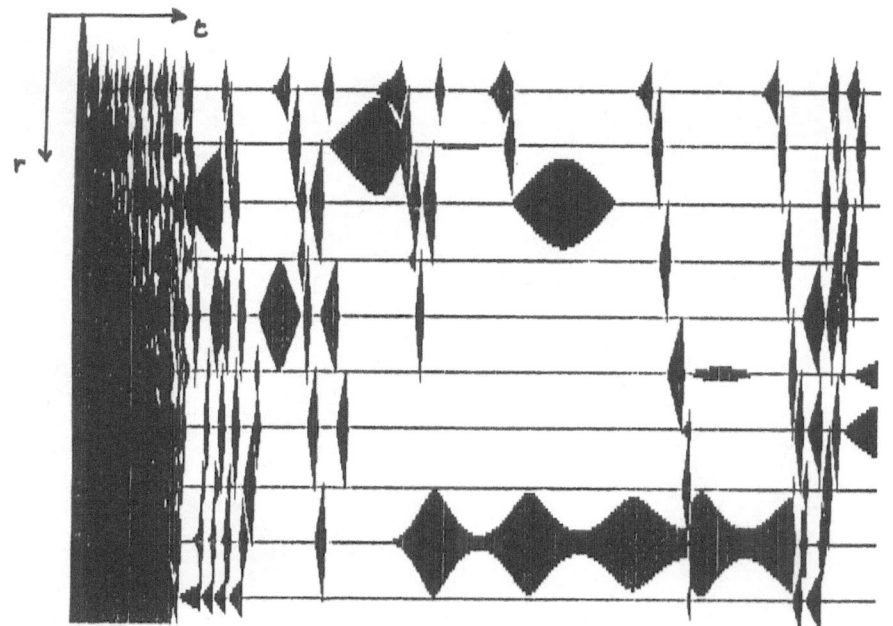

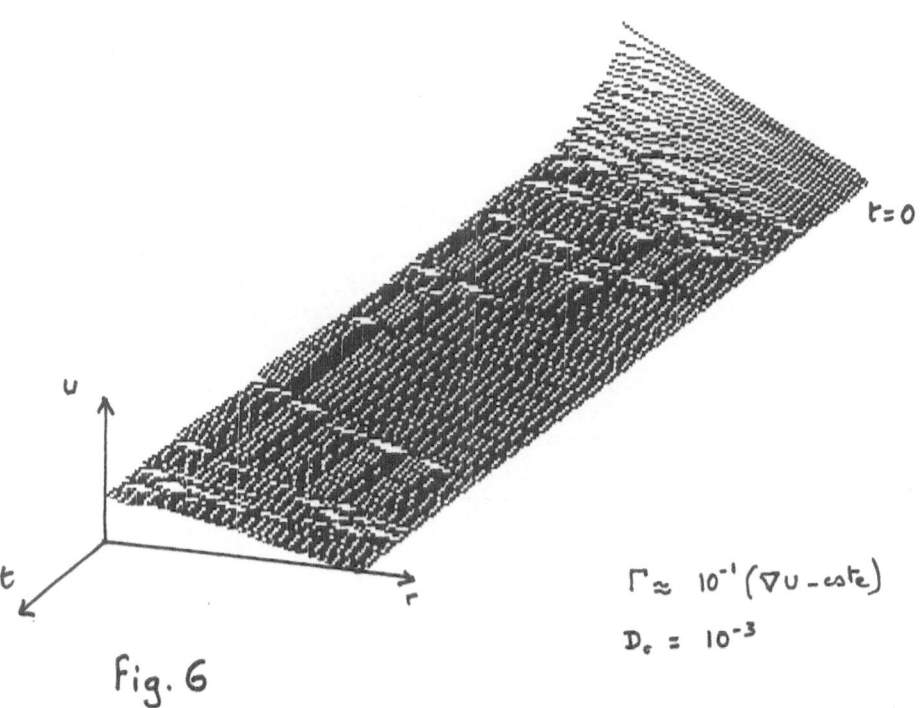

$\Gamma \approx 10^{-1} (\nabla u - cste)$

$D_c = 10^{-3}$

Fig. 6

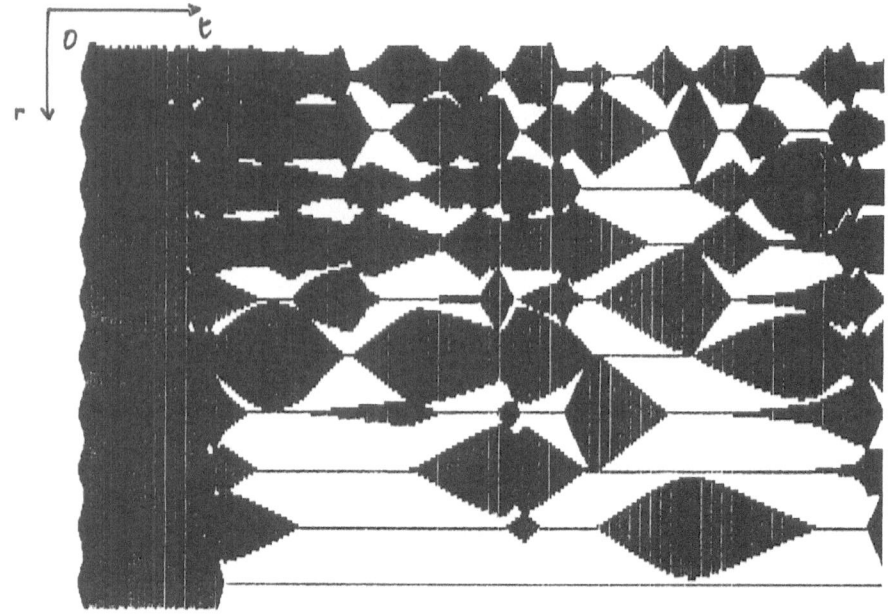

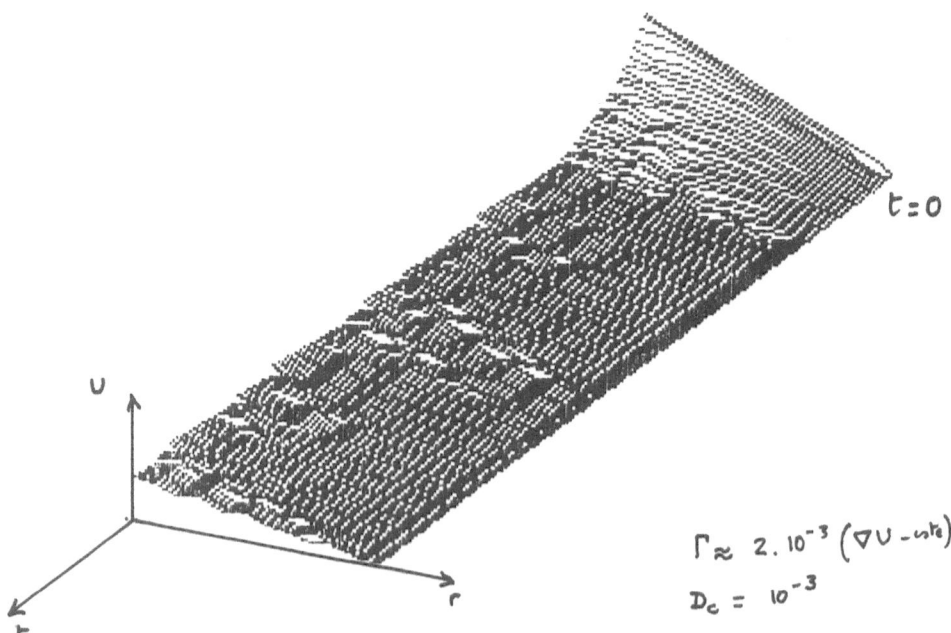

$$\Gamma \approx 2 \cdot 10^{-3} \left(\nabla U - u^{1/2} \right)$$
$$D_c = 10^{-3}$$

fig. 7

123

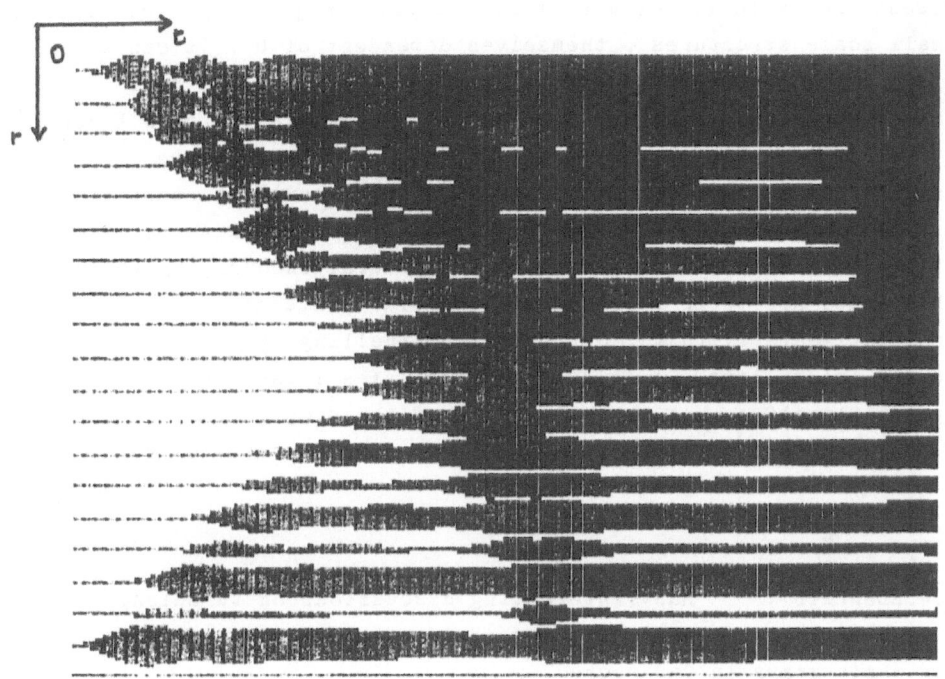

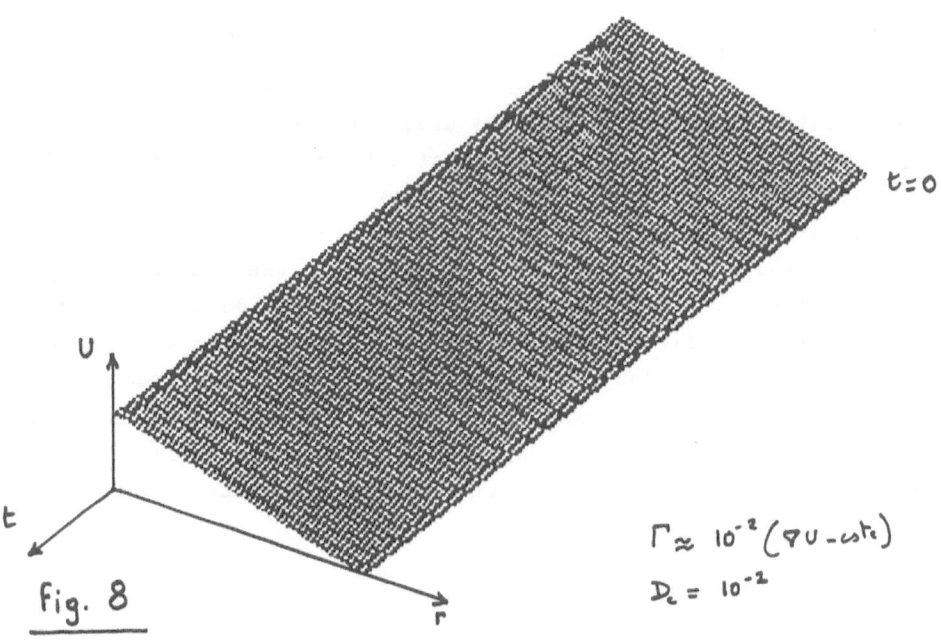

fig. 8

$\Gamma \approx 10^{-2} (\nabla U - \omega t_c)$

$D_c = 10^{-2}$

$t = 0$

represent the evolution of a spatially dependent quantity U controlled
by small scale structures , themselves dependent of U . It can be
adapted to study several transport phenomena (e.g. radial mass
transfert across an accretion disk around a compact star or a black
hole [4]) . Its most obvious application , to tokamak energy
transport , needs careful developments , in particular to take into
account a real particle distribution instead of a monokinetic
population . Such a study is under way . Several encouraging features
of this model are to be stressed already ,for instance it has been
shown experimentally that , during plasma fuelling by injection of
deuterium ice pellets , a cold front propagates inwards faster than
the pellet itself : this is reminiscent of the propagation mechanism
mentionned above , although numerical comparisons would be premature .

ACKNOWLEDGMENTS :   Many interesting discussions on turbulence and
------------------   transport were held with T.Schep and J.Kuijpers
during their stays in Cadarache .

APPENDIX : Transport when some KAM torii still exist

For this first experiment , we have used a naive and probably
uncorrect approach of the dependency of the transverse diffusion
coefficient on the stochasticity parameter : namely we have done the
calculations by attributing a value $\approx \epsilon^2$, ill-named $D_{\varrho\perp}$ to the
coefficient whenever the overlap was exceeded , whereas for values of
$\epsilon < 1$ we took D_ϱ /(1.-s) ; this specific question of the transport
across partially destroyed KAM-torii , in presence of (collisionnal)
diffusion has been the object of several studies , and especially of
[5] . The choice of the grossly simplified dependance used here was
done because the goal of this paper is rather to demonstrate the model
properties than to use it systematically for a realistic application.

REFERENCES :
----

[1]   A.Samain , Ann.Phys.,1979,[4],395

      P.C.Liewer : Nucl.Fusion vol.25 , n° 5 (1985) p.543

[2]   "Turbulence & anomalous transport in magnetised plasmas "
      Cargese workshop ,1986 , D.Gresillon et al. editors , Editions de
      Physique , BP112 , F91944 Les Ulis Cedex , France

      T.Stringer : JET report P (85) 17

[3]   W.M.Manheimer : J.Phys. (Paris) vol.40 (1979) C7-269

[4]   J.Kuijpers : private communication

[5]   M.A.Dubois , M.S.Benkadda : to be published

White Noise and Stochastic Variational Calculus for

Gaussian Random Fields

Takeyuki HIDA

Department of Mathematics

Nagoya University

Nagoya, 464-01, Japan

§0. Introduction

The purpose of this paper is two fold.   Namely,

1) White Noise Analysis, revisited, and

2) a proposal of stochastic variational calculus for Gaussian random
   fields.

Concerning the first subject, we should like to emphasize the
important and, in fact, more basic roles played by a white noise in
the theory of infinite dimensional calculus.   As is well-known, the
history of white noise analysis dates back to 1970 odd and since then
it has developed steadily and successfully as one of the main streams
of infinite dimensional calculus, though to some extent the coverage
is a matter of taste.   We have seen rapid development of the theory
made particularly during the past several years and its various kind
of applications in quantum dynamics.   Now it seems to be time to have
a state-of-the-art survey of the white noise theory.

Let us start with the complex Hilbert space $(L^2) \equiv L^2(E^*, \mu)$ of
white noise functionals, where E^* is a space of generalized functions
and μ is the white noise measure.   To carry out the causal calculus,
which has been prescribed in [4]-[6], we have introduced differential
operators ∂_t, $t \in T$ (the parameter set taken to be a manifold).   Then,
we are naturally led to introduce reasonably larger classes of generalized white noise functionals, where the ∂_t's, their adjoints and
Laplacians have rich domains.   Perhaps it would be better to develop

the theory, in this note, in a logical order from Section 1 to Section 4, rather than tracing the heuristical development.

The second topic is provided to propose a new method of investigation of Gaussian random fields. Our basic idea is that the way of dependcency of the fields as the parameter moves around should be circumstanciated by considering the geometrical structure, namely the symmetry, of the parameter space. The interesting properties can be observed when the parameter spaces or their restrictions are taken to a symmetric spaces like spheres S^d or set of circles, since we can use the symmetry group to describe the geometric structure. This will be discussed in Section 5 and Section 6.

It is also noted that there are some particular cases where the parameter set has no symmetry, but classical theory can be applied. For example, the parameter set is chosen as the collection of all C^∞-contours and the variation of Green's function depending on a contour can be discussed. A stochastic version of this theory gives us some interesting probabilistic meaning.

§1. Background

We start with, as usual, a Gel'fand triple

$$E \subset L^2(T, d\sigma) \subset E^*,$$

where T is a Riemannian manifold and where $d\sigma$ is the volume element derived from the riemannian metric. The space E is usually taken to be a σ-Hilbert nuclear space which is dense in $L^2(T, d\sigma)$. In the case where the symmetry group $G(T)$ for T is given, then the measure $d\sigma$ is assumed to be invariant under the action of the group $G(T)$.

Let a characteristic functional

$$(1.1) \qquad C(\xi) = \exp[\ -\frac{1}{2}\|\xi\|^2\ ]\ , \qquad \xi \in E, \qquad \|\ \| \text{ the } L^2(T, d\sigma)\text{-norm},$$

be given. Then, we are given a probability measure μ on E^* such that

$$(1.2) \qquad C(\xi) = \int_{E^*} \exp[i<x,\xi>]\ d\mu(x).$$

The measure space (E^*, μ) obtained above is called a (T-parameter) white noise, which is a realization of a stationary Gaussian random field $\{W(\xi); \xi \in E\}$ having independent values at every point of T, with the characteristic functional $C(\xi) = E\{\exp[iW(\xi)]\}$ of the form given by (1.1). The measure μ is called a white noise measure with parameter space T or a T-parameter white noise measure. It is also considered as the probability distribution of a white noise.

As soon as we are given a T-parameter white noise (E^*, μ), we can form a complex Hilbert space $L^2(E^*, \mu)$, which is often denoted by (L^2). A member $\varphi(x)$ of (L^2) is called a white noise functional, or a Brownian functional.

Let $\{\xi_n\}$ be a complete orthonormal system (c.o.n.s.) in $L^2(T, d\sigma)$. Then $\{<x, \xi_n>\}$ forms a system of independent standard Gaussian random variables defined on the probability space (E^*, μ). With this system we can form the so-called Fourier-Hermite polynomials based on $\{\xi_n\}$, which are given by the following formula

(1.3) $\quad h_{\{n_k\}}(x) = c \cdot \prod_k H_{n_k}(<x, \xi_k>/\sqrt{2})$ (finite product),

where c is a constant. The sum $\sum_k n_k$ is called the degree of the polynomial.

Let $\mathcal{H}_n$ ($n \geq 0$) be the subspace of (L^2) spanned by all the Fourier-Hermite polynomials of degree n.

A direct-sum decomposition of (L^2) is now established.

Theorem 2.1. (Wiener-Itô decomposition). The Hilbert space (L^2) admits the following decomposition:

(1.4) $\quad (L^2) = \bigoplus_{n=0}^{\infty} \mathcal{H}_n$.

Proof. Since those Fourier-Hermite polynomials are mutually orthogonal, so are the subspaces $\mathcal{H}_n$. Noting that those polynomials form a complete orthonormal system with a suitable choice of constant we can prove the formula (2.2).

The $\mathcal{G}$-transform defined by

(1.5) $(\mathscr{G}\Phi)(\xi) = \int \Phi(x + \xi) \, d\mu(x),$ $\Phi \in (L^2),$

gives us the following isomorphisms:

(1.6) $\mathscr{K}_n \simeq \sqrt{n!} \, \widehat{L^2(T, d\sigma)^{n\otimes}}$ $d\sigma$: the volume element on T,

                                                                $\otimes$: symmetric tensor product,

in the sense that

(1.7) $(\mathscr{G}\Phi)(\xi) = (f, \xi^{n\otimes}) = \int_{T^n} \cdots \int f(u_1, \cdots, u_n) \xi(u_1) \cdots \xi(u_n) \, d\sigma^n.$

where $(\, , \,)$ is the inner product in $L^2(T, d\sigma)^{n\otimes}$.

In view of (1.7), the representation of Φ by the $\mathscr{G}$-transform is called the <u>integral representation</u>.

§2. <u>Generalized white noise functionals</u>.

It is easy to see that the isomorphism (1.6) can be extended to

(2.1) $\mathscr{K}_n^{(-n)} \simeq \sqrt{n!} \, \widehat{H^{-(n+1)/2}(T^n, d\sigma^n)}$

to introduce a class $\mathscr{K}_n^{(-n)}$ of generalized white noise functionals of degree n, where $H^m(T^n)$, $m \in \mathbb{Z}$, is the Sobolev space of order m over T^n.

Set, in particular, n = 1. Then, we consider a functional Φ in $\mathscr{K}_1^{(-1)}$ such as

(2.2) $(\mathscr{G}\Phi)(\xi) = \int_C f(s) \xi(s) \, d\sigma_C(s) ,$

where $d\sigma_C$ is the measure on C derived the Riemannian metric on T. Note that f(s) should be viewed as a function on T concentrated on C. Such a functional Φ defines the restriction of the parameter to C.

Returning to the space $\mathscr{K}_n^{(-n)}$, we should further note that it is our big advantage to be able to have individual $\mathscr{K}_n$ extended to much larger space $\mathscr{K}_n^{(-n)}$. Indeed, such an extension is well established because of the actual function space structure of $L^2(T, d\sigma)$. Also, we can employ the usual technique used when we introduce generalized functions on the circle in terms of the Fourier series. Intuitively speaking, we form a weighted sum of $\mathscr{K}_n^{(-n)}$, $n \geq 0$:

(2.3) $(L^2)^- = \bigoplus_{n=0}^{\infty} c_n \mathscr{K}_n^{(-n)} .$

where $\{c_n\}$ is a decreasing sequence of positive numbers. The space

$(L^2)^-$ is a Hilbert space such that

$$(L^2)^- = \{ \Phi = \sum \Phi_n \ ; \ \Phi_n \in \mathcal{H}_n^{(-n)}, \ \sum c_n^2 \|\Phi_n\|_{-n}^2 < \infty \},$$

where $\| \ \|_{-n}$ is the $\mathcal{H}_n^{(-n)}$-norm.

A member of the $(L^2)^-$ is called a <u>generalized white noise</u> <u>functional</u>. Here, the test functional space $(L^2)^+$ must have been introduced in a usual manner so that a test functional - generalized functional pairing is established. Indeed, we are given a Gel'fand triple:

$$(2.4) \qquad (L^2)^+ \subset (L^2) \subset (L^2)^-.$$

(See [5] for a heuristic interest.)

Another class of generalized functionals looks like an infinite dimensional analogue of the Schwartz space $\mathcal{G}$ over R^d. Take a self-adjoint operator H on $\overset{2}{L}(T,\nu)$, and use the second quantization technique to obtain the test functional space $(\mathcal{G})$. Again form a Gel'fand triple

$$(2.5) \qquad (\mathcal{G}) \subset (L^2) \subset (\mathcal{G})^*.$$

The space $(\mathcal{G})$ forms an algebra dense in (L^2). Exponential functions of the form $\exp[<x,\xi>]$, $x \in E^*$, $\xi \in E$, are members of $(\mathcal{G})$.

In the spaces $(L^2)^-$ and $(\mathcal{G})^*$ just established above, we enjoy much freedom in order to carry out the causal calculus using the differential operators ∂_t's, Laplacians and so forth.

§3. <u>Rotation group</u>.

A rotation of the space E means a linear homeomorphism g of E such that

$$(3.1) \qquad \|g\xi\| = \|\xi\| \qquad \text{for every } \xi \in E.$$

The collection of such g's will be denoted by $O(E)$. The original idea concerning this notion is due to H. Yoshizawa (1969). With the usual product $O(E)$ forms a group. The group $O(E)$ is called a <u>rotation group of</u> E. If necessary, it can be topologized by, for example, the compact-open topology.

The adjoint operator g^*, acting on E^*, of the rotation g can be defined by

$$<x,\ g\xi> = <g^*x,\ \xi>, \qquad x \in E^*,\ \xi \in E.$$

The collection $0^*(E^*) = \{g^*;\ g \in 0(E)\}$ again forms a group, called a <u>rotation group of</u>· E^*.

The two groups $0(E)$ and $0^*(E^*)$ are isomorphic under the correspondence

$$g <\text{———}> g^*, \qquad g \in 0(E),\ g^* \in 0^*(E^*).$$

The following assertion is well known.

<u>Proposition</u> 3.1. The white noise measure μ is $0^*(E^*)$-invariant :

(3.2) $\qquad g^*\mu = \mu \qquad$ for any $g^* \in 0^*(E^*)$.

We are going to introduce three important subgroups of $0(E)$; namely, classes I, II and III. They have different characteristic properties and share their roles in both probability theory and functional analysis. The first two classes may be defined by choosing a c.o.n.s. $\{\xi_n\}$ in $L^2(T,\nu)$ such that $\xi_n \in E$ for every n.

I. <u>Finite dimensional rotations</u>.

Let E_n be the subspace of E spanned by the ξ_i, $1 \le i \le n$. The collection of all rotations g such that $g\big|_{E_n^\perp} = $ identity forms a subgroup, denoted by G_n, of $0(E)$. The group G_n is obviously isomorphic to the n-dimensional rotation group $SO(n)$. The inductive limit

(3.3) $\qquad G_\infty \equiv \underset{n}{\vee}\ G_n$

is called the subgroup of finite dimensional rotations.

II. <u>The Lévy group</u>.

Let π denote an automorphism of the set N of positive integers. We set

$$\rho(\pi) = \underset{N \to \infty}{\lim \sup} \frac{1}{N} \#\{n \le N\ ;\ \pi(n) > N\},$$

where $\#\{\ \cdot\ \}$ means the cardinal number of the set inside of the $\{\ \}$. For $\xi = \underset{n}{\sum} a_n\xi_n$ we define g_π by

$$g_\pi\xi = \underset{n}{\sum} a_n \xi_{\pi(n)} \quad ,$$

and define $\mathscr{G}$ by

$$\mathscr{G} = \{g_\pi \in O(E);\ \pi \text{ is an automorphism of } N,\ \rho(\pi) = 0\}.$$

The collection $\mathscr{G}$ is a subgroup of $O(E)$ called the Lévy group (see [3],IIIème Part.). We often consider a larger group like $\tilde{\mathscr{G}} \equiv \mathscr{G} \vee G_\infty$ when we discuss harmonic analysis.

    III. Whiskers.

The third class depends not on $\{\xi_n\}$ but does heavily depend on the geometric structure of the manifold T. Take a diffeomorphism ψ of T such that g_ψ defined by

(3.4) $\qquad (g_\psi \xi)(u) = \xi(\psi(u)) |\frac{d\psi}{d\nu}(u)|^{1/2}$

is a member of $O(E)$. We are pArticularly interested in a continuous one-parameter subgroup $\{g_t\}$ of $O(E)$, each member of which is given by a diffeomoephism ψ_t of T as in (3.4).   The group property

$$g_t g_s = g_{t+s}$$

necessarily requires

(3.5) $\qquad\qquad \psi_t \circ \psi_s = \psi_{t+s}, \qquad t,s \in R^1.$

Such a one-parameter subgroup is called a whisker.

    Since each $\psi_t(u)$ is a diffeomorphis of T onto itself, the relation (3.5) imples that there exists a diffeomorphism $f(u)$ of T such that

(3.6) $\qquad \psi_t(u) = f[f^{-1}(u) + t].$

With this expression of ψ_t the following assertion can be proved.

    Proposition 3.2. If the manifold T is either an abelian group or a symmetric space, a whisker is uniquely determined by its infinitesimal generator

(3.7) $\qquad \alpha \equiv \frac{d}{dt} g_t \big|_{t=0} ,$

which is expressed in the form

(3.8) $\qquad \alpha = a(u)\frac{d}{du} + \frac{1}{2} a'(u),$

where $a(u) = f'[f^{-1}(u)].$

    General theory tells us that commutation relations of whiskers can be described in terms of those generators.

There are several subgroups of O(E) consisting of whiskers which are isomorphic to classical linear groups, and which have their own probabilistic meanings.   For details, see [4] Chapt.5, [7], [8].

§4. Stationary random fields.

For further concrete discussion we need to specify the parameter set   T.   Namely, if the manifold is either an abelian group like R^d, or else a symmetric space like $S^d \simeq SO(d+1)/SO(d)$, then we establish the following results.

i)   The case where T is an ablean group.   The   measure ν on T is taken to be the Haar measure.   Now let g_t, t ∈ T, be defined by

(4.1) $g_t \xi(u) = \xi(ut^{-1})$.

Then, we are given a one-parameter group $\{g_t\} \subset O(E)$.   The adjoint operators g_t^* also form a one-parameter group, and the U_t defined by

(4.2) $(U_t \Phi)(x) = \Phi(g_t^* x)$,   t ∈ T,

is a unitary operator acting on (L^2), since μ is invariant under g_t^*. The operator U_t extends to a continuous linear operator on $(L^2)^-$. The collection $\{U_t;\ t \in T\}$ is a   continuous   one-parameter unitary group which defines $(L^2)^-$-valued   stationary random field $\{X(t)$; t ∈ T}   in such a way that

                    $X(t) = U_t \Phi$,   t ∈ T,

for a given $\Phi \in (L^2)^-$.

Examples of stationary random fields.

1) White noise.

Let   T   be R^d.   The white noise (E^*, μ), given in Section 1, defines a generalized Gaussian random field   X(t)   by setting

(4.3) $X(t) \equiv X(t,x) = x(t)$, $x \in E^*$.

It is a stationary random field, since it is expressible as

          $X(t) = U_t X(0)$,

where U_t comes from the shift operator as in (3.7):

(4.4) $S_t :\ \xi(u) \ \longrightarrow\ \xi(u - t)$.

This means that the white noise is stationary, where the group in question is R^d itself.

2) <u>Lévy's Brownian motion with T = S^1 or S^d</u> (unit sphere).

For simplicity, T is taken to be the unit circle S^1 and the measure ν is the uniform measure $d\theta$ on S^1. The white noise measure μ is introduced on E^* with the characteristic functional

(4.5) $$C(\xi) = \exp[-\tfrac{1}{2} \|\xi\|^2],$$

where $\| \ \|$ is the $L^2(S^1, d\theta)$-norm.

The Lévy Brownian motion $\{X(\theta); \theta \in S^1\}$ with parameter space S^1 is a Gaussian system with $E(X(\theta)) = 0$ satisfying

(4.6) $$E\{|X(\theta) - X(\theta')|^2\} = \rho(\theta, \theta'),$$

where $\rho(\theta, \theta')$ is the Riemannian distance on S^1 between θ and θ' (see P. Lévy [3]). Such a process may be realized by white noise integral as is given below:

(4.7) $$X(\theta) = 2^{-1/2} [\int_{s(\theta)} x(\alpha)d\alpha - \int_{S^1 - s(\theta)} x(\alpha)d\alpha],$$

where $s(\theta)$ is the semi-circle : $s(\theta) = \{\varphi \ ; \ |\theta - \varphi| \leq \pi/2 \}$ with center θ. Obviously $\{X(\theta)\}$ is stationary with respect to the group of rotations of the circle.

In a similar manner, we can form a Lévy Brownian motion with with parameter space S^d : $\{X(\theta) \ ; \ \theta \in S^d\}$, where $\theta = (\theta_1, \theta_2, \ldots, \theta_d)$ with $0 \leq \theta_1 < 2\pi$, $0 \leq \theta_2, \theta_3, \ldots, \theta_d \leq \pi$. Namely, we start with an S^d-parameter white noise (E^*, μ), where E^* is the space of generalized functions on S^d. Then, form $X(\theta)$ by the white noise integral in a similar manner to (4.7):

(4.8) $$X(\theta) = c\{\int_{s(\theta)} x(\theta)d\sigma(\theta) - \int_{S^d - s(\theta)} x(\theta)d\sigma(\theta)\} \ ,$$

where $s(\theta)$ is the semi-sphere of S^d with center θ, $d\sigma$ is the surface element and c is a positive constant such that $c^2 = 2^{-1}\Gamma(\tfrac{d+1}{2})\pi^{(d-1)/2}$. With this expression it is easy to prove the following equality:

$$E|X(\theta) - X(\theta')|^2\} = \rho(\theta, \theta'), \qquad \rho : \text{Riemannian distance.}$$

This implies that $\{X(\theta)\}$ is a Lévy's Brownian motion.

<u>Remark</u>. Restriction of the parameter θ.

i) For Lévy's Brownian motion the restriction of parameter is easily done by the mapping

(4.9) $\pi : (\theta_1, \theta_2, \ldots, \theta_d) \longrightarrow (\theta_1, \theta_2, \ldots, \theta_{d-1}, \frac{\pi}{2})$.

However, it is not straight forward, as we have seen in [6], in the case of white noise.   Actually, particular generalized white noise functionals can serve to obtain the restriction, thereby we are given a white noise concentrated on lower dimensional submanifold.

In both cases, the Brownian motion and the white noise, we can form them with parameter space of one dimension lower due to natural mapping of restrictions.

§5. Gaussian random fields depending on a manifold.

We are now ready to discuss a Gaussian random field $\{X(C); C \in \mathbf{C}\}$ depending on a Riemannian manifold C in the Euclidean space R^d. We assume that $\mathbf{C}$ consists of C^∞-manifold homeomorphic to the sphere S^{d-1} and that each $X(C) = X(C,x)$, $x \in E^*$, is a generalized white noise functional that is linear in x.   Namely, $X(C)$ always lives in the space $\mathcal{H}_n^{(-n)}$.

Here are some illustrative examples.

Example 1. The Lévy Brownian motion $\{X(a); a \in R^d\}$ represented as a white noise integral (McKean's representation).   Let $\mathbf{C} = \{ C_a; a \in R^d\}$, C_a being the (d-1)-dimensional sphere with diameter $\overline{oa}$ where o denotes the origin.   Each $X(a)$, which may be written as $X(C_a)$, is expressed, on the probability space (E^*, μ), in the form

(5.1) $X(C_a) = c(d) \int_{[C_a]} |u|^{-(d-1)/2} x(u) \, du^d$, $x \in E^*$,

where $[C_a]$ is the ball with boundary C_a, and where $c(d)$ is a constant given by

$$c(d) = \{2^{d-2}(d-1)|S^{d-1}|^{-1} \cdot B(\frac{d-1}{2}, \frac{d-1}{2})\}^{1/2}.$$

Thus the system $\{X(C_a); a \in R^d\}$ is a version of the R^d-parameter Lévy Brownian motion.

Example 2 (Si Si [11]).   Take the Lévy Brownian motion $\{X(a); a \in R^2\}$.   Let C be a C^∞-curve homeomorphic to a circle.   For a fixed point p the conditional expectation

(5.2)   Y(C) = E{X(p) / X(a), a ∈ C}

is a random variable depending on a curve C.   We therefore have a

Gaussian random field {Y(C); C ∈ **C**}, where **C** = {C ; C$^{\infty}$-curve, home-

omorphic to a circle}.

If, in particular, **C** is replaced by the class **C**$_0$ of circles that

pass through the origin, then the explicit form of Y(C) expressed as

an integral of X(a) over C can be obtained (see [11]).

Example 3.   Applications of the Dirichlet problem ([8]) and the

Neumann problem.   Let D be a domain in R^d, and assume that the

boundary C = ∂D is smooth enough.   Take an ordinary d-dimensional

Laplacian operator Δ, and let G(u,v;C) be the Green's function for

the domain D (with boundary C) and the Laplace equation   Δ f = 0.

Then, a random variable X(C)   on (E*, μ) is deefined by

(5.3) $X(u,C) = \int_D G(u,v;C) \, x(v) \, d\sigma(v)$, $x \in E^*$,

dσ : Lebesgue measure on R^d.

If C is fixed, the X(u,C) is a random field with parameter space R^d

and it holds that

(5.4)   Δ X(u,C) = x(u).

This can rigorously be proved by applying the 𝒮-transform,   although

X(u,C) is not an ordinary function of u but a random function.

Example 4.   Under the same situation as in the last example, we

can even define

(5.5) $Z(u,C) = \int_C N(u,v;C) \, Y(v) \, d\sigma(v)$,   u ∈ D,

by choosing a suitable Gaussian random field {Y(v); v ∈ C}.

In parallel with the Neumann problem for the partial differen-

tial equations, we can discuss harmonic property and boundary value

for the random field { Z(u,C) ; C ∈ **C**}.   Our interest lies however

in the variation of Z(u,C) in C for a fixed u, since the classical

theory tells us the explicit form of the variation of N(u,v;C) like

in the case of the Green's function.

§6. Variational calculus for Gaussian random fields.

Given a Gaussian random field $\{ X(C) ; C \in \mathbf{C} \}$, where $\mathbf{C}$ is a collection of Riemannian manifolds in a Euclidian space. Then, we are interested in the way of dependency of $X(C)$ when C moves and deforms within the class $\mathbf{C}$. What we are going to discuss in this note is, of course, far from the general theory, however some special cases can be discussed by using their proper techniques, so that one can observe the stochastic character and even hidden symmetry.

Two particular cases will be discussed.

[1] The class of manifolds is chosen to be $\mathbf{C}_0$ the collection of all $(d-1)$-dimensional spheres in R^d. In an obvious manner $\mathbf{C}_0$ may be identified with $R^d \times R_+$ as a topological space.

We remind that there is a conformal group, denoted by $\tilde{C}(d)$, that is acting on R^d, and that it consists of the following i) – iv): Let u denote the variable running through R^d. Then,

i) shifts $u \longrightarrow u - t$, $t \in R^d$,

ii) isotropic dilation $u \longrightarrow ue^t$, $t \in R^1$,

iii) rotations group $SO(d)$

iv) special conformal transformations = conjugates to the shifts

with respect to w,

where w is the reflection: $u \longrightarrow \dfrac{u}{|u|^2}$.

Put the transformations i) – iv) together. And one is given the conformal group which is $\frac{1}{2}(d+1)(d+2)$ dimensional.

The following assertion can easily proved.

<u>Proposition</u> 6.1. The class $\mathbf{C}_0$ of spheres is invariant under the action of the conformal group $\tilde{C}(d)$, and the action of the group on the space $\mathbf{C}_0$ is continuous and transitive.

With this property of the conformal group, we can speak of the variation of a random field depending on a sphere. Set

(6.2) $X(C) = \int_C F(s) X(s) \, dv(s)$, $C \in \mathbf{C}$,

where $\{X(s); s \in C\}$ is a continuous Gaussian random field, $F(s)$ is continuous and dv(s) is the surface element over the sphere C.

Infinitesimal deformation δC of C is induced by infinitesimal change of members in C̃(d) and eventually it gives us the variation of X(C). Hence, we have to consider the action of the Lie algebra of C̃(d). Let C(d) be the unitary representation of C̃(d) on E ; namely for g̃ ∈ C̃(d)

$$g\xi(u) = \xi(\tilde{g}u)|J|^{1/2}, \qquad u \in E, \quad J :\text{Jacobian.}$$

We can take a base { α_j; $1 \le j \le \frac{1}{2}$ (d+1)(d+2)} of the Lie algebra of of the group C(d). Members of the base may come from one-parameter subgroups (whiskers) of O(E) by taking infinitesimal generators as in the formula (3.7). With these notations we establish

Theorem 6.1. Let X(C) be given by (6.1) with X(s) in $\mathcal{K}_1$, and assume that C runs only through C_0. Then, the variation δX(C) of X(C) is expressed in the form

$$(6.2) \qquad \delta X(C) = \sum_j dt_j \int_C \{\alpha_j(FX)(s)\delta_j(s)dv(s) + (FX)(s)\delta_j(dv(s))\},$$

where $\delta_j(s)$ denotes the difference between C and C + δC, and where $\delta_j(dv(s))$ stands for the infinitesimal difference of the surface element dv at s.

Proof. First apply $\mathcal{G}$-transform to the expression (6.2) so that we obtain an ordinary functional of ξ and C. Then, we appeal to the classical theory of calculus of variations (see e.g. Lévy [2]), where we see a formula for a functional I = $\int_C$ u ds, C :contour in R^2,

$$(6.3) \qquad \delta I = \int_C (\delta u\ ds + u\delta ds),$$

where ds is the line element along the curve. The conclusion (6.2) can be proved by paraphrasing the above formula, and by extending the result to the case of higher dimensional manifold. (See [6] for more interpretation.)

We then consider a white noise integral

$$(6.4) \qquad X(x) = \int_{C_0} F(s)\ x(s)\ dv(s), \qquad x \in E^*.$$

where C_0 passes through the origin. The diameter of C_0 is denoted as $\overline{oa}$.

Consider now the subgroup of O(E) which leaves the C_0 invariant. Such a group, denoted by G_a, involves a subgroup of the group generated by special conformal transformations, the isotropic dilation and the isotropy group at a, which is isomorphic to SO(d-1).

Let H denote the Hilbert space $L^2(C_0,dv)$ and define U_g by

(6.5) $\qquad (U_g f)(v) = f(gv)|J|^{1/2}$, $\qquad f \in H$, $\qquad$ J : Jacobnian.

Then, we can easily prove the following proposition by applying the reflection with respect to the unit sphere to show that the group G_a is isomorphic to the homothety group acting on R^{d-1}.

Proposition 6. 2. The unitary reoresentation $U = \{U_g; g \in G_a\}$ of the group G_a is irreducible.

Note that U is identified with a subgroup of O(E).

Theorem 6.2. Let $X(x)$ be defined by (6.4). Then, the space spanned by $\{X(g^*x); g \in G_a\}$ coincides with the space spanned by the system $\{<x,\xi> : \xi \in \mathcal{Y}(C_a)\}$.

Proof. Observe the expression of $X(x)$ in (6.4) and apply g^* to x. Then we have

$$X(g^*x) = \int_{C_a} (gF)(s)\, x(s)\, dv(s).$$

Since gF, $g \in G$, generates dense subset of $L^2(C_a, dv)$, $x(s)$ can be recovered, and the theorem has been proved.

[2] Let $\mathbb{C}$ be the class of all possible C^∞-manifolds isomorphic to a sphere, while the random fields with parameter set $\mathbb{C}$ is very much restricted.

Theorem 6.3. ([9]) Let $X(u, C)$ be the field given by (5.4). Then, we have

(6.6) $\quad \delta X(u,C) = \int_D \delta G(u,v;C)\, x(v)d\sigma(v) + \int_C G(u,s;C)x(s)\delta n(s)dv(s).$

Proof. The $\mathcal{Y}$-transform of the random variable X(u,C) is given by

$$\{\mathcal{Y}X(u,C)\}(\xi) = \int_D G(u,v;C)\xi(v)d\sigma(v).$$

Take its variation when C changes by δC. Then, we have (see P. Lévy, [2])

$$\int_D \delta G(u,v;C)\xi(v)d\sigma(v) \; + \; \int_C G(u,s;C)\xi(s)\delta n(s)dv(s).$$

Applying the $\mathcal{Y}^{-1}$-transform, we obtain (6.6), where the second term, of the above expression correspons to a generalized white noise functional.

Remark. 1) The formula of the variation of $G(u,v;C)$ may be given by the Hadamard equation

$$\delta G(u,v;C) = -\frac{1}{2\pi} \int_C \frac{\partial}{\partial n} G(u,m;C) \frac{\partial}{\partial n} G(m,v;C) \delta n(s)dv(s).$$

Remark. 2) The first and the second terms of the right hand side of (6.6) can be discriminated, since they have different order in the mean square.

To close this section we should like to note an important remark concerning the concept of the innovation in the generalized sense, although we do not intend to give a definition in the case of random fields.

Consider the case where the variation is taken around a circle. We know many concrete examples where a white noise integral over the circle arises and the term is discriminated from others, like in the case of $X(u,C)$ as in (6.6). We can also see interesting examples in [11-12] with this property. What we should claim is that the white noise defining the X does come out from the variation. In terms of the above $X(u,C)$ as an example, we can form the original white noise $x(u)$ by taking the variation not by using the formula (5.,4). Such a situation is well illustrated also in the paper [10].

[REFERENCES]

[1] P. Lévy, Sur la variation de la distribution de l'électricité sur un conducteur dont la surface se déforme. Bull. Soc. math. France, 46 (1918), 35 - 68.

[2] —————, Problèmes concrets d'analyse fonctionnelle. Gauthier-Villars, 1951.

[3] —————, Le mouvement bnrownien fonction d'un point de la sph-

ère de Riemann. Rendiconti del Circolo Mat. di Palermo. ser. II 8 (1959) 297 - 310.

[4] T. Hida, Brownian motion. Iwanami 1975; english ed. Springer-Verlag, 1980.

[5] ——— , Analysis of Brownian functionals. Carleton Math. Lec. Notes no.13, 1975.

[6] ——— , White noise analysis and Gaussian random fields. Proc. 24th Winter School of Theoretical Physics, Karpacz, 1988.

[7] T. Hida, K.-S. Lee and S.-S. Lee, Conformal invariance of white noise. Nagoya Math. J. 98 (1985), 87 - 98.

[8] T. Hida and Si Si, Variational calculus for Gaussian random fields. Proc. 1988 Warsaw Conference.

[10] K. -S. Lee, White noise approach to Gaussian random fields. (to appear).

[11] Si Si, A note on Lévy's Brownian motion. Nagoya Math. J. 108 (1887), 121 - 130.

[12] ——— , A note on Lévy's Brownian motion, II. Nagoya Math. J. 114 (1989), 165 - 172.

[13] ——— , Gaussian processes and conditional expectations. BiBoS Notes Nr. 292/87.

CHAOS IN VIBROTRANSPORTATION.

Max-Olivier HONGLER

Institut de Microtechnique

Département de Mécanique

Ecole Polytechnique Fédérale de Lausanne

CH-1015 LAUSANNE.

Abstract : A vibro-impact device commonly used in automated assembly lines is discussed in the light of recent developments of non-linear dynamics. Specifically, the existence of non-linear phenomena such as cascade of bifurcations and chaotic solutions are examined. The illustration presented here, namely a vibratory transporter has, besides its own engineering interest, the merit to exhibit a dynamics described by a well known 2-dimensional, dissipative mapping.

1. INTRODUCTION.

In 1961, the eminent Professor of Mechanical Engineering, R. M. Rosenberg concluded an article devoted to non-linear oscillations in the following terms [1] : "*The outlook regarding progress in non-linear oscillations is bright for those who like to do research, and bleak for those who like to see results. (....). At the present time, no hope exists for a unified theory or body of knowledge regarding the solutions of non-linear problems. The basic reason behind this statement is the failure to define the field* ." Since this remark was written, great progresses has been achieved in the field of non-linear dynamics. Recently, the engineering community has started to explore more systematically the implications of these new mathematical developments [2,3]. Illustrations ranging from magnetically levitated vehicles, chaos in elastic continua, impact print head, non-linear electric circuits, etc...are reported ; (see further references in [3]). The aim of the present paper is to bring a contribution to this exploration in the domain of automated assembly systems.

We shall discuss here a dissipative, non-linear system driven by external impulsive forces. The dynamics of this system is described by a discrete mapping which now stands as one corner-stone of the studies in non-linear science. Let us here emphasize that the dynamics of impulsively driven systems is exactly described by non-linear mappings ; whereas, in the study of differential equations, mappings often result either from approximations or modelizations of the original equations of the motion. Besides its own engineering interest, the device discussed in this paper, presents the advantage to admit dynamical evolutions equations already encountered in the mathematical literature.

Before, we introduce our particular device, let us first formally exhibit the type of equations of motion we will have to deal with. These have the recurrent form :

$$\tau_{n+1} = f_1(\tau_n, \psi_n) \tag{1a}$$

$$\psi_{n+1} = f_2(\tau_n, \psi_n), \tag{1b}$$

where f_1 and (or) f_2 are non-linear functions . The mappings to be derived are dissipative i.e. the Jacobian of Eqs. (1a,b) is less than one.. Among the infinitely rich variety of possible choices for the function f_1 and f_2 , let us mention here :

$$f_1 = \tau_n + \alpha \psi_n \tag{2}$$

$$f_2 = \varepsilon \psi_n + (1+\varepsilon)\cos(\tau_{n+1}), \tag{3}$$

where α is an external parameters and $0 < \varepsilon < 1$ relates the dissipation ; the Jacobian of this transformation equals ε. Eqs. (2) and (3) describe the so-called dissipative standard mapping which is discussed in [3].

Our paper is organised as follows : In section 2, we introduce the problem of vibro-transportation. It is observed that Eqs. (2) and (3) are embedded in the dynamics of this system. In section 3, we report results of numerical investigations performed for a set of parameters which occur in actual situations. Finally, section 4 is devoted to conclusions and remarks..

2. VIBRO-TRANSPORTATION.

One of the difficulties in the realization of automatic assembly lines is to convey parts to the ad-hoc locations in the chain. A solution, commonly adopted, is the use of vibratory transporters (also called vibratory feeders). Basically, a vibratory feeder is constituted by an oscillating track on which the parts to be conveyed are disposed. When the track is set into motion, the mobile, lying on it, is itself set in movement. Since the pioneering work A.H. Redford & G. Boothroyd [6], theoretical and experimental aspects of vibro-transportation have been abundantly studied, (a selection of articles is given in [7]). This important activity of research clearly reflects the difficulties which the constructors of feeders have to deal with.

Schematically, the device is represented in Fig.1 and 2 where the notations to be used are introduced. The reference frame xOy is mobile and attached to the track.

In actual applications, the vibratory transporter is either a bowl or a linear track. Here, we shall restrict our discussion to the linear case for which the centripetal and Coriolis accelerations are absent- (the dynamics for the bowl shape case presents, in its essence, identical features as locally it reduces to the case Fg. 2).

In view of Fig. 2, the general equations of the motion have the form :

$$m\ddot{x}(t) = ma\omega^2 \sin(\omega t) - mg\sin(\alpha) + F \qquad (4a)$$

$$m\ddot{y}(t) = mb\omega^2 \sin(\omega t + \gamma) - mg\cos(\alpha) + N, \qquad (4b)$$

where dots denote the derivatives with respect to the time, F and N stand respectively for the friction and the constraints forces, α is the slope of the track, g the gravitational acceleration and γ the phase shift between the parallel and perpendicular components of the excitation force.

Depending on the external parameters, various types of motions exist and a detailed analysis of the possible periodic motions is given in [8]. Here, we shall confine our attention to the pure jumping regimes (i.e. sticking to the track is neglected). In these ballistic regimes, the dynamics between the impacts with the feeder, simply reduces to free flight equations, namely :

$$\ddot{u}(\tau) = \sin(\tau) - k \qquad (5a)$$

$$\ddot{v}(\tau) = \eta\sin(\tau + \gamma) - ktg(\alpha) \qquad (5b)$$

where τ, $u(\tau)$, $v(\tau)$, k, η are dimensionless quantities defined by :

$$k = g\cos(\alpha)/b\omega^2 \quad ; \quad \eta = a/b \quad ; \quad \tau = \omega t$$

$$u = y/b \quad ; \quad v = x/a . \tag{6}$$

In the ballistic regimes, the parameter k <1 ; typically for the feeder presented in this talk k $\cong$ 0.5 to 0.8.

The dynamics at the nth impact time $\tau = \tau_n$ is specified in the form :

$$\frac{\partial}{\partial \tau}u(\tau)\Big|_{\tau = \tau_n + \varepsilon} = -R_\perp \frac{\partial}{\partial \tau}u(\tau)\Big|_{\tau = \tau_n - \varepsilon} \tag{7a}$$

$$\frac{\partial}{\partial \tau}v(\tau)\Big|_{\tau = \tau_n + \varepsilon} = R_{//} \frac{\partial}{\partial \tau}v(\tau)\Big|_{\tau = \tau_n - \varepsilon} \tag{7b}$$

where the coefficients of the perpendicular, (parallel) restitution are denoted respectively by $R_\perp$ and $R_{//}$ and ε is a infinitesimal quantity which relates times just before and after the impact time $\tau = \tau_n$. Obviously we have :

$$0 < R_\perp < 1 \quad \text{and} \quad 0 < R_{//} < 1. \tag{8}$$

Now , let us introduce the notations :

$$\frac{\partial}{\partial \tau}u(\tau)\Big|_{\tau = \tau_{n+\varepsilon}} = \Psi_n \quad \text{and} \quad \frac{\partial}{\partial \tau}v(\tau)\Big|_{\tau = \tau_{n+\varepsilon}} = \Phi_n \tag{9}$$

Using Eqs. (7-9), we can renew the initial conditions each time an impact has occured. Hence, the direct integration of the free flight Eqs. (5a, b) yields the set of non-linear mappings :

$$-k/2 \; (\tau_{n+1}-\tau_n)^2 + (\cos(\tau_n)+\Psi_n)(\tau_{n+1}-\tau_n) =$$
$$= \sin(\tau_{n+1})-\sin(\tau_n) \qquad (10)$$

$$\Psi_{n+1} = R_\perp(\; k(\tau_{n+1}-\tau_n)+\cos(\tau_{n+1})-\cos(\tau_n)-\Psi_n) \qquad (11)$$

$$\Phi_{n+1} = R_{//}(-\eta \; (\cos(\tau_{n+1}+ \gamma) - \cos(\tau_n+\gamma)) -$$
$$-ktg(\alpha)(\tau_{n+1}-\tau_n) + \Phi_n). \quad (12)$$

The transport rate itself can be calculated with the mean velocity W_n (in the parallel direction) attained between successive impacts ; thus we obtain :

$$W_n = \frac{1}{\tau_{n+1}-\tau_n} \int_{\tau_n}^{\tau_{n+1}} \left(\frac{\partial}{\partial\tau'} v(\tau')\right) d\tau' = \qquad (13)$$

$$W_n = \left\{\eta\cos(\tau_n+\gamma)+\Phi_n\right\}-\frac{k}{2}tg(\alpha)\left(\tau_{n+1}- \tau_n\right) - \left(\frac{\eta}{\tau_{n+1}-\tau_n}\left(\sin(\tau_{n+1}+\gamma)-\sin(\tau_n+\gamma)\right)\right)$$

The dynamics of the model is now completely characterized by Eqs. (10-13).

Let us discuss the solutions. First of all, one has to remark that Eqs. (10) and (11) can be discussed independently of Eq. (12). Eqs. (10) and (11) are precisely the mapping recently studied in [9,10]. Let us briefly recall the results obtained in [4,5,9,10]. The mappings Eqs. (10) and (11) exhibits the cascade of bifurcations (here, the control parameter is k), discovered by Myrberg and Feigenbaum [11]. The period one solutions (i.e. $\tau_n = \tau_0 + (2\pi n)r$, $\mathbf{N} \ni r$) are immediately found in the form:

$$\Psi_{n+1} = \Psi_n = \Psi = \frac{(2\pi r k)\, R_\perp}{1+R_\perp} \tag{14a}$$

$$\tau_0 = A\cos\left\{\pi r k\left(\frac{1-R_\perp}{1+R_\perp}\right)\right\} \tag{14b}$$

Once $R_\perp$ is fixed and hence the mobile to convey selected, the unique control parameter of the problem is k. The stability intervals for the period one solutions are obtained by a linearization procedure. The result reads [4,9,12] :

$$k_{1,r} > k > k_{2,r} \tag{15}$$

where $k_{1,r}$ and $k_{2,r}$ read :

$$k_{1,r} = \left\{ \pi^2 r^2 \left(\frac{1-R_\perp}{1+R_\perp}\right)^2 + 4\frac{\left(1+R_\perp^2\right)^2}{\left(1+R_\perp\right)^4} \right\}^{-\frac{1}{2}}$$

and

$$k_{0,r} = \frac{1}{\pi r}\left(\frac{1+R_\perp}{1-R_\perp}\right)$$

When k is decreased below $k_{1,r}$ a stable period two orbit is found. This behaviour is observed until a new critical value, say $k_{2,r}$, is reached, where a new period doubling occurs.... and so on until $k_{\infty,r}$, where the chaotic regime is attained [4,5,11,13,14]. A sketch of this situation is summarized in Fig. 3. The succession of the critical values approaches the accumulation point $k_{\infty,r}$ according to the the equation [4,5,11,13,14] :

$$\lim \frac{k_{\eta+1,r} - k_{\eta,r}}{k_{\eta+2,r} - k_{\eta+1,r}} = 4.6992... \qquad \eta = 1, 2, 3, ... \tag{16}$$

Using Eq. (13), the transport rate in the simple periodic regimes defined by Eqs (14a,b) takes the form :

$$W_n = W = \pi r k \left\{ \eta \cos(\gamma) \left(\frac{1 - R_\perp}{1 + R_\perp} \right) - tg(\alpha) \left(\frac{1 + R_{//}}{1 - R_{//}} \right) \right\} -$$

$$- \eta \sin(\gamma) \left(1 - \pi^2 k^2 r^2 \left(\frac{1 - R_\perp}{1 + R_\perp} \right)^2 \right)^{\frac{1}{2}} \qquad (17)$$

While it is relatively obvious to obtain Eq. (17), the estimation of the transport is far less trivial in the case of chaotic regimes. To simplify the expressions without lost of generality, let us confine ourselves now to the case $\gamma = 0$. With the use of Eq. (10), Eq. (13) can be written in the form :

$$W_n = \left(\Phi_n - \eta \Psi_n \right) + \left(\tau_{n+1} - \tau_n \right) \left(\frac{\eta k}{2} - \frac{k}{2} tg(\alpha) \right) \qquad (18)$$

In the chaotic regime, the quantities Φ_n, Ψ_n, and τ_n form pseudo-stochastic sequences which statistical properties are unknown. To calculate the average transport rate, one would in fact need the probability densities governing these quantities. Analytical results which give such invariant measures are not yet available. Hence, one has to resort to numerical exploration. Let us distinguish between two regimes.

1) For relatively large restitution parameter, namely $0.8 < R_\perp < 1$, the mapping Eqs. (10) and (11) can be approximated [4,16]. This is achieved by observing that the quantity :

$$(\sin(\tau_{n+1})-\sin(\tau_n))/(\tau_{n+1}-\tau_n)\cong 0 \tag{19}$$

is a vanishingly small quantity for appropriately choosen initial conditions [4,16]. Using this approximation, the mapping Eqs. (10-11) can be revritten in the form :

$$\tau_{n+1} = \tau_n + \frac{2}{k}\left(V_n\right) \tag{20a}$$

$$V_{n+1} = R_\perp V_n + \left(1 + R_\perp\right)\cos\left(\tau_{n+1}\right) \tag{20b}$$

where we have introduced the notation :

$$V_{n+1} = \cos(\tau_n) + \Psi_n$$

The mapping Eqs. (20a,b) exhibits precisely the form of Eqs. (2) and (3).

2) In actual realizations, the restitution coefficients are of the order $R_\perp \cong 1/3$ and $R_{//} \cong 0.2$. In this case the approximation Eq. (19) is not valid and the implicit mapping Eq. (10) has to be solved numerically.

Such calculations have been performed in [15]. Typical results are sketched in Fig. 4, where, for a fixed $\eta = 4$, the quantity : $<W> = 1/N \sum_n(W_n)$ is plotted against the external control parameter k. The period one regimes defined for $k_{1,r} < k < k_{2,r}$; r=1, 2, 3, 4 in Eq. (15) lead to high transport rates $<W>$. In the chaotic regions, $<W>$ exhibits a (positive definite) random looking behavior with a mean lower than the values obtained in the periodic cases. Remark that the statistical nature of $<W>$ in the chaotic regime looks to be independent on the parameter k ; i.e. no net tendancy for the mean emerges. Let us further devote a special attention to the fine structure of the curve $<W>$ in chaotic regions. We can ask whether this fine structure remains unchanged under (reasonably small) perturbations and how it is affected by the precision used to iterate the mapping Eqs. (10) and (11) ; (remember, Eq.(10) is implicit and has therefore to be solved by succesive increments) ? To answer these questions, using identical τ_0, Ψ_0 and Φ_0 , we have calculated $<W>$, with different increments ; (the increments chosen range from 0.016 to 0.026 by step of size 0.001). We clearly observe that the period-one regions are unchanged. In the random regimes howewer, the $<W>$ are indeed dependent on the choice of the increments. In Fig 5, we show the mean $<<W>>$ obtained from these different calculations. The fine structure of <<W>> exhibited in Fig. 5 therefore is more robust under small perturbations than the fine structure of Fig. 4. Observe how the details drawn in Fig. 4 has been smoothed in Fig. 5

In actual feeders, the reliability of the mean transport rate is one of the crucial property. Indeed, the set of external control parameters (i.e. amplitude of the excitations, frequency etc....) are always subject to variations due industrial environment. Hence, the problem is to determine transport regimes which are not too sensitive to variations of these external parameters. It seems intuitively clear, that beside the simplest periodic solutions (which are obtained for k in relatively large bands), the chaotic regime is also likely to favour a mean transport rate relatively insensitive to external parameters. In the subharmonic perodic regimes, very tiny changes in the operating conditions are sufficient to induce a change of period in the Feigenbaum cascade. On the other hand, we have to stress that external noise is always present in the system. This in turn has the effect of truncating the original cascade of bifurcations [9,10]. From the conceptual point of view, the role played by the chaotic solutions is interesting. Indeed, here the chaos would appear as a useful behaviour in contrary to most situations where random solutions are considered as a nuisance.

4. CONCLUSIONS AND PERSPECTIVES.

We have obtained the dynamical equation of motions of two common mechanical devices in the form of non-linear, dissipative sets of mappings Eqs. (2) and (3). These mappings are among the simplest models discussed in non-linear dynamics. In particular, cascade of bifurcations and chaotic solutions are present. In the devices presented here, we point out that not only the periodic behaviour is interesting for actual applications. Indeed, it might well happen that the chaotic regime is precisely the one to be tuned for the requested task. This situation can be intuitively expected when, for instance, the behaviour of a non-linear dynamical system has to be relatively insensitive to variations of the external parameters which govern the equations of the motions. Indeed, once in the chaotic regime, the details of the motion become almost irrelevant ; only the invariant distributions of the pseudo-stochastic variables contain the relevant informations. An other class of mechanical devices where a simple non-linear mapping plays an important role are the gearboxes models [18,19,20,21]. In this case the relevant dissipative mapping is the Fermi map which originally has been derived in the context of cosmic ray acceleration.

Although its ubiquous presence, the influence of external noise has been omitted in this paper. Its presence has a tendancy to smooth the invariant measures and hence, to favour the property of chaotic regimes to be less sensitive to small variations external parameters.

ACKNOWLEDGEMENTS.

Prof. Dr. C.W. Burckhardt is warmly thanked for his hospitality at the Institut de Microtechnique. I am indebted to Prof. J. Figour who introduced me to the problem of vibratory feeding.

REFERENCES.

1. R. M. ROSENBERG. "Nonlinear oscillations". App. Mech. Rev. 14 ,(1961), 837.

2. L.O. CHUA. "Special issue on chaotic systems". Proc. of IEEE. 75,

3. F.C. MOON. "Chaotic vibrations. An introduction for applied scientists". (1987, John Wiley.

4. J. GUCKENHEIMER & P.J. HOLMES. "Nonlinear oscillations, dynamical systems and bifurcations of vector fields" . App. Math. Sc. 42 , (1983), Springer Verlag.

5. A.J. LICHTENBERG & M.A. LIEBERMAN. "Regular and stochastic motion". App. Math. Sc. 38, (1983), Springer Verlag.

6. A. H. REDFORD & G. BOOTHROYD. "Vibratory feeding". Proc Instn. Mech. Engrs. 182, (1967-68), 135.

7. M.-O. HONGLER & J. FIGOUR. "Periodic versus chaotic motion in vibratory feeders". Helv. Phys. Acta. to appear.

8. O. TANIGUCHI, M. SAKATA, Y. SUZUKI & Y. OSANAI. "Studies on vibratory feeders". Bull. of the JSME. 6, (1963), 37.

9. N. B. TUFILLARO, T.M. MELLO, Y.M. CHOI & A.M. ASLBANO. "Period doubling boundaries of a bouncing ball". J. Physique **47**, (1986), 173.

10. N.B. TUFILLARO, & A.M. ALBANO. " Chaotic dynamics of a bouncing ball". Am J. of Phys. **54**, (1986), 939.

11. C. MIRA . "Chaotic dynamics". (1987),World Scientific, Singapoore, NewJersey & Hong-Kong.

12. J. INOUE, S. MIYAURA & A. NISHIYAMA. " On the vibrotansportation and vibroseparation". Bull. of the JSME. **11**, (1968), 167.

13. M.J. FEIGENBAUM. "Qualitative universality for a class of transformations". J. Stat. Phys. **19**, (1978), 25.

14. P. Collet & J.-P. ECKMANN. "Iterated map on the interval as dynamical systems". (1980), Birckhaüser, Basel.

15. M.-O. HONGLER, P. CARTIER & P. FLURY. " Numerical study of a non-linear mapping describing vibrotransporation",.Preprint (1988)).

16. C. N. BAPAT,S. SANKAR & N. POPPLEWELL. " Repeated impacts on a sinusoidally vibrating table, reappraised". J. Sound & Vib. **108**, (1986), 1477.

17. Ya. F. VAYNKOF & S.V. INOSOV. "Non-periodic motion in vibratory conveyors". Mechanical Sc. Maschinovedeniye. **5**, (1976), 1.

18. F. PFEIFER & F. KUCUKAY. "Eine erweiterte Theorie mechanische Stosstheorie und uhre Anwendung in der Getriebdynamik". VDI-Zeitschr Bd. 127, (1985), 341.

19. F. KUCUKAY & F. PFEIFER. " Uber Rasselschwingungen in KFZ-Schallgetrieben". Ing . Archiv. 56, (1986), 25.

20. M.-O. HONGLER & L. STREIT. "on the origin of chaos in gearbox models". Physica 29D, (1988), 402.

21. K. KARAGIANNIS . "Chaotic motion in gearboxes". These proceedings.

FIGURE CAPTIONS.

Figure 1:   Vibratory feeder.

Figure 2:   Modelization of a vibratory feeder.

Figure 3:   Scenario of the dynamical behavior. $R_{\perp} = 1/3$

Figure 4.   Mean transport rate <W> as a function of the excitation parameter k, (the increment on k is 0.0025). $R_{\perp}=1/3$; $\alpha = 0.0$; $\gamma = 0.0$. For each values of k, we perform 600 iterations of the mapping.

Figure 5:   Mean of <W> over a selection of 9 different values of the increments used to solve Eq. (7). For each values of k, (the increment on k is 0.0025), we perform 400 iterations.

160

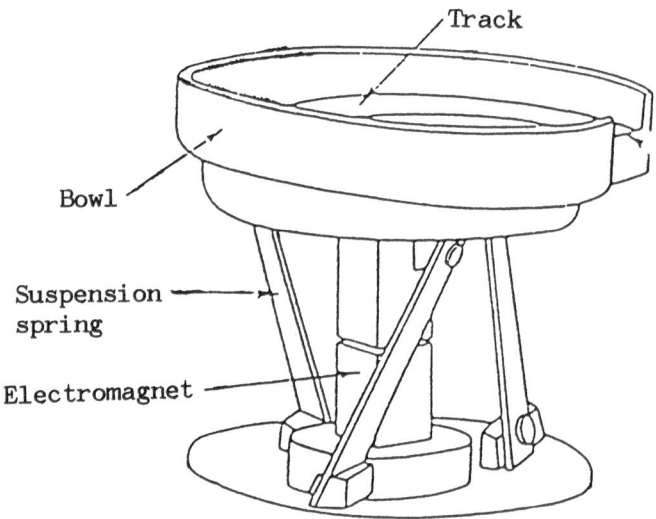

Fig 1

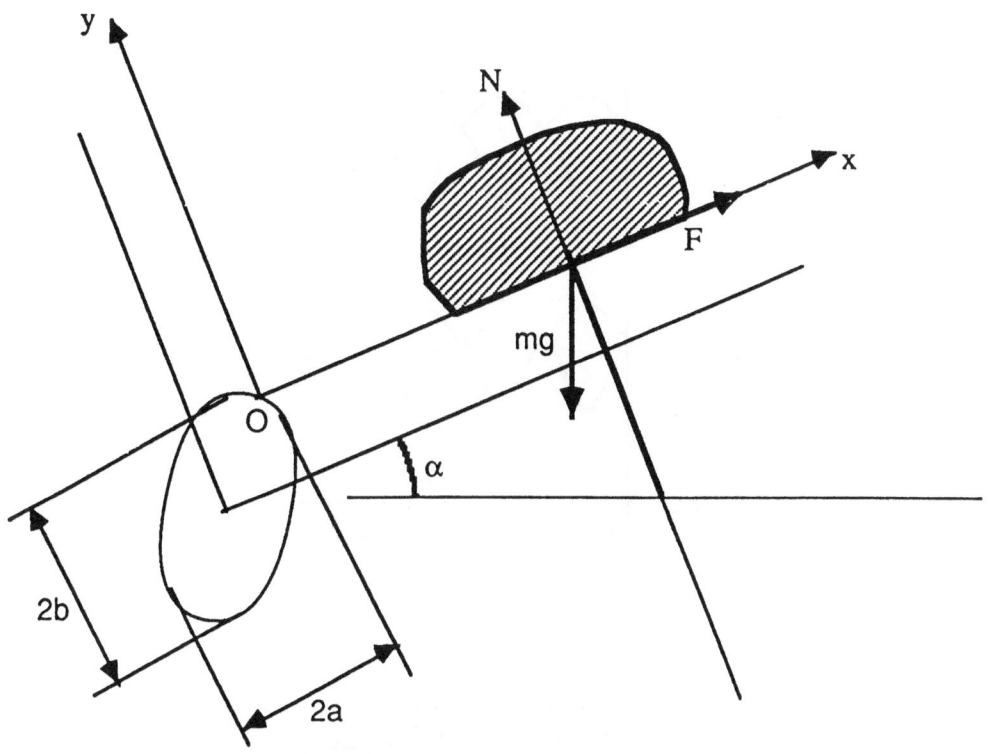

Figure 2

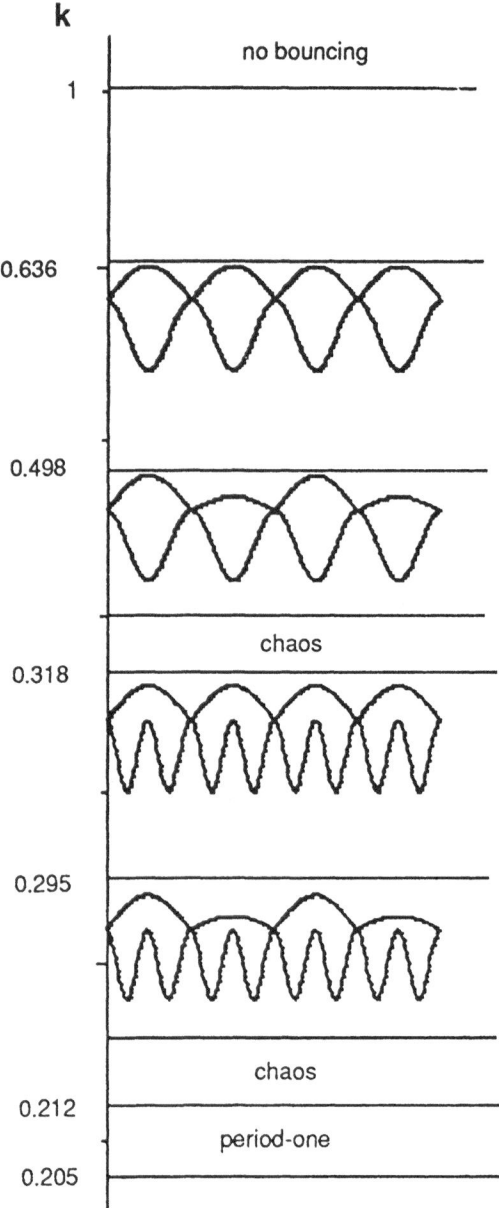

Fig. 3

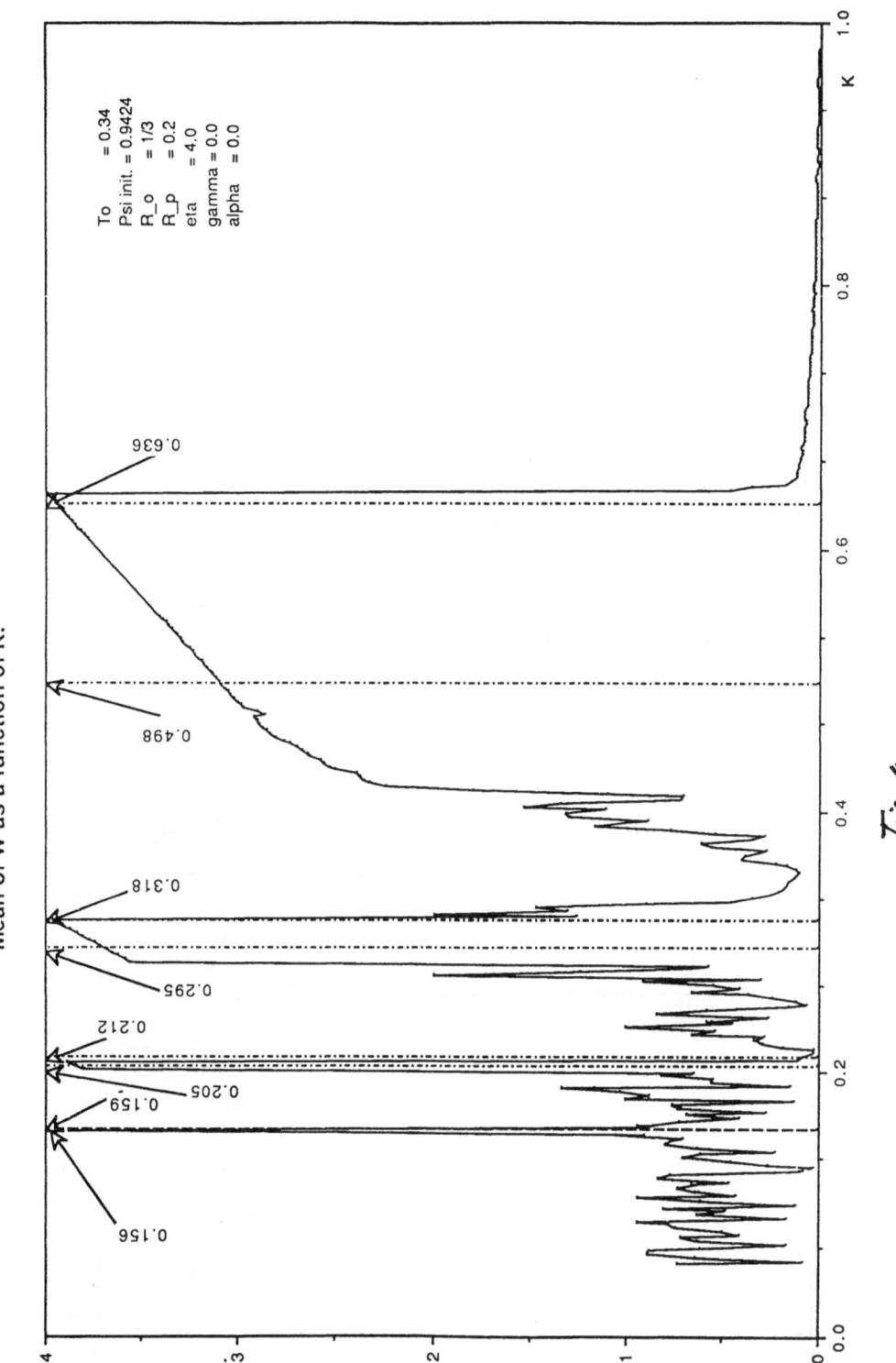

Mean of W as a function of K.

To = 0.34
Psi init. = 0.9424
R_o = 1/3
R_p = 0.2
eta = 4.0
gamma = 0.0
alpha = 0.0

Fig. 4

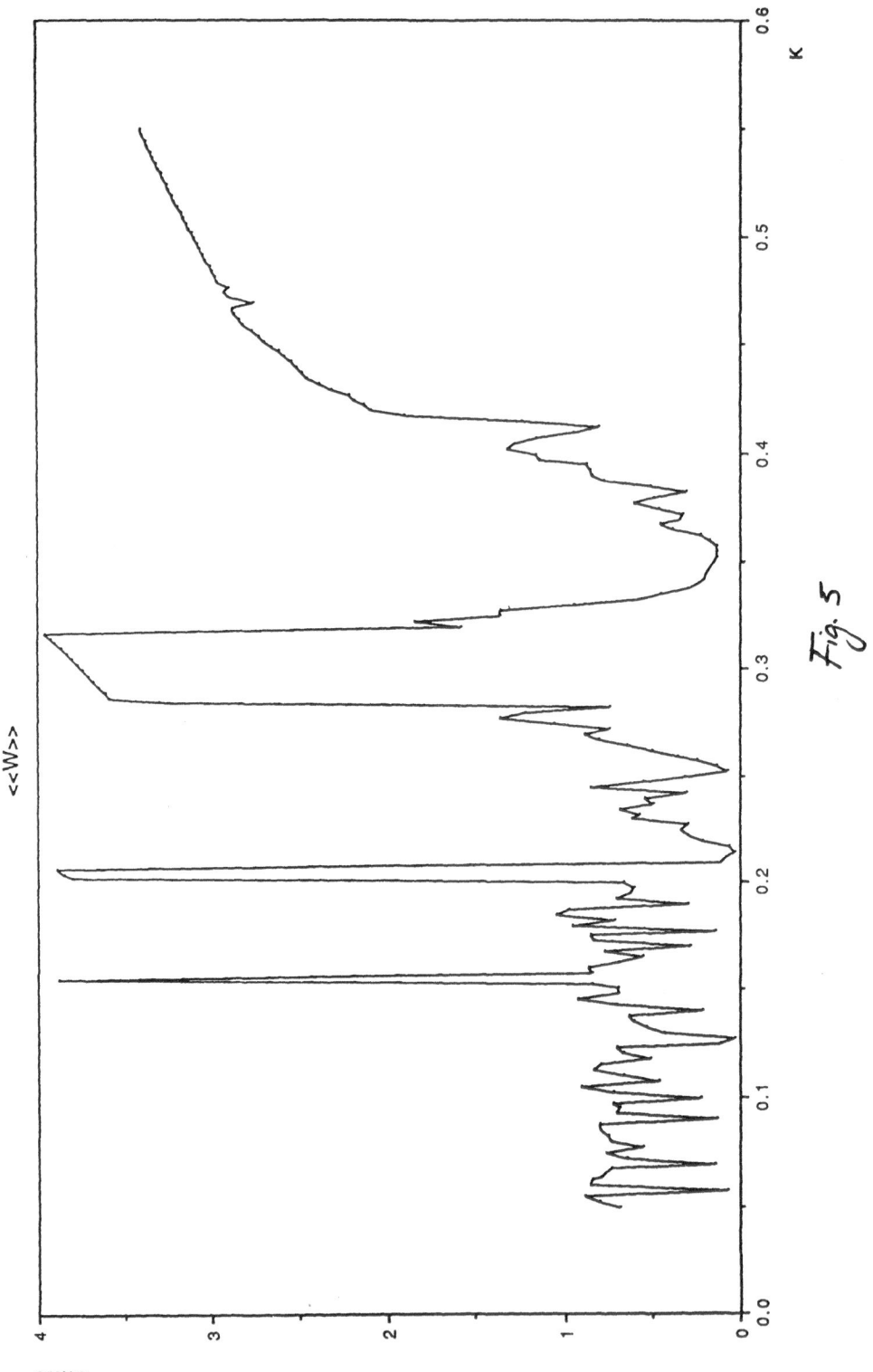

Fig. 5

Random perturbation and its application
to simulated annealing

Chii–Ruey Hwang

Institute of Mathematics, Academia Sinica

Taipei, TAIWAN 11529

1.   **Introduction.**   This is a brief expository report on the mathematical theory of simulated annealing.   The mathematical modelling and some results are stated in Section 2, examples in Section 3, remarks in Section 4.

Annealing is a physical process of lowering temperature slowly in order to reach the global minimum energy states.   By simulating such a process, one may find solutions for global minimization problems.

Of course, this is not a cure–all method.   In practice one has to build into the optimization problem a local structure in order to implement simulated annealing effectively.   Usually this might be problem–dependent and sometimes even very difficult.   We will not discuss this here.   For applications one may consult [1] and references listed there.

The common difficulty encountered in global optimization problem is: The cardinality of the state space is too large such that a direct search is not feasible. Travelling salesman problem is such an example.   Or when the state space is of continuum, one usually finds local minima instead.

The idea of simulated annealing introduced by Kirkpatrick et al [20] and Cerny [4] is to mimic the Metropolis method [23] but, in the meantime, decrease the "temperature" slowly.   Randomness induced by the "thermal" perturbations enables the process to escape from being trapped in the local minima.   Time is used to exchange for space and the annealing procedure will converge to global minima. Hence, in practice nearly global–minimum solutions can be found by this approach.

Simulated annealing was used in image restorations by Geman and Geman [10] where they also proved the first convergence result.

Grenander [13] used stochastic differential equations to describe a continuous version of simulated annealing. This approach reveals the close relation between simulated annealing and random perturbations.

Hajek [14] gave a general mathematical model for the discrete state space case. He also introduced some very useful concepts, e.g. weak reversibility, cups, bottoms, critical constant for convergence in probability etc.

Mathematically simulated annealing is to study the large time behaviors of certain nonhomogeneous Markov process $X(t)$, with a proper annealing rate $T(t)$, which is a "diagonalization" of a family of homogeneous (more or less) Markov processes. Theoretical results can be found in [5, 6, 7, 8, 11, 14, 17, 18, 19, 20, 22, 24, 25, 26, 27]. The basic questions are: Where does the process $X(t)$ go and how does it distribute as t goes to infinity? What are the critical constants for the annealing rate $T(t)$ and the convergence rate for $X(t)$? Can these results really help in doing the real implementation of simulated annealing?

2.   Mathematical modelling and some results

Let U be a given function from S to $\mathbb{R}$ and assume that the minimum of U is attainable in S. The goal is to find the global minima of U.

We will restrict ourselves to $S = \mathbb{R}^d$ or S being a finite set. Let

$$\underline{S} := \{x: U(x) = \min_y U(y)\}.$$

Let us consider the usual simulated annealing set up first. Let π_T denote the Gibbs distribution with density

$$\frac{1}{Z_T} \exp \frac{-U(x)}{T},$$

where Z_T is the normalizing constant.

Note that under mild conditions, π_T converges weakly to a unique probability π_0 concentrating on $\underline{S}$ as $T \downarrow 0$ [16]. So if $X_T(t)$ is a process with π_T as the equilibrium distribution and let $X(t) = X_{T(t)}(t)$ with $T(t)$ going to zero at a proper rate, one would expect that $X(t)$ converges in certain sense to π_0 or the set $\underline{S}$.

Note that what we described in the above paragraph is just a diagonalization procedure. Hence, actually π_T can be more general than the Gibbs distribution as long as π_T converges to $\underline{S}$ in certain sense. The difficult part is to determine the right cooling schedule $T(t)$. We call $X(t)$ a simulated annealing process.

For $S = \mathbb{R}^d$, consider for each $T > 0$,

(1) $$dX_T(t) = -\nabla U(X_T(t))dt + \sqrt{2T}\ dW(t);$$

(2) $$dX(t) = -\nabla U(X(t))dt + \sqrt{2T(t)}\ dW(t),$$

where $W(t)$ is a d–dimensional Brownian motion.

Note that π_T is the equilibrium distribution of X_T. One may regard (1) as random perturbations of the dynamic system

$$\frac{dX(t)}{dt} = -\nabla U(X(t)).$$

Corresponding to (1) and (2), one may consider a more general random perturbation problem, $\epsilon = \sqrt{2T}$

(3) $$dX^\epsilon(t) = b(X^\epsilon(t))dt + \epsilon\sigma(X^\epsilon(t))dW(t);$$

(4) $$dX(t) = b(X(t))dt + \epsilon(t)\sigma(X(t))dW(t).$$

The motivation of considering this general setup will become clear when we consider the examples for discrete case in the next section.

From another point of view (3) and (4) are closely related to a singular perturbation problem:

(5) $$L^\epsilon = \frac{\epsilon^2}{2} \sum_{ij} a_{ij} \frac{\partial^2}{\partial x_i \partial x_j} + b\cdot\nabla\ ,$$

where $a = \sigma\sigma^*$.

Under mild condition, one can prove the following results.

(I)   Let μ^ϵ be the invariant measure of the process X^ϵ in (3). Then for any $\alpha > 0$ there exists $\delta > 0$ such that for any compact set F in $\mathbb{R}^d$ there is $\epsilon_0 > 0$, the following holds

$$|E_x[f(X^\epsilon(T))] - \int f(y)d\mu^\epsilon(y)| \leq \|f\|_\infty \exp(-\frac{\delta}{\epsilon^2}),$$

where x is in F, f is a bounded continuous function, $\|f\|_\infty$ is the supnorm, $\epsilon \leq \epsilon_0$, $T = \exp\frac{\Lambda+2}{\epsilon^2}$. Λ here is a critical constant defined explicitly by a and b. [17, I].

This result gives the relationship between the proper time scale in terms of the perturbation and how far away the process is from its corresponding equilibrium.

Again under suitable conditions we have

(II)   For $c \geq \Lambda$ and bounded continuous f, the process defined by (2) satisfies: for $T(t) = \frac{c}{\log(t+2)}$,

$$E_y f(X(t)) \to \int f(x) d\pi_0(x) \text{ as } t \to \infty,$$

uniformly for y in a compact set. Here π_0 is the weak limit of π_T. [17 III, 25].

(III)   There exists a constant Λ_H such that the process defined by (2) satisfies: for any $\epsilon > 0$

(6) $P_x\{X(t) \text{ in an } \epsilon \text{ neighborhood of } \underline{S}\} \to 1, \ t \to \infty,$

uniformly over the starting point x in a compact set, if $T(t) = \frac{c}{\log(t+2)}$ with $c \geq \Lambda_H$. If $c < \Lambda_H$, (6) fails. [17 III].

Λ_H here is the constant defined by Hajek [14]. $\Lambda_H \leq \Lambda$. Λ is the critical constant for weak convergence, Λ_H for convergence in probability.

Note that Λ is also closely connected to the limiting behavior of the second eigenvalues of (5) in the case $-b = \nabla U$, $a \equiv 1$ [17 II].

Now we turn to the finite state space case. Here we only consider the discrete time situation and state a general setup.

Let $X_T(n)$ be a discrete time Markov chain with state space S and transition probability $P_T(x,y)$ defined by

(7) $P_T(x,y) = Q_T(x,y) \exp(-\frac{\alpha_{xy}}{T})$, if $x \neq y$,

$P_T(x,x) = 1 - \sum_{y \neq x} P_T(x,y),$

where $Q_T(x,y)$ is a transition matrix, $\alpha_{xy} \geq 0$. We will assume that there exists a transition matrix $Q(x,y)$ and positive constants c_1, c_2 such that

$$c_1 Q(x,y) \leq Q_T(x,y) \leq c_2 Q(x,y).$$

If $\alpha_{xy} = [U(y) - U(x)]^+$, $Q_T(x,y) = Q_T(y,x)$, then π_T is an invariant measure for X_T.

A simulated annealing process (discrete space, discrete time) is an inhomogeneous Markov chain $X(n) = X_{T(n)}(n)$ with a proper annealing function $T(n)$.

We consider the case when (7) is ergodic for $T > 0$. Note that $T = 0$, (7) is not ergodic in general. This is a singular perturbation situation.

Using α_{ij}, one can define another set S_0 similar to the set $\underline{S}$ introduced before. If (7) is weak reversible and $\alpha_{ij} = (U(j) - U(i))^+$ then $\underline{S} = S_0$ [18].

Let $T(n) = \dfrac{c}{\log(n+2)}$ and under mild condition

(IV)   For each $c > 0$, $X(n)$ converges weakly to a limiting distribution which may depend on $X(0)$.

For $c \geq \Lambda$, the limiting distribution is independent of the starting point $X(0) = i$,

$$\lim_{n \to \infty} P\{X(n) = j| \ X(0) = i\} = \mu_j \quad \text{if} \quad j \in S_0,$$

$$= 0 \quad \text{otherwise.}$$

Moreover,

$$\lim_{n \to \infty} \frac{P\{X(n) = j| \ X(0) = i\}}{\epsilon(n)^{w(i)}} = \mu_j, \quad i,j \in S,$$

where $\epsilon(n) = \exp(-\dfrac{1}{T(n)})$, μ_j and w_j are define by α_{ij}'s, $\sum\limits_{j \in S_0} \mu_j = 1$. [19].

Actually $W(i) = U(i) - \min\limits_{j \in S} U(j)$ if $\alpha_{ij} = \{U(j) - U(i)]^+$.

(V)   If $c \geq \Lambda_H$, then for all $i \in S$

$$\lim_{n \to \infty} P\{X(n) \in S_0| \ X(0) = i\} = 1.$$

If $c < \Lambda_H$, then for all $i \in S$,

$$\overline{\lim_{n \to \infty}} \ P\{X(n) \in S_0| \ X(0) = i\} < 1. \ [19].$$

(IV) gives the rate of convergence and the limiting distribution explicitly if the annealing rate is slower than the critical constant. (V) demonstrates that Λ_H is the critical constant for convergence is probability.

3.   Examples

Travelling Salesman Problem

Let m cities be labelled by $1, ..., m$. A tour of these m cities is a permutation of $\{1, ..., m\}$, i.e. the tour follows the order $\sigma(1), \sigma(2), ..., \sigma(m), \sigma(1)$. Define

$$U(\sigma) = \sum_{i=1}^{m} c(\sigma(i), \sigma(i+1)),$$

where $\sigma(m+1) = \sigma(1)$, $c(j,k)$ denotes the travelling cost of σ from city j to city k and S has $m!$ elements.

The generating mechanism is as follows: if at time n the chain is at state σ, a state σ' is generated by picking $i < j$ at random and reverse the order of travelling between i and j of the tour σ, i.e.

(8) $\quad \sigma'(k) = \sigma(k) \qquad$ if $k < i$ or $k > j$,

$\qquad\qquad = \sigma(j - (k-i)) \quad$ otherwise.

Then the probability of accepting σ' is $\exp - \dfrac{[U(\sigma') - U(\sigma)]^+}{T(n)}$. More precisely, the corresponding transition matrix (7) has

$$Q_{\sigma\sigma'} = \frac{m(m-1)}{2}, \quad \text{if there exist } i < j \text{ such that (8) holds,}$$

$$= 0 \qquad\qquad \text{otherwise;}$$

and $\alpha_{\sigma\sigma'} = [U(\sigma') - U(\sigma)]^+$.

Note that (8) defines a neighborhood structure in S.

The following two examples are commonly used in image processing and statistical physics. They will be stated in a quite general way.

We assume that there is an underlying graph L of size N. For each $i \in L$, there associates a level set L_i of size m_i, $m_i > 1$. W.l.o.g. one may assume that L is a lattice and all m_i's are equal.

The state space S is defined by

$$\{ x \mid x(i) \in L_i , i \in L \}.$$

Gibbs Sampler

For a fixed site i, a transition matrix G_i describing the transitions at that fixed site i by using local Gibbs distribution is defined by

$$G_i(x,y) = \frac{\exp\left(-\frac{U(y)}{T}\right)}{\sum\limits_{z \in N(i,x)} \exp\left(-\frac{U(z)}{T}\right)} \quad , \text{ if } y \in N(i,x);$$

$$= 0 \quad , \quad \text{otherwise};$$

where $N(i,x) = \{ z \mid x(j) = z(j) \ \forall \ j \neq i \}$.

So if we pick a site at random and then use the Gibbs sampler, the transition matrix (7) can be written as

$$RG = \frac{1}{N} (G_1 + G_2 + \cdots + G_N),$$

and the corresponding $Q(x,y)$ and α_{xy} are:

For $x \neq y$,

$$\left[\begin{array}{l} \text{if } y \in N(i,x) \text{ for some i, then } Q(x,y) = \frac{1}{N \cdot n(i,x)} , \\[2mm] \qquad\qquad \alpha_{xy} = U(y) - \min\limits_{z \in N(i,x)} U(z) ; \\[2mm] \text{otherwise,} \quad Q(x,y) = 0. \end{array} \right.$$

Here $n(i,x)$ is the number of z such that $U(z) = \min\limits_{y \in N(i,x)} U(y)$.

If the sites are swept systematically, say according to the order of their indices, then the corresponding transition matrix (2.1) can be written as

$$SG = G_1 G_2 \cdots G_N$$

Note that it is very easy to implement SG, but the corresponding α_{xy}'s are too messy to write down.

Metropolis Sampler

For a fixed site i, M_i represents the transition matrix by using local Metropolis method, i.e. the transition is defined as follows: Pick a level from $L_i - \{x(i)\}$ at random, say $y(i)$, and define $y(j) = x(j)$ for $j \neq i$, the acceptance probability of y is $\exp{\frac{-[U(y)-U(x)]^+}{T}}$. More precisely, for $x \neq y$

$$M_i(x,y) = \frac{1}{m_i - 1} \exp{\frac{-[U(y) - U(x)]^+}{T}} ,$$

if y differs from x at only one site i;

$$= 0 , \qquad \text{otherwise};$$

and $M_i(x,x) = 1 - \sum_{y \neq x} M_i(x,y).$

So if we pick a site at random, and then use Metropolis sampler

$$RM = \frac{1}{N} (M_1 + M_2 + \cdots + M_N).$$

The corresponding transition matrix (7) has $Q_T(x,y)$ and α_{xy}:

For $x \neq y$

$$Q_T(x,y) = \frac{1}{N} \frac{1}{m_i - 1} \text{ , if } y \in N(i,x) \text{ for some } i;$$

$$= 0 , \qquad \text{otherwise.}$$

$$\alpha_{xy} = [U(y) - U(x)]^+ = U(y) - \min (U(x), U(y)).$$

The systematic sweep $SM = M_1 M_2 \cdots M_N.$

Again, α_{xy} are too messy to write down, but it is easy to implement such a procedure.

4.   Remarks

1.   The Metropolis and Gibbs samples described in the previous section are very easy to implement and hence are used widely. But the comparison among RM, SM, RG, SG   seems not easy. There are results in a recent work [20], but still not quite complete.

2.   How to incorporate the idea of "parallelism" into the modelling? [1, 10, 26].   The α_{ij}'s will be very messy if one tries to update several sites simultaneously.   And the corresponding equilibrium distribution for fixed   T   is not very clear either.   Of course, one has to avoid the check board effect.

3.   If the annealing schedule depends on the state too, i.e. $T(t) = T(t, X(t))$,   will this speed up the convergence? This might change the nature of the problem. E.g.   if $T(t, X(t)) = T(t)U(X(t))$, then we might have to study the singular perturbation of degenerate diffusion processes or Markov chains.

4.   In the continuous state space case, when the drift b is $-\nabla U$ and σ is I, it looks like the Metropolis sampler. What is the corresponding Gibbs sampler?

5.   What is the waiting time for the global minimum? It is not hard to cook up an example of three states such that the expected waiting time is infinity.

References

[1]   P.J.M. van Laarhoven and E.M.L. Aarts (1987), Simulated Annealing: Theory and Applications, D.Reidel, Dordrecht.

[2]   K. Binder (1978), Monte Carlo Methods in Statistical Physics, Springer–Verlag, N.Y.

[3]   O. Catoni (1988), Grandes déviations et décroissance de la température dans les algorithmes de recuit, C.R.Acad.Sci.Paris, t.307, Série I, 535–539.

[4]   V. Cerny (1982), A thermodynamical approach to the travelling salesman problem: an efficient simulation algorithm, preprint. Inst. Phys. & Biophysics, Comenius Univ., Bratislava.

[5]   T.–S. Chiang, Y. Chow (1987), On the convergence rate of annealing processes. To appear in SIAM J. Control and Optimization.

[6]   T.–S. Chiang, Y. Chow (1987), A limit theorem for a class of inhomogeneous Markov processes, Technical Report, Institute of Math., Academia Sinica.

[7]   T.–S. Chiang, C.–R. Hwang, S.–J. Sheu (1987), Diffusion for global optimization in $\mathbb{R}^n$, SIAM J. Control and Optimization.

[8]   D.P. Connors, P.R.Kumar (1987), Balance of recurrence order in time inhomogeneous Markov chains with application to simulated annealing, preprint.

[9]   M.I. Freidlin, A.D. Wentzell (1984), Random Perturbations of Dynamical Systems, Springer–Verlag, New York.

[10] S. Geman, D. Geman (1984), Stochastic relaxation, Gibbs distribution, and the Baysian restoration of images, IEEE Trans. Pattern Analysis and Machine Intelligence, vol.6, 721–741.

[11] S. Geman, C.–R. Hwang (1986), Diffusion for global optimization, SIAM J. Control and Optimization, vol.24, No.5, 1031–1043.

[12] S. Geman, D.E. McClure (1987), Statistical methods for tomographic image reconstruction, to appear in Proceedings of the 46th Session of the ISI, Bulletin of the ISI, Vol.52.

[13] U. Grenander (1984), Tutorial in Pattern Theory, Lecture Notes Volume, Div.Appl.Math. Brown Univ.

[14] B. Hajek (1985), Cooling schedules for optimal annealing, preprint.

[15] B. Hajek (1985), A tutorial survey of theory and applications of simulated annealing, Proceedings of the 24th IEEE Conference on Decision and Control, vol.2, 755–760.

[16] C.–R. Hwang (1980), Laplace's method revisited, weak convergence of probability measures, Ann. Probab. 8, 1177–1182.

[17] C.–R. Hwang, S.–J. Sheu (1986), Large time behaviors for perturbed diffusion Markov processes with applications I, II, III Technical Report, Institute of Math., Academia Sinica.

[18]   C.–R. Hwang, S.–J. Sheu (1988), On the weak reversibility condition in simulated annealing, To appear in Soochow J. of Math.

[19]   C.–R. Hwang, S.–J. Sheu (1988), Singular perturbed Markov chains and exact behaviors of simulated annealing process, Technical Report, Institute of Math., Academia Sinica.

[20]   C.–R. Hwang, S.–J. Sheu (1989), Remarks on Gibbs sampler and Metropolis sampler, Technical Report, Institute of Math., Academia Sinica.

[21]   S. Kirkpatrick, C.D. Gelatt, and M.P. Vecchi (1983), Optimization by simulated annealing, Science 220, 671–680.

[22]   H.J. Kushner (1985), Asymptotic global behavior for stochastic approximations and diffusion with slowly decreasing noise effects: global minimization via Monte Carlo, preprint, Div. Appl. Math., Brown Univ.

[23]   N. Metropolis, A.Rosenbluth, M.Rosenbluth, A. Teller, E. Teller, (1953), Equation of state calculations by fast computing machines, J. of Chem. Physics, 21, 1087–1092.

[24]   D. Mitra, F. Romeo, A Sangiovanni–Vincentelli (1986), Convergence and finite time behavior of simulated annealing, Adv. Appl. Prob. 18, 747–771.

[25]   G. Royer (1988), A remark on simulated annealing of diffusion processes, Tech. Report, Département de Mathématiques et d'informatique, Univ. d'Orléans.

[26]   A. Trouvé (1988), Problèmes de convergence et d'ergodicité pour les algorithmes de recuit parallélisés, C.R.Acad.Sci.Paris, t.307, Série I, 161–164.

[27]   J. Tsitsiklis (1985), Markov chains with rare transitions and simulated annealing, preprint, Laboratory for Information and Decision Systems, Massachusetts Institute of Techonology.

RATTLING VIBRATIONS IN GEARBOXES

K. Karagiannis
Lehrstuhl B fuer Mechanik, TU Muenchen
Postfach 202420, D-8000 Muenchen 2, BRD

1. Introduction

The coexistence of bachlash, excitation and low level load in machinery is a common feature and often leads to undesirable rattling vibrations and noise. This paper will deal with rattling in passenger car gear boxes.

Fig. 1 shows the design of a five speed gear box. Principally, it consists of the input shaft, countershaft and the output shaft as well as the gear wheels that correspond to the different speeds. When a certain speed is activated the specific wheels are meshing whereas the other ones are running without load. On the one hand there always exists some backlash between those loose gear wheels on the output shaft and the corresponding wheels on the countershaft. On the other hand the angular velocity of the input shaft is fluctuating due to the variations of the engine torque. This leads to rattling vibrations and gear box noise that affects the comfort of the passenger. The mechanism as such is characterized by the fact that the free flight of the gear wheels is constrained by the backlash limits where partly elastic impacts occur. These impacts represent an excitation mechanism for the gear box which leads to the mentioned undesirable noise emission. Under specific conditions such as no-load operation and the forth speed the noise level is especially annoying. Therefore the automobile industry shows increasing interest in solving this problem.

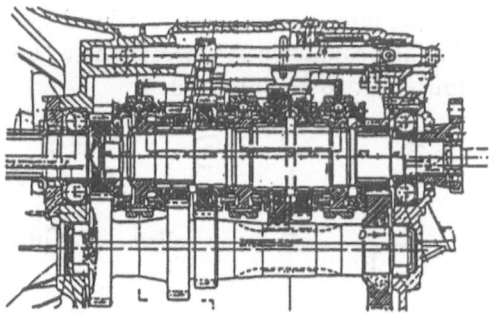

Fig. 1 Five speed gear box

The literature that exists about this topic may be split up into two different categories. Elaborate multibody system modelling methods are very common today including numerical integration procedures for the nonlinear differential equations of motion. The interpretation of the results, though, is a difficult task since most often they have the form of time series, spectral density functions or statistical moments that are functions of a large number of parameters. On the other hand there are a number of recent works that handle simple models using the tools of modern dynamical system theory. The principal behavior of the system can be investigated and information about the structure of the resulting motions be achieved. Nevertheless, for systems with many degrees of freedom the interpretation of the results itself becomes very difficult.

Papers [1] and [2] develop sophisticated modelling techniques for the simulation of gear box rattling. The contacts at the backlash constraints are modelled as impacts and the integration of the equations of motions requires an extensive amount of computation time. In [3] the simulation of the gear motions is performed on an analog computer. The constraints are modelled by nonlinear stiffness characteristics. [4] and [5] introduce a theoretical approach and numerical investigations of a lumped mass within rigid constraints. In [6] mapping procedures are applied to one and two stage gear mechanisms. The authors of [7] and [8] analytically and experimentally examine the behavior of a jumping ball on a fortuitously moving plate.

This paper deals with the investigation of the mechanisms, their origins and the relations between the noise levels and the system parameters. In the following a multibody model for the gear box will be presented, the equations of motion will shortly be mentioned and a discretization of the motion be derived. For the discrete models of a one stage and a two stage gear mechanism the results of the numerical simultions are presented that are achieved using mapping methods. An approximation of the statistical functions will be shown. The results are then compared to those of experimental measurements.

2. Gear box model

2.1. Equations of motion

The physical model for the investigations is displayed in Fig. 2. A gear mechanism with n stages is considered taking into account only rotational motions. The bodies are enumerated in such a way that the countershaft corresponds to body number 1 and the loose gear wheels on the output shaft to bodies 2...n.

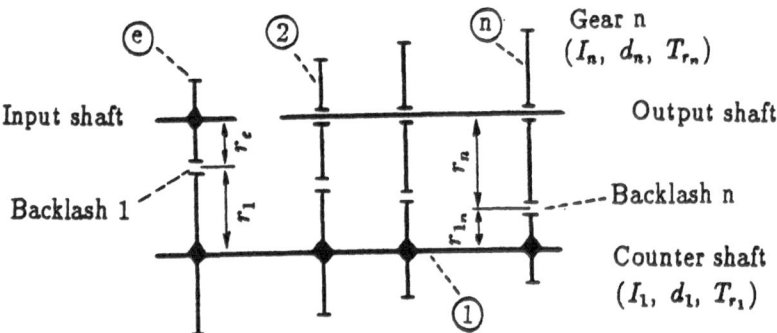

Fig. 2 Model of n stage gear mechanism

If there was no exciting fluctuation in the input shaft angular velocity the gear wheels would rotate with their respective nominal constant speeds. Therefore it obviously is usfull to describe the system state using the deviations from these nominal velocities.

Mathematically we use the angle coordinates

$$\varphi = [\varphi_1, \ldots, \varphi_n]^T \in \Re^n$$

as generalized coordinates.

According to the above mentioned subdivision of the motion we distinguish into free flight where the wheels don't have any contact and the impact phase when one or more constraints are hit. During free flight only the moments transmitted by the gear lubricant oil appear. They are modelled as linear damping and constant moments.

The equations of motion have the form

$$\mathbf{I}\ddot{\varphi} + \mathbf{D}\dot{\varphi} + \mathbf{T}_r = 0 \tag{1}$$

$$\text{if} \quad s_j \in (-v_j, 0), \quad j = 1, \ldots, n$$

with
$\mathbf{I} = diag\{I_1, I_2, \ldots, I_n\} \in \Re^{n,n}$ the matrix of the moments of inertia,
$\mathbf{D} = diag\{d_1, d_2, \ldots, d_n\} \in \Re^{n,n}$ the damping matrix,
$\mathbf{T_r} = \{T_{r_1}, T_{r_2}, \ldots, T_{r_n}\} \in \Re^n$ the vector of constant moments and
v_j the backlash within the n-th gear mesh.

The vector of relative distances in the backlashes $\mathbf{S} = [s_1, s_2, \ldots, s_n]^T \in \Re^n$ in the following will be called the vector of play coordinates. These coordinates will have values in the ranges $(-v_j, 0)$, $j = 1, \ldots, n$. They are related to the angle coordinates by the regular transformation

$$\mathbf{S} = \mathbf{Y}\varphi + v\, e \tag{2}$$

where the regular matrix $\mathbf{Y} \in \Re^{n,n}$ and the vector $v \in \Re^n$ depend on the geometric properties of the gear box and $e(t)=r_e\varphi_e(t)$ represents the excitation of the input gear. r_e is the basic radius of the input gear wheel and φ_e is the fluctuation of the motion of the input shaft. Mearsurements of the excitation φ_e showed that it may be sufficiently approximated by a single harmonic function $\varphi_e(t) = A\sin(\omega t)$ with amplitude A of the fluctuations of the angular motion.

The solution of the equations of motion is

$$\varphi\,(t) = \varphi\,(t_0) + \mathbf{B}^{-1}[\mathbf{E} - exp[-\mathbf{B}(t-t_0)]](\dot{\varphi}\,(t_0) + \mathbf{B}^{-1}\mathbf{c}) - \mathbf{B}^{-1}\mathbf{c}(t-t_0) \tag{3}$$

$$\dot{\varphi}\,(t) = exp[-\mathbf{B}(t-t_0)](\dot{\varphi}\,(t_0) + \mathbf{B}^{-1}\mathbf{c}) - \mathbf{B}^{-1}\mathbf{c} \tag{4}$$

$$\text{if} \quad s_j \in (-v_j, 0), \, j = 1, \ldots, n$$

with the abbreviatios $\mathbf{B} = \mathbf{I}^{-1}\mathbf{D}$ and $\mathbf{c} = \mathbf{I}^{-1}\mathbf{T}_r$.

The contact phase is modelled as partly elastic impact since the gear wheels are made of hardened steel and the load is very low. Using impact theory one gets transition equations from the system state before an impact to the state after it. For backlash i the transition equations are

$$
\begin{aligned}
t^+ &= t^- \\
\varphi^+ &= \varphi^- \\
\dot{\varphi}^+ &= \mathbf{U}_i \dot{\varphi}^- + \zeta_i \dot{e}
\end{aligned}
\tag{5}
$$

with - specifying values before and + specifying those after the impact. $\mathbf{U}_i \in \Re^{n,n}$ and $\zeta_i \in \Re n$ are the transition matrix and a transition vector, respectively, that depend on the geometry of the gear box, the number of restitution and the impacting pair of gear wheels.

In order to achieve a more evident mathematical form the equations of motion are transformed into the play coordinates :

$$
\tilde{\mathbf{S}} + \mathbf{YBY}^{-1}\dot{\mathbf{S}} = \boldsymbol{v}\,\tilde{e} + \mathbf{YBY}^{-1}\boldsymbol{v}\,\dot{e} - \mathbf{Yc}
\tag{6}
$$

$$
\text{if } \ s_j \in (-v_j, 0) \ j = 1, \dots, n
$$

and

$$
\dot{\mathbf{S}}^+ = \mathbf{YU}_i\mathbf{Y}^{-1}\dot{\mathbf{S}}^- + (\mathbf{E} - \mathbf{YU}_i\mathbf{Y}^{-1})\boldsymbol{v}\,\dot{e}
\tag{7}
$$

$$
\text{if } \ s_i \in \{-v_i, 0\}, \ s_j \in (-v_j, 0), \ j = 1, \dots, n, \ j \neq i
$$

It is obvious that this is a non-holonomic, self-excited, nonlinear mechanical system. Due to the impulsive parameter excitation it is not possible to use the approximation methods of nonlinear dynamics.

2.2 Discrete model

From equations 6 and 7 it is obvious that the time should be introduced into the system state space in order to have the dynamical behavior of the system completely described. Now $\mathbf{q} = [t, \varphi^\top, \dot{\varphi}^\top]^\top \in [0, \frac{2\pi}{\omega}) \times \Re^{2n}$ stands for the state vector of the angle coordinates whereas $\mathbf{z} = [t, \mathbf{S}^\top, \dot{\mathbf{S}}^\top]^\top \in [0, \frac{2\pi}{\omega}) \times \Re^{2n}$ is the state vector in play coordinates. As the excitation is periodic the vector field of the dynamical system has the same period and therefore the time t may be normed within the interval of the excitation period.

The motion is exactly defined if the sequence of system states before or after impacts is known. In addition, the trajectories between impacts don't have any influence on the noise emission that is caused only by the impacts themselves. With regard to these considerations it is preferable to use a time discrete description instead of the time continuous one. This is easily done by the definition of point mappings onto Poincare sections (P.S.).

For the system states before impacts the P.S. is

$$\Sigma^- = \{ \text{ q rsp. z })/ \ (s_1 = -v_1 \wedge \dot{s}_1 < 0) \vee (s_1 = 0 \wedge \dot{s}_1 > 0)$$
$$\vee \ldots \vee (s_n = -v_n \wedge \dot{s}_n < 0) \vee (s_n = 0 \wedge \dot{s}_n > 0) \}$$

whereas for the states after impacts it is defined as

$$\Sigma^+ = \{ \text{ q rsp. z })/ \ (s_1 = -v_1 \wedge \dot{s}_1 \geq 0) \vee (s_1 = 0 \wedge \dot{s}_1 \leq 0)$$
$$\vee \ldots \vee (s_n = -v_n \wedge \dot{s}_n \geq 0) \vee (s_n = 0 \wedge \dot{s}_n \leq 0) \}$$

Two mappings are defined : the mapping describing the impact

$$\mathbf{F}_s \ : \ \Sigma^- \to \Sigma^+, \in \ \Re^{2n+1}$$

is defined by equation 5 and

$$\mathbf{F}_f \ : \ \Sigma^+ \to \Sigma^-, \in \ \Re^{2n+1}$$

represents the free flight. The latter one results from the solution of the equations of motion during free flight and an additional equation determining the position at the next impact. Now the whole motion may be written in the form

$$\ldots \mathbf{q}_i^- \to \mathbf{q}_i^+ \to \mathbf{q}_{i+1}^- \to \mathbf{q}_{i+1}^+ \to \mathbf{q}_{i+2}^- \ldots$$

The stability of the process may be judged by the functional matrices of the mappings. The functional matrix of the whole motion is achieved by sequential multiplication of the matrices corresponding to the separate transitions. The form of the functional matrices for the k-th impact and the following free flight, respectively, is

$$\mathbf{M}_{s_k} = \frac{\partial \mathbf{q}_k^+}{\partial \mathbf{q}_k^-} \qquad \mathbf{M}_{f_k} = \frac{\partial \mathbf{q}_{k+1}^-}{\partial \mathbf{q}_k^+}$$

It is impractibale to judge on the stability of periodic solutions by the investigation of eigenvalues of the functional matrix in the near of fix points since these points, in general, can not be determined analytically. In this case the motion is calculated by numerical evaluation of the mappings so that the stability of periodic solutions becomes obvious. Nevertheless, we determine the greatest Ljapunov exponent σ as a measure for the regularity of the motion :

$$\sigma = lim_{n \to \infty} \frac{1}{n} \ln(\frac{|\mathbf{w}_n|}{|\mathbf{w}_0|}) \tag{8}$$

with

$$\mathbf{w}_n = \prod_{k=1}^{n} \mathbf{M}_{f_k} \mathbf{M}_{s_k} \mathbf{w}_0$$

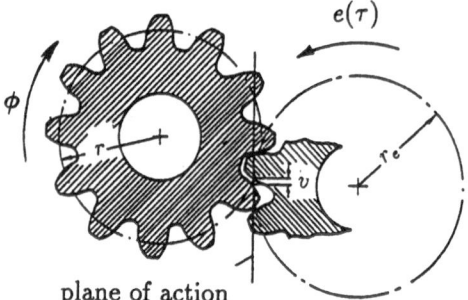

plane of action

Fig. 3 One stage gear model

3. Results

3.1. Results for the one stage model

In the forth speed and during no-load operation all gear wheels as well as the countershaft are running freely without load whereas in other speeds the momentum flux passes from the input shaft through the countershaft via the engaged gear wheel to the output shaft. In this case only the gear wheels not under load may be considered as one stage gears neglecting their influence on the motion of the gear train under load. Fig. 3 shows the single stage model. The parameter space of the one stage gear has small dimension. Even more, using similarity rules it can be further reduced by three. The equations of motion for the normalized play coordinates are

$$\left. \begin{array}{rll} \ddot{s} + \beta \dot{s} &=& \ddot{e} + \beta \dot{e} + \gamma \quad \text{if } s \in (-1,0) \\ \dot{s}^+ &=& -\epsilon \dot{s}^- \quad\quad \text{if } s \in \{-1,0\} \end{array} \right\} \tag{9}$$

$$\text{with} \quad e = \alpha \sin(\tau)$$

where $s = s/v$, $\tau = \omega t$, $\beta = d/(I\omega)$, $\gamma = (T_r r)/(I\omega^2 v)$, $\alpha = (r_e A)/v$ are the normalized play coordinates themselves, the time, damping, constant moments and amplitudes, respectively. Derivatives in equation 9 are with regard to the normalized time τ.

As mentioned above the vector $\mathbf{z} = (\tau, \mathbf{s}, \dot{\mathbf{s}})^\mathsf{T} \in [0, 2\pi) \times \Re^2$ is used within the description of the discrete system. The mapping $\mathbf{F}_{s_k}$ of the k-th impact is given by

$$\left. \begin{array}{rll} \tau_k^+ - \tau_k^- &=& 0 \\ \mathbf{s}_k^+ - \mathbf{s}_k^- &=& 0 \\ \dot{\mathbf{s}}_k^+ - \epsilon \dot{\mathbf{s}}_k^- &=& 0 \end{array} \right\} \tag{10}$$

whereas for the k-th free flight phase the appropriate mapping $\mathbf{F}_{f_k}$ is

$$\left. \begin{array}{l} -\mathbf{s}_{k+1}^- + \mathbf{s}_k^+ + \alpha(\sin(\tau_{k+1}^-) - \sin(\tau_k^+)) + \frac{1}{\beta}(1 - exp(-\beta(\tau_{k+1}^- - \tau_k^+))) \\ \quad (\dot{\mathbf{s}}_k^+ - \alpha\cos(\tau_k^+) - \frac{\gamma}{\beta}) + \frac{\gamma}{\beta}(\tau_{k+1}^- - \tau_k^+) = 0 \\ \\ \alpha\cos(\tau_{k+1}^-) + (\dot{\mathbf{s}}_k^+ - \alpha\cos(\tau_k^+) - \frac{\gamma}{\beta})exp(-\beta(\tau_{k+1}^- - \tau_k^+))\frac{\gamma}{\beta} - \dot{\mathbf{s}}_{k+1}^- = 0 \\ \\ \mathbf{s}_{k+1}^- - \mathbf{s}_k^+ \in \{-1, 1, 0\} \end{array} \right\} \tag{11}$$

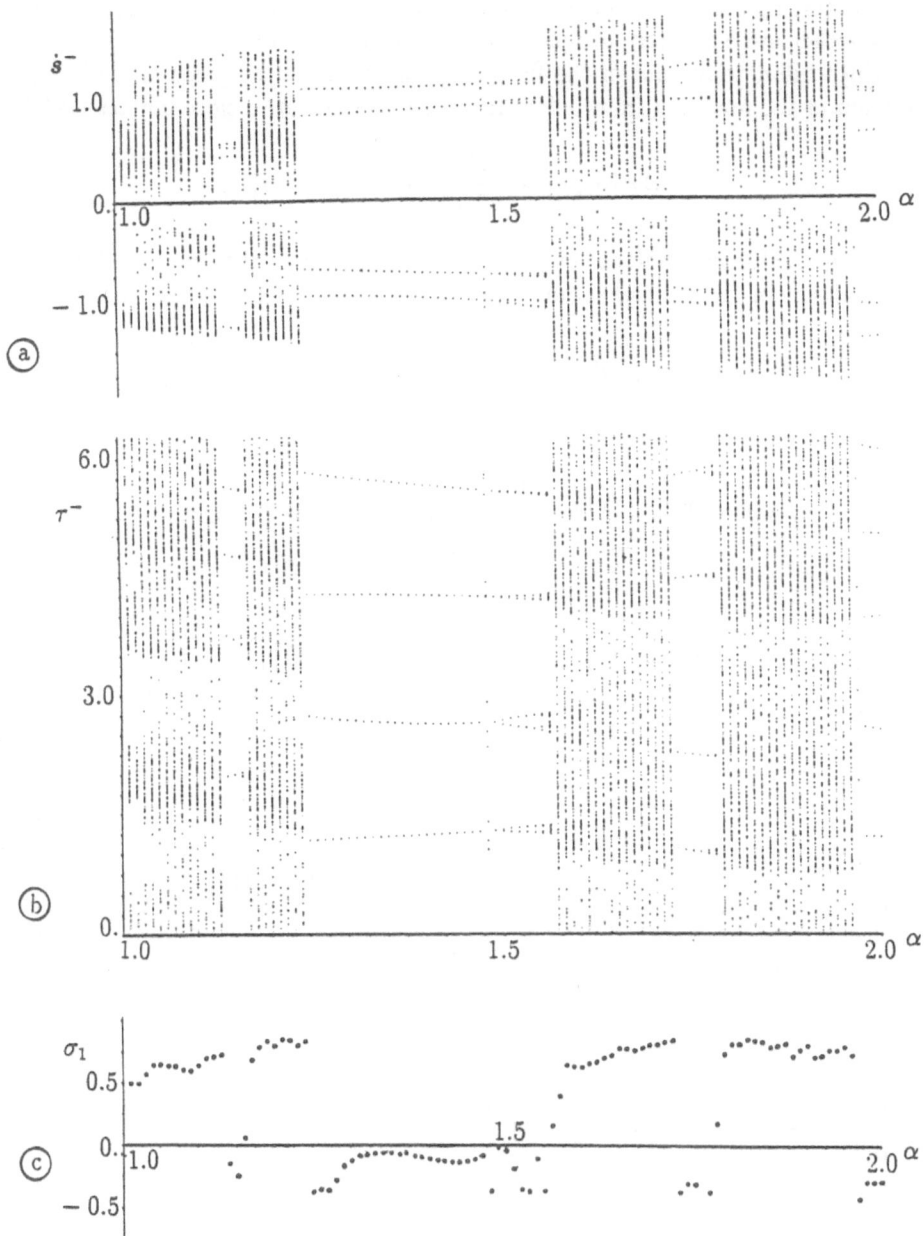

Fig. 1 Bifurcation diagram of one stage gear
a : bifurcations of relative velocities within backlash
b : bifurcations of normalized time
c : greatest Ljapunov exponent
the bifurcation parameter is α and the other parameters are
$\beta = 0.1, \ \gamma = 0.1, \ \epsilon = 0.9$

The greatest Ljapunov exponent can be calculated by equation 8. Although the mappings altogether are extremely complex the sum of the Ljapunov exponents may be calculated since the determinant of the functional matrices can be given analytically. The determinant of the mapping from before impact k to the state before impact k+1 is

$$det(\mathbf{M}_{f_k}\mathbf{M}_{s_k}) = \epsilon^2 \frac{\dot{s}_k^-}{\dot{s}_{k+1}^-} \exp(-\beta(\tau_{k+1}^- - \tau_k^+))$$

Consequently, the determinant of the mapping from before the first to before the n-th impact results from the multiplication of the determinants of elementary mappings.

$$det(\prod_{k=1}^{n} \mathbf{M}_{f_k}\mathbf{M}_{s_k}) = \prod_{k=1}^{n} det(\mathbf{M}_{f_k}\mathbf{M}_{s_k}) = \epsilon^{2n}\frac{\dot{s}_1^-}{\dot{s}_{n+1}^-}\exp(-\beta\sum_{k=1}^{n}(\tau_{k+1}^- - \tau_k^+)) \qquad (12)$$

Using these equations the sum of the exponents can be calculated :

$$\sigma_1 + \sigma_2 + \sigma_3 = \lim_{n\to\infty}\frac{1}{n}\ln|det(\prod_{k=1}^{n}\mathbf{M}_{f_k}\mathbf{M}_{s_k})| = 2\ln(\epsilon) - \beta < \tau_{k+1}^- - \tau_k^+ > \qquad (13)$$

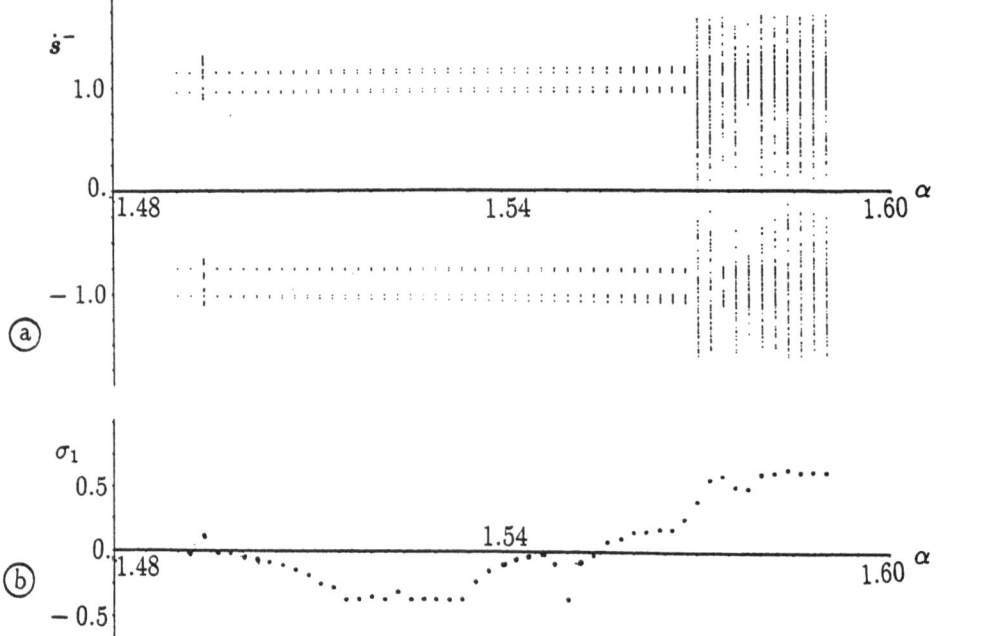

Fig. 5 Details of bifurcation diagram fig. 4
a : Detail of fig. 4 a
b : Detail of fig. 4 c

One of the exponents is identically zero as it corresponds to the defining mapping within the P.S. The sum of the other two is negative which is not a surprising result since the mappings have a dissipative character resulting in a shrinking phase space. Note, that in equation 13 only dissipative term appear, i.e. the number of restitution ϵ and the

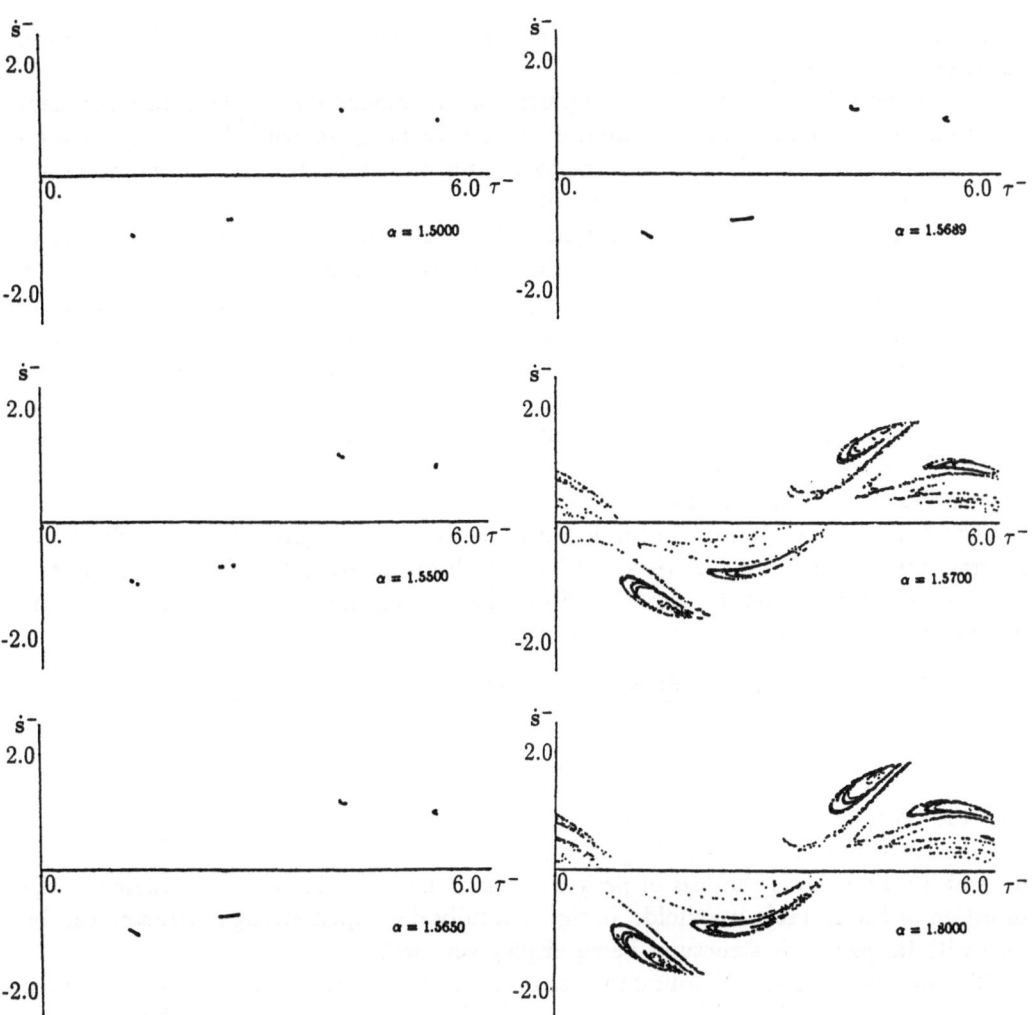

Fig. 6 Change in the form of the attractor in the P.S.
Parameters like in fig. 4

normalized damping multiplied by the mean flight time which is a measure for the mean shrikage during the flight phase.

The numerical evaluation of the mappings and the calculation of the greatest exponent has been performed on a digital computer the results being shown below. In fig. 4 a bifurcation diagram is displayed where the parameter α (amplitude of excitation/backlash) is varied. (In order to get a stationary situation the mapping is performed 5000 times of which the last 300 results are displayed.) Fig. 4 a shows the diagram for the relative velocities within the backlash, fig. 4 b is the respective diagram for the time points τ^- in the range of $(0, 2\pi]$ and fig. 4 c shows the greatest Ljapunov exponent. The qualitative properties of the motion strongly depends on the parameters of the system. In certain parameter regions the solution is periodic and through a sequence of socalled bifurcations becomes chaotic. This feature repeats itself over and over. The bifurcation itself is the destabilization of a periodic solution resulting in another solution with twice the period order. The distance between two bifurcations decreases rapidly so that the bifurcation sequence soon results in chaotic behavior.

This feature is shown in detail in fig. 5 which is magnifies parts of fig. 4. From this diagram the parameters $\alpha_1, \alpha_2, \alpha_3$ are extracted which correspond to the first, the second and the third bifurcation, respectively. From these a Feigenbaum number as well as the parameter, for which chaos appears, can be calculated. The values are

$$\delta = \frac{\alpha_2 - \alpha_1}{\alpha_3 - \alpha_2} = 5 \quad (4.669)$$

$$\alpha_\infty = \alpha_1 + \frac{1}{1 - 1/\delta}(\alpha_2 - \alpha_1) = 1.556 \quad (1.557)$$

with the exact theoretical results given for comparison in parentheses.

The development of the corresponding strange attractor in the P.S. is given in fig. 6 , which nicely shows the loss of periodicity and the spreading of the attractor in the direction of the instable manifold. In fig. 7 a fully developed strange attractor can be seen with its geometric structures being display very well.

The specific behavior according to changes in the parameter α can also been observed when changing the other system parameters. Besides the mentioned type of bifurcations one can find socalled "inner" ones for which the impacts at one of the backlash limits bifurcates whereas at the opposite limit the impact is stable. The subdivision into these two categories depends on the choice of the initial conditions but has no specific influence on mean and other statistical values. Therefore it has not been investigated in detail.

As the fix points are determined numerically it is not meaningful to try to exactly investigate their specific behavior.

3.2 Results for the two stage model

The modelling has already been presented and the complete description of the motion is possible only in $\Re^5$. As the dimension of the parameter space is very high irrespective of a possible reduction by three by means of similarity rules. Therefore in the following we will use unnormalized parameters.

Qualitatively the results are the same as in the above case of the one stage gear, i.e. periodic as well as chaotic motion with a sequence of bifurcations in between. Since the phase space has more dimensions the representation of results becomes more difficult.

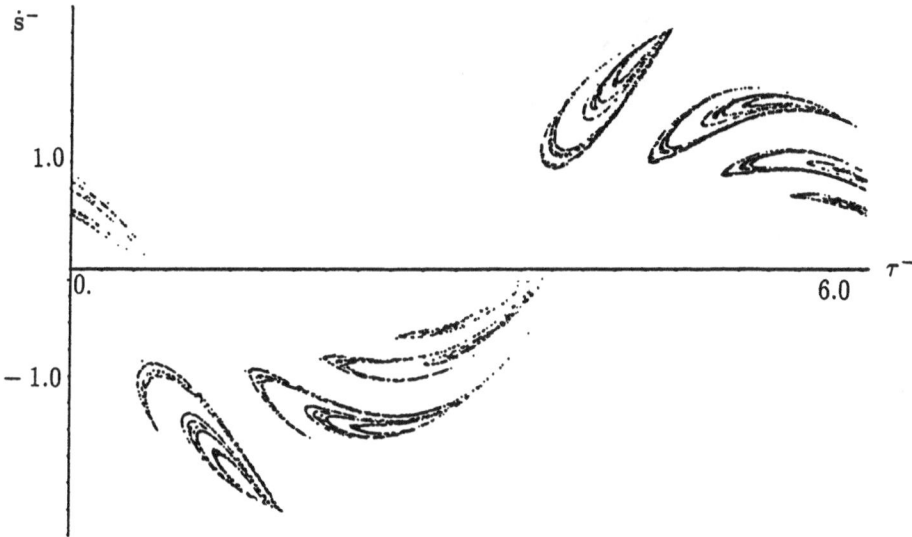

Fig. 7 Strange attractor
Parameters are : $\alpha = 3$, $\beta = 0.1$, $\gamma = 0.1$, $\epsilon = 0.9$ and 10.000 iterations

Here only projections of this space onto certain planes will be shown.

Fig. 8, for example, shows different projections for a chaotic motion for a specific parameter combination. The first diagram (a) is the projection into the plane of velocities, the second (b) a projection onto normalized time and the velocities in backlash 1 after impacts and the third (c) a similar graph for backlash 2. The inner structure of the attractor as compared to the single stage model has vanished as a property of the projection, and the interpretation of the results becomes more difficult.

Fig. 9 shows the same projections as fig. 8 for a periodic motion of high order.

4. Parameter dependence of mean values

4.1. Parameter dependence for one stage model

In order to optimize the design of the gear box the influence of the parameters on noise emission has to be investigated. The mechanisms of sound transmission and radiation are very sophisticated and not a topic of this paper. It is assumed that the squared means of the relative velocities in the backlashes at impacts can be used as a measure for the radiated noise. The derivation of the parameter dependence of the mean values is certainly possible numerically but at this point we would like to introduce a different modelling approach to achieve analytical approximations.

Here we investigate a one stage gear with neglection of damping and constant moments. The mappings are written with respect to normalized values and the angle will be used as a generalized coordinate. The resulting form of the mappings is

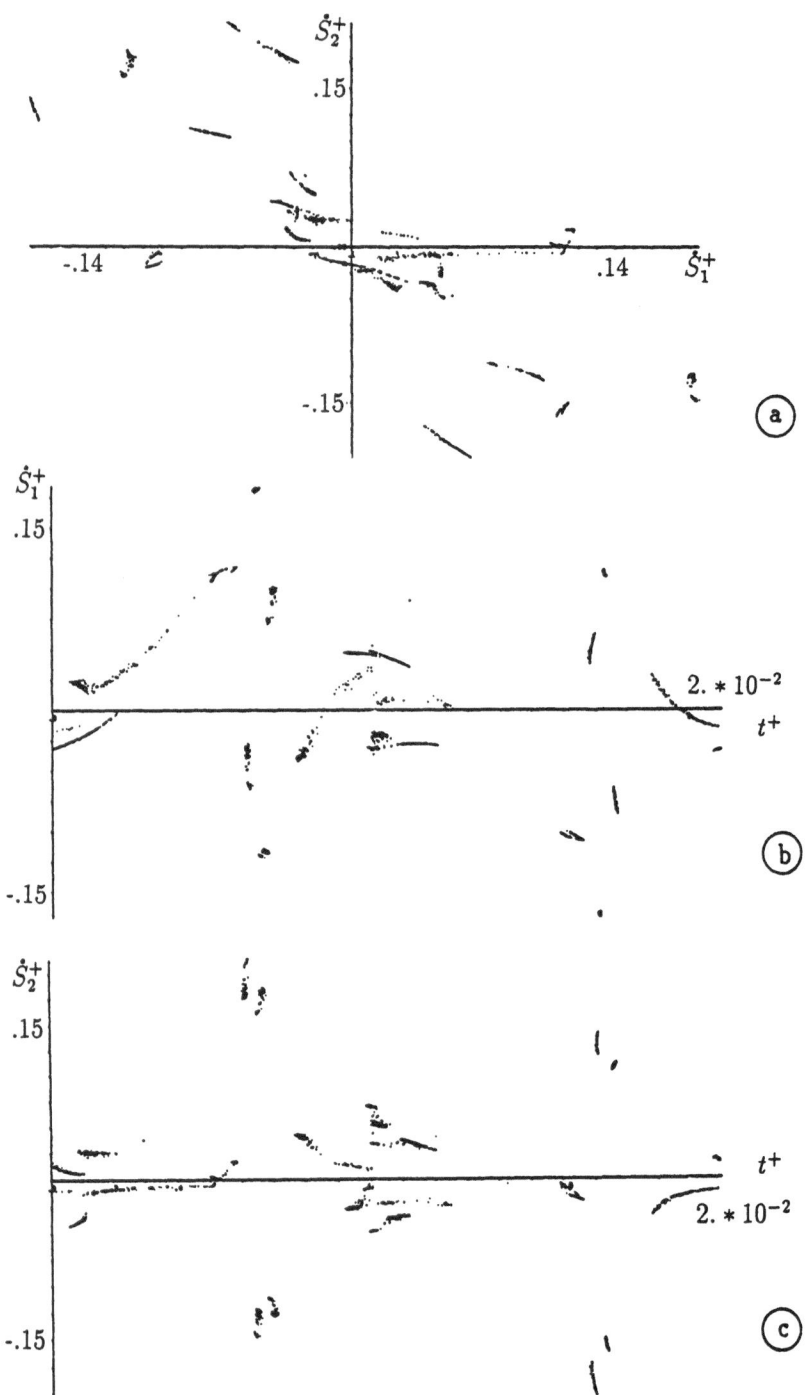

Fig. 8 Chaotic motion of the two stage model
a : plane of velocities
b/c : plane of normalized time and velocity in one backlash

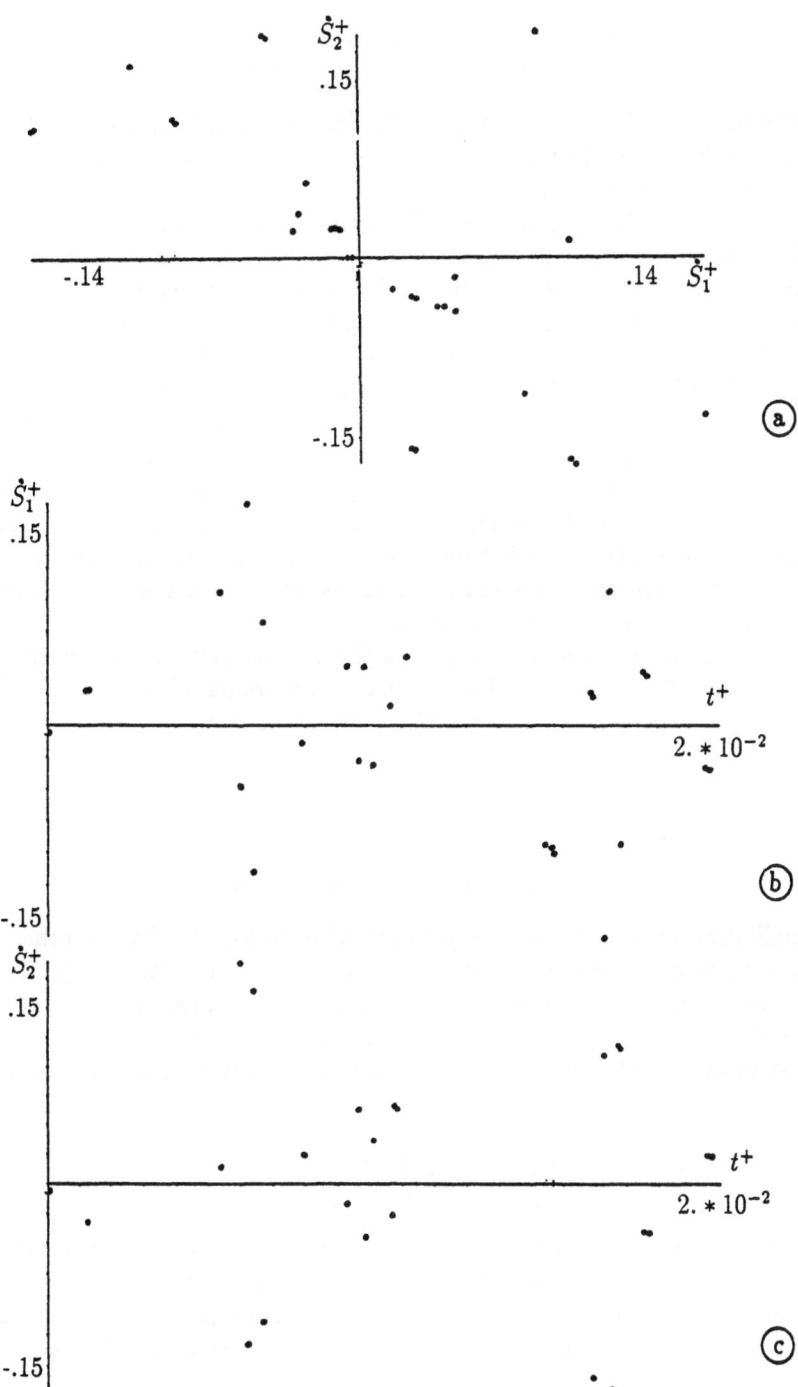

Fig. 9 Periodic motion of the two stage model
a : plane of velocities
b/c : plane of normalized time and velocity in one backlash

$$\left.\begin{array}{l}\dot{\phi}_{k+1} = -\epsilon\dot{\phi}_k + (1+\epsilon)\alpha\cos(\tau_{k+1}) \\ \dot{\phi}_k(\tau_{k+1}-\tau_k) - \alpha(\sin\tau_{k+1}-\sin\tau_k) + s_{k+1} - s_k = 0\end{array}\right\} \quad (14)$$

with the normalized angle $\phi = (r\varphi/v)$ and the normalized angular velocity $\dot{\phi}_k$ during the k-th flight phase. The superscripts + and - need not be used as the angular velocity does not change during the flight.

From the analysis of vibrations it is well known that for the system of equation 14 there exist periodic as well as chaotic solutions. Irrespective of the qualitative characteristics of the motion it always appears in the attractor basin. Infinitely repeating the mapping very densely fills the attractor manifold even for chaotic motions. If the distribution in this region was known this would mean a complete description of the stationary motion. For the chaotic case this tends to be a continuous function in the attractor manifold whereas for the periodic case it reduces to a number of Dirac functions. It corresponds to the invariant distribution in the ergodic theory of dynamical systems.

It even would be sufficient to have the marginal distribution of the normalized time points of the impacts to be able to achieve characteristic statistical results. This marginal distribution also depends on the behavior of the system and in the same manner as the distribution in the attractor manifold tends to become continuous for the chaotic and a number of Dirac impulses for the periodic case.

Fig. 6 shows that the periodic as well as the chaotic motions appear in the vicinity of the fix points. This results in the fact that the marginal distribution in the near of the corresponding points also shows significant values compared to the rest of the region. Since the equations 14 are invariant under the transformation

$$\begin{array}{rcl}\tau & \to & \tau + \pi \\ \dot{\phi} & \to & -\dot{\phi} \\ (s_{k+1} - s_k) & \to & -(s_{k+1} - s_k)\end{array} \quad (15)$$

every distribution of impacts will be periodic with period π. On the other hand, if a motion is not symmetric within the backlash transforming the motion by 15 will always yield in a correct solution. Taking the mean of the transformed and the un-transformed motion regains the mentioned periodicity of the marginal distribution.

Now let us assume the marginal distribution being known. Equations 14 are replaced by

$$\left.\begin{array}{l}\dot{\phi}_{k+1} = -\epsilon\dot{\phi}_k + (1+\epsilon)\alpha\cos(\tau) \\ \rho(\tau)\,,\ \tau\in[0,2\pi)\end{array}\right\} \quad (16)$$

which is a discrete stochastic equation. Time τ is asumed a random variable with some probability density $\rho(\tau)$ corresponding to the distribution found above.

From equation 16 the mean values can be determined analytically. Calculating the expected value for equation 16 yields in the following recursive formula for the first order moment

$$m_{1_{k+1}} = -\epsilon\, m_{1_k} + (1+\epsilon)\alpha J_1$$

where $J_1 = \int_0^{2\pi}\rho(\tau)\cos\tau d\tau$ and $m_{1_k} = E[\dot{\phi}_k]$

Here m_{1_k} is the expected value of the angular velocities after k repetitions. The term J_1 in this case vanishes due to the mentioned periodicity in the marginal distribution $\rho(\tau)$
The above iteration converges to zero if $\epsilon < 1$:

$$m_1 = \lim_{k \to \infty} m_{1_k} = 0$$

The derivation of the second moment is possible by quadrature of equation 16 and determination of the expected values :

$$m_{2_{k+1}} = \epsilon^2 \, m_{2_k} + (1 + \epsilon\,)^2 \alpha^2 J_2 - 2\epsilon(1 + \epsilon)\alpha E[\dot\phi_k \cos \tau]$$

$$\text{wobei} \quad J_2 = \int_0^{2\pi} \rho(\tau) \cos^2 \tau d\tau \quad \text{and} \quad m_{2_k} = E[\dot\phi_k^2]$$

In this formula m_{2_k} stands for the expected value of the squared angular velocities after k iterations. Obviously

$$E[\dot\phi_k \cos \tau] \approx 0$$

holds since the values $\dot\phi$ are symmetric and the normalized time points of impacts appear in the vicinity of $\pi/2$ und $3\pi/2$. This approximation results in the following iteration formula for the second moment

$$m_{2_{k+1}} = \epsilon^2 \, m_{2_k} + (1 + \epsilon\,)^2 \alpha^2 J_2$$

m_{2_k} can explicitly be given as a function of the initial conditions :

$$m_{2_k} = \frac{1 - \epsilon^{2k}}{1 - \epsilon}(1 + \epsilon)\alpha^2 J_2 + \epsilon^{2k} m_{2_0} \tag{17}$$

Here m_{2_0} represents the influence of the initial conditions. Equation 17 converges very rapidly as $\epsilon < 1$. In parallel, the influence of the initial conditions — or of the asumed distribution of the initial conditions — deminishes. For $k \to \infty$ equation 17 converges to the second moment m_2 :

$$m_2 = \lim_{k \to \infty} m_{2_k} = J_2 \frac{1 + \epsilon}{1 - \epsilon}\alpha^2 \tag{18}$$

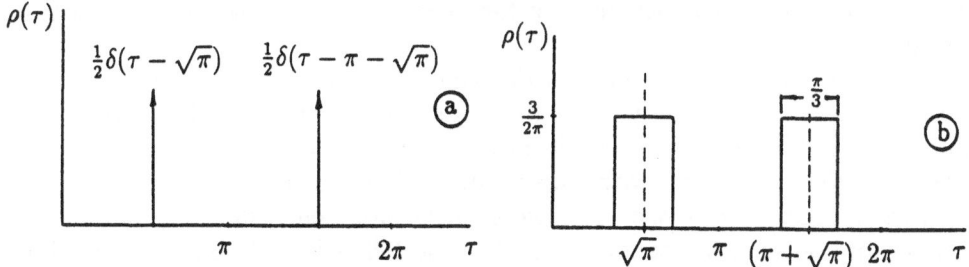

Fig. 10 Assumed marginal distribution $\rho(\tau)$

The faktor J_2 in the above equation could be determined by numerical simulation but instead is only estimated since as a proportionality factor its influence is small. Let us assume marginal distributions $\rho(\tau)$, one for a periodic and the other one for a chaotic case

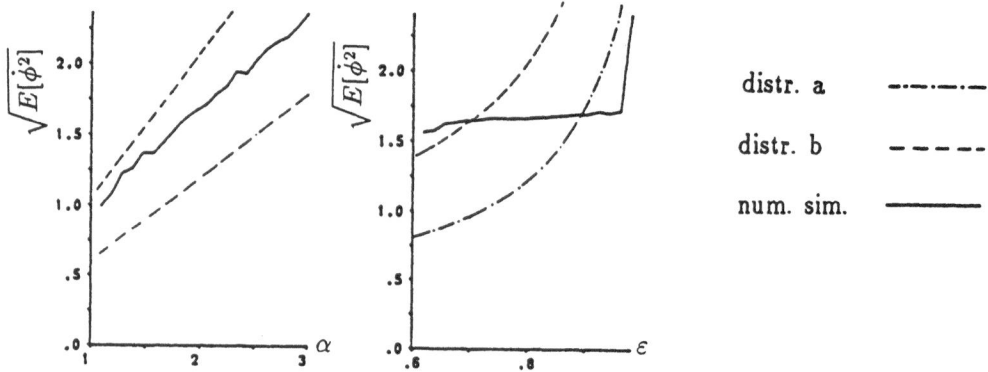

Fig. 11 Second moments from approximation (a and b) and simulation (c)

(see fig. 10). From these distributions the integral J_2 is calculated and introduced into equation 18 . Fig. 11 shows the second moments resulting from this approximation for both of the assumed marginal distributions on the one hand and the numerical simulation on the other.

In spite of the simplicity of the applied method the qualitative agreement is sufficient. Resetting the normalized parameters into unnormalized forms yields in the following relation between the second moment and the system parameters :

$$E[\dot{\varphi}^{\,2}] = J_2 \frac{A^2\omega^2}{(r/r_e)^2} \frac{(1+\epsilon)}{(1-\epsilon)} \tag{19}$$

This form clearly shows the influence of the single parameters. The intensity of the appearing velocities rises quadratically with the excitation frequency and amplitude, decreases quadratically with the transmission factor but for $\epsilon < .8$ seems to be independent of the number of restitution.

4.2. Parameter dependece for the two stage model

Also in the two stage case let us investigate a simplified model neglecting damping and constant moments. As mentioned in subsection 3.2 we will use unnormalized parameters trying to find analytical relations between the expected mean values and the system parameters.

As in the one stage case depending on the parameters there appear periodic as well as chaotic solutions in the vicinity of the fix points (compare fig. 8 and 9). Again the system can be described completely by the invariant distribution.

Here we consider only the marginal distribution of the impact times within the excitation period. Compared to the one stage case the situation is more complicated as impacts may appear in either backlash 1 or 2 . The P.S. achieved from the numerical simulation according to subsection 3.2 shows that the impacts in the two backlashes are correlated. This suggests a simple approximate relation between the distributions of the impacts in backlash 1 and 2. It is assumed that after l_1 impacts in backlash 1 l_2 impacts in backlash 2 appear. The value of the quotient l_1/l_2 depends on the backlashes and moments of inertia and should be identified numerically. For the case of approximately equal moments of inertia of the loose gear wheels and if the quotient of the backlashes is in the range of 0.5–2 then $l_1 \approx l_2 \approx 1$. This means that after an impact in backlash 1 an impact in backlash 2 may be expected. Assuming this to be exactly the case yields in an approximation of the real motion which still will sufficiently approximate the attractor.

Let us now assume the marginal distribution of impact times in backlash 1 was known a priori. Instead of the mappings we will use the following discrete stochastic equations

$$\left. \begin{array}{l} \dot{\varphi}_{2_{k+2}} = \mathbf{U}_2 \dot{\varphi}_{1_{k+1}} \\ \dot{\varphi}_{1_{k+1}} = \mathbf{U}_1 \dot{\varphi}_{2_k} + \zeta \, \cos(\omega t) \\ \rho(t) \, , \ t \ \in \ [0, \frac{2\pi}{\omega}) \end{array} \right\} \qquad (20)$$

$\dot{\varphi}_{2_k} = (\dot{\varphi}_1, \dot{\varphi}_2)^{\mathsf{T}} \ \in \ \Re^2$ is the vector of angular velocities of the two gear wheels after the k-th impact in backlash 2 and $\dot{\varphi}_{1_{k+1}} = (\dot{\varphi}_1, \dot{\varphi}_2)^{\mathsf{T}} \ \in \ \Re^2$ are the angular velocities after the *(k+1)*-th impact in backlash 1 . The vektor $\zeta \ = \omega \zeta_1$ is introduced as an abbreviation. The transfer matrices $\mathbf{U}_1$, $\mathbf{U}_1$ and the vektor ζ have already been defined in section 2, subscripts 1 and 2 represent the stage in which the impact occurs. Equations 20 assume a probability density for the impacts in backlash 1. After each of these impacts one in backlash 2 follows.

Using the stochastic model of equations 20 one can calculate the expected values of the appearing velocities. For this purpose we formulate the equations in the following way :

$$\left. \begin{array}{l} \dot{\varphi}_{2_{k+2}} = \mathbf{U}_2 \mathbf{U}_1 \dot{\varphi}_{2_k} + \mathbf{U}_2 \zeta \, \cos(\omega t) \\ \dot{\varphi}_{1_{k+1}} = \mathbf{U}_2^{-1} \dot{\varphi}_{2_k} \\ \rho(t) \, , \ t \ \in \ [0, \frac{2\pi}{\omega}) \end{array} \right\} \qquad (21)$$

The first equation contains only relations between system states in backlash 2 and can easily be processed further. Determination of the expected values yields

$$\left. \begin{array}{l} \mu_{2_{k+2}} = \mathbf{U}_2 \mathbf{U}_1 \mu_{2_k} + \mathbf{U}_2 \zeta \, J_1 \\ \mu_{1_{k+1}} = \mathbf{U}_2^{-1} \mu_{2_k} \end{array} \right\} \qquad (22)$$

$$\text{with} J_n = \int_0^{2\pi/\omega} \rho(\omega t) \cos^n(\omega t) dt$$

$\mu_{2_k} = E(\dot{\varphi}_1, \dot{\varphi}_2)^{\mathsf{T}}$, is the expected value of the velocities after the k-th impact in backlash 2 and $\mu_{2_{k+1}} = E(\dot{\varphi}_1, \dot{\varphi}_2)^{\mathsf{T}}$ represents the expected value of the velocities after the *(k+1)*-th impact in backlash 1 .

Equation 22 converges if the absolute eigenvalues of the matrix $\mathbf{U}_2 \mathbf{U}_1$ are less than one. For the interesting parameter region this holds true. The expected values can explicitly be found as functions of the initial conditions and their limits can be given :

$$\left. \begin{array}{l} \mu_2 = \lim_{k \to \infty} \mu_{2_k} = (\mathbf{E} - \mathbf{U}_2 \mathbf{U}_1)^{-1} \mathbf{U}_2 \zeta \, J_1 \\ \mu_1 = \lim_{k \to \infty} \mu_{1_{k+1}} = \mathbf{U}_2^{-1} \mu_2 \end{array} \right\} \qquad (23)$$

Evaluating this form with respect to the parameters of the model yields

$$\mu_1 = \mu_2 = J_1 \, A\omega(1/i_1, 1/(i_1 i_2))^{\mathsf{T}} \qquad (24)$$

with $i_1 = \frac{r_{i_1}}{r_e}$, $i_2 = \frac{r_2}{r_{i_2}}$ being the transmission factors of the two stages. This result was to be expected as the excitation is amplified by the transmission factors weighted by the value J_1 which depends on the marginal distribution of the impacts.

As in calculating the first order moments we will distinguish between the two back-lashes for the calculation of expected values of the squares of the velocities. $C_{2_k} = [\dot{\varphi}_1 \dot{\varphi}_2^T] \in \Re^{2,2}$ is the correlation matrix of velocities after impacts in backlash 2 and k iterations of the process, and $C_{1_{k+1}} = [\dot{\varphi}_1 \dot{\varphi}_2^T] \in \Re^{2,2}$ is the correlation matrix of the velocities after impacts in backlash 1 and k iterations of the process. The evaluation of C_2 is done by multiplying the first of the equations 21 with its transposed and extracting the expected values.

$$
\begin{aligned}
C_{2_{k+2}} = \ & U_2 U_1 C_{2_k} (U_2 U_1)^T + \\
& [U_2 \zeta \, \mu_{2_k}^T (U_2 U_1)^T + U_2 U_1 \mu_{2_k} (U_2 \zeta)^T] J_1 + \\
& U_2 \zeta \, (U_2 \zeta)^T J_2
\end{aligned}
\tag{25}
$$

This equation is a recursive matrix equation which converges under the condition that the absolute eigenvalues of the matrix $U_2 U_1$ are less than one. The limit is a solution of the above equation and so a matrix equation for its determination appears :

$$
\begin{aligned}
C_2 = \ & U_2 U_1 C_2 (U_2 U_1)^T + \\
& [U_2 \zeta \, \mu_2^T (U_2 U_1)^T + U_2 U_1 \mu_2 (U_2 \zeta)^T] J_1 + \\
& U_2 \zeta \, (U_2 \zeta)^T J_2
\end{aligned}
\tag{26}
$$

$$
\text{where} \quad C_2 = \lim_{k \to \infty} C_{2_k}
$$

Similarly, a corresponding matrix equation can be derived for the correlation matrix of the velocities in backlash 1 :

$$
\begin{aligned}
C_1 = \ & U_1 U_2 C_1 (U_1 U_2)^T + \\
& [\zeta \, \mu_1 (U_1 U_2)^T + U_1 U_2 \mu_1 \zeta^T] J_1 + \\
& \zeta \, \zeta^T J_2
\end{aligned}
\tag{27}
$$

$$
\text{where} \quad C_1 = \lim_{k \to \infty} C_{1_{k+1}}
$$

The correlation matrices can be determined analytically with the results being a little bit lengthy. Like in the one stage case the integral J_1 vanishes due to the symmetry of the attraction region if all possible solutions are summed up. If only a single attractor is considered the integral should still have a small value. The elements of the correlation matrix are

$$
C_{1_{11}} = \frac{A^2 \omega^2}{i_1^2 (1 - \epsilon)} \frac{g_1(\epsilon, x)}{g(\epsilon, x)} J_2
\qquad
C_{1_{22}} = \frac{A^2 \omega^2}{i_1^2 i_2^2 (1 - \epsilon)} \frac{g_2(\epsilon, x)}{g(\epsilon, x)} J_2
$$

$$
C_{1_{12}} = \frac{A^2 \omega^2}{i_1^2 i_2} \frac{g_3(\epsilon, x)}{g(\epsilon, x)} J_2
$$

and

$$C_{2_{11}} = \frac{A^2\omega^2}{i_1^2(1-\epsilon)} \frac{f_1(\epsilon, x)}{f(\epsilon, x)} J_2 \qquad C_{2_{22}} = \frac{A^2\omega^2}{i_1^2 i_2^2(1-\epsilon)} \frac{f_2(\epsilon, x)}{f(\epsilon, x)} J_2$$

$$C_{2_{12}} = \frac{A^2\omega^2}{i_1^2 i_2} \frac{f_3(\epsilon, x)}{f(\epsilon, x)} J_2$$

$$\text{with} x = \frac{I_2/I_1}{i_2^2}$$

The functions $f_1, f_2, f_3, f, g_1, g_2, g_3, g$ are high order polynomials of ϵ and x . The properties of the proposed formulas agree well with experimental results. From the mentioned expected values other characteristics can be derived such as energy dissipation by the impacts or the intensity of the impacts. At this point we will not go into this subject more detailed.

5. Experimental results

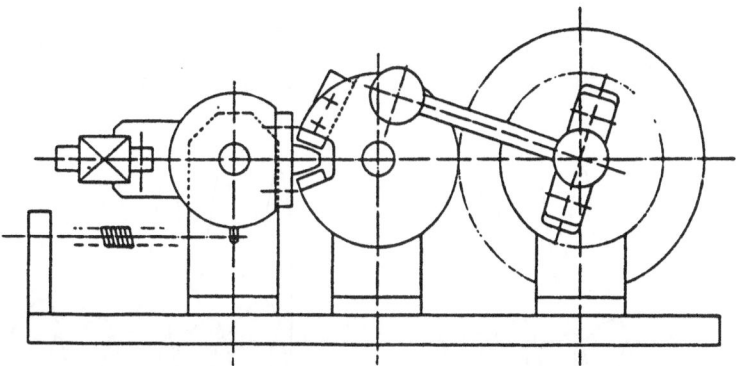

Fig. 12 Experimental set-up

The theoretical results are to be verified at an experimental set-up. A model of a one stage gear has been constructed as displayed in fig. 12 . A wheel with a single involute tooth is excited by vibrating constraints. In addition, a eddy current brake simulates linear damping forces whereas a week spring adds an almost constant momentum. The vibrating constraints themselves are driven via an excenter mechanism by a direct current motor.

All parameters that are important for the problem can be varied such as the excitation frequency and amplitude, the backlash, the damping factor and the constant moment. Measurement results are the motions of the two wheels, the tooth wheel and the one with the vibrating constraints. The signals are then further processed on a digital computer.

In fig. 13 three signals are shown which correspond to different damping values; the other parameters are constant. The frequency of the ecxitation is comparatively low and the backlash great because for these conditions the signals are more suitable. The same behavior as in the experiment is found by numerical integrations of the one stage gear

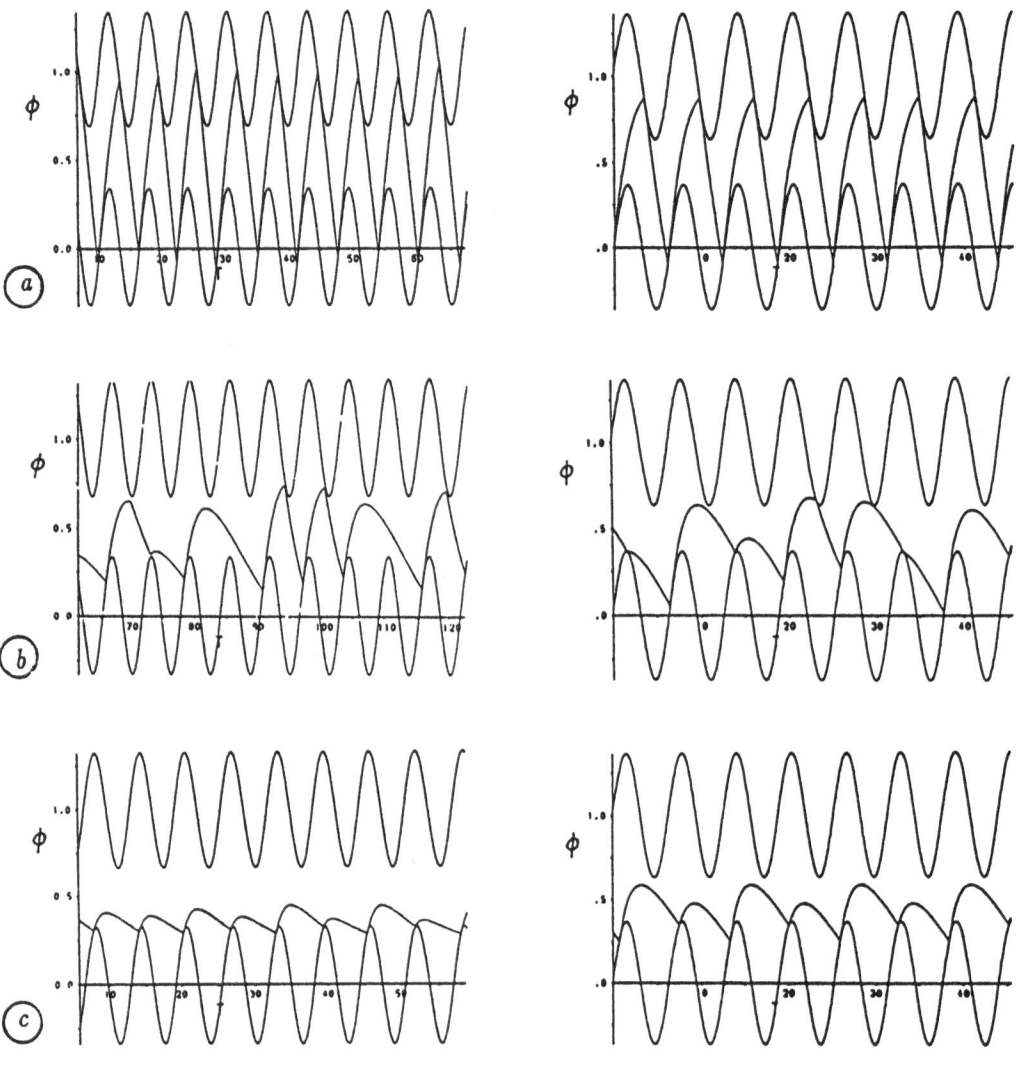

Fig. 13 Comparison measurement (left) - simulation (right)
a : periodic motion b=0.30
b : chaotic motion b=0.50
c : subperiodic motion b=0.55
Parameters are : $\alpha = 0.324$ $\gamma = 0.043$ $\epsilon = 0.4$

model. The number of retstitution is low (≈ 0.4) but it seems to be due to the design. In real gearboxes we expect a number of restitution of about 0.8.

ACKNOWLEDGMENTS
Support of this work by the DAAD (German Academic Exchange Service) is gratefully appreciated.

References

[1] F. Pfeiffer und F. Kücükay : Eine erweiterte Stoßtheorie und ihre Anwendung in der Getriebedynamik. VDI-Z Bd. 127 (1985) Nr. 9 - Mai(1)

[2] F. Kücükay und F. Pfeiffer : Über Rasselschwingungen in Kfz-Schaltgetrieben. Ingenieur-Archiv 56 (1986) 25-37

[3] Toshimitsu Sakai, Yuhji Doi, Kenichi Yamamoto : Theoretical and Experimental Analysis of Rattling Noise of Automotive Gearbox. SAE Passenger Car Meeting Dearborn, Michigan June 1981

[4] S. W. Shaw : The dynamics of a Harmonically Excidet System Having Rigid Amplitude Constraints. Part 1 : Subharmonic and Local Bifurcations. Journal of Applied Mechanics June 1985, Vol. 52 453-458

[5] S. W. Shaw : The dynamics of a Harmonically Excidet System Having Rigid Amplitude Constraints. Part 2 : Chaotic Motions and global Bifurcations. Journal of Applied Mechanics June 1985, Vol. 52 459-464

[6] F. Pfeiffer : Seltsame Attraktoren in Zahnradgetrieben. Ingenieur-Archiv 58 (1988) 113-125

[7] L. A. Wood and K. P. Byrne : Analysis of a random repeated impact process. Journal of Sound and Vibration (1981) 78(3), 329-345

[8] L. A. Wood and K. P. Byrne : Experimental investigation of a random repeated impact process. Journal of Sound and Vibration (1982) 85(1), 53-69

FEEDBACK CONTROL OF RESISTIVE MODES IN TOKAMAKS

M.F.F.Nave

Laboratorio Nacional de Engenharia e Tecnologia Industrial, Sacavem,
Portugal
present address: JET Joint Undertaking, Abingdon, Oxon, U.K.

ABSTRACT

The control of the amplitude of rotating resistive modes is important to
prevent disruptive conditions, thereby extending the operating space of
tokamaks. To gain information for the design of practical feedback
schemes, it is useful to analyse the response of non-linear modes to time
dependent boundary conditions imposed by external conductors and a
resistive vessel. Here it is shown how a dynamic system describing the
evolution of the amplitude and the phase of the modes is obtained. The
system presents regimes of both phase and frequency instabilities and
shows clearly that the problem of amplitude stabilisation is strictly
related to the stabilisation of the phase.

1- INTRODUCTION

Among a large number of devices designed for achieving the conditions for

fusion reactions, the Tokamak is the most extensively studied. It consists

of a toroidal vessel where a plasma, an ionised gas of hydrogen isotopes,

is confined away from the walls by a system of magnetic fields. A

schematic view of a tokamak is shown in figure 1. The main component of

the magnetic field, created by coils surrounding the vessel, is in the toroidal direction. Also in the toroidal direction runs a plasma current, produced by an external transformer, which has the function of heating the plasma by Joule dissipation and creates a small poloidal magnetic field essential for magnetic equilibrium. The resultant magnetic field is helical.

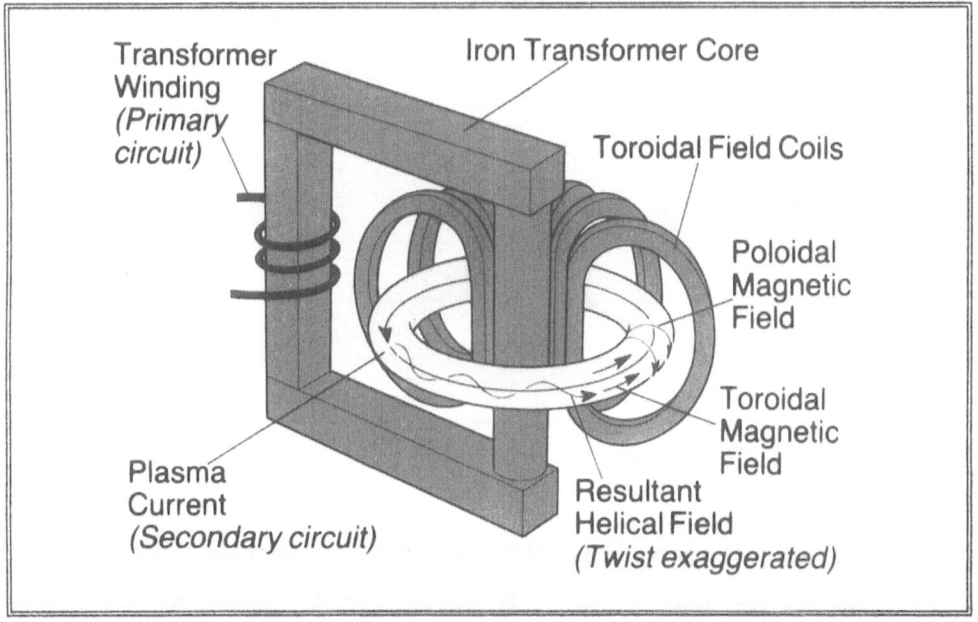

Figure 1. Tokamak magnetic field configuration

To achieve a net energy gain from the nuclear fusion reactions certain conditions have to be met. The temperature must be sufficiently high to overcome the Coulomb repulsion between two positively charged nuclei which implies that the plasma must be thermally isolated from the walls for a sufficient time, and to guarantee a sufficient number of fusion reactions the density must also be high. In order to obtain those conditions bigger and bigger tokamaks have been built in the past four decades. An example of a large tokamak, whose operating conditions are close to those required for a net energy gain, is the European tokamak JET /1/, shown in figure 2.

198

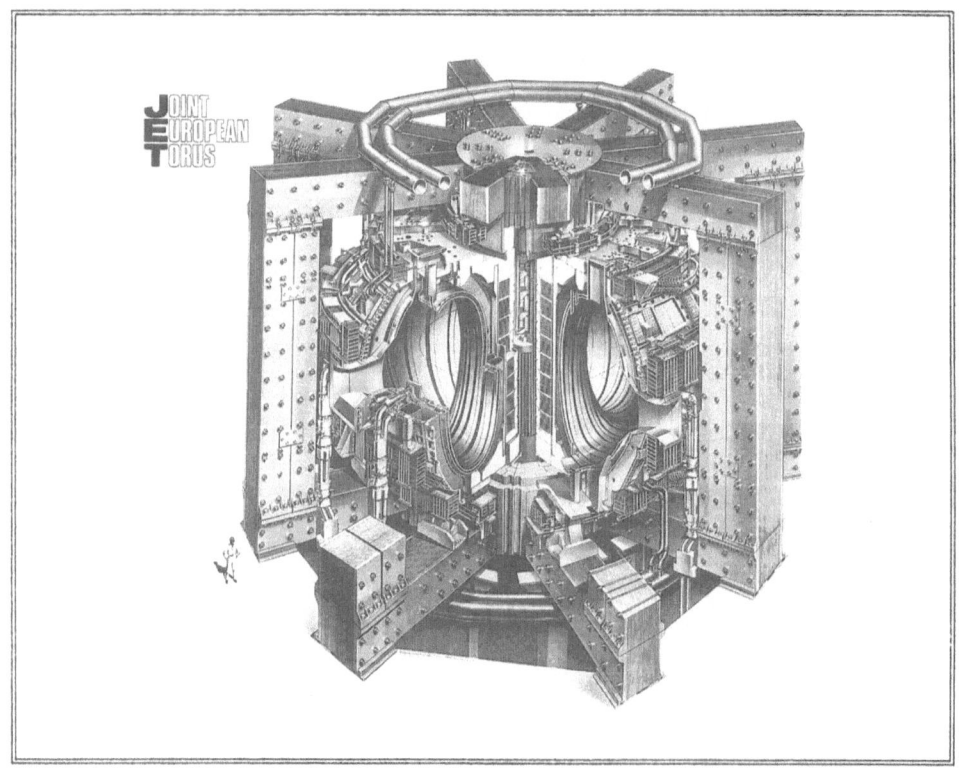

Figure 2. The JET tokamak.

However, no matter how large the machine, considerable losses of both particle and energy confinement is caused by a variety of macroscopic instabilities. The most dangerous of these instabilities known as 'disruption' /2/, causes a rapid cooling of the plasma and a collapse of the plasma current. Although in some cases the plasma recovers, major disruptions literally 'kill' the plasma, limiting the operation of tokamaks at high currents and high densities. Moreover, disruptions lead to large mechanical stresses which may damage the machine.

A considerable effort, both in experimental and theoretical studies, has been devoted to understanding the disruptions. The aim is to avoid them. Nevertheless, many questions remain unanswered and to add complexity to the problem, disruptions occurring in different operating regimes appear ultimately to have different causes and may have to be described by different models /3/. However, some of the events preceding them are commom to all types of disruptions. Usually, before a disruption occurs growing perturbations of the magnetic field are observed. Thus, even if one does not understand all the physics involved, one could try to control a disruption by controlling the appearance or the growth of the magnetic precursors.

These magnetic perturbations are believed to be related to the nonlinear evolution of 'tearing' modes, which are resistive magnetohydrodynamic (mhd) modes driven unstable by unfavourable current gradients /4-8/. In their non-linear stage tearing modes would change the magnetic topology with the appearance of rotating magnetic islands .

Several methods have been suggested in the literature in order to control the tearing modes. Here one is concerned with the electrodynamic techniques which are particularly suitable to control the boundary conditions of the stability problem. In these techniques, a magnetic field generated by an arrangement of external coils produces a magnetic field equivalent to and which counteracts the magnetic field perturbation produced by currents within the plasma. Three classes of control have been considered :

a)   stabilisation by means of static helical currents /9-12/,

b)   amplitude- and phase-tuned oscillating currents linked to the measured signals by a feedback loop /13-17/, and

c)   high-frequency currents to achieve dynamic stabilisation /18,19/.

The first technique can only have a limited success. As will be shown below, the control of the amplitude cannot be considered independently from the control of the phase, and therefore the second technique should be more suitable to stabilise rotating modes. Here we concentrate on this second technique which at the moment is being attempted in the DITE tokamak /20/. A feedback stabilisation system shall also be in operation in JET at the end of 1990 /21/.

The following text starts with a brief review of the tearing mode stabilisation problem in the standard large aspect ratio tokamak approximation (i.e., the torus is approximated by a cylinder). Experimental evidence for tearing modes is illustrated with JET data. Then, the equations governing the non-linear evolution of the mode amplitude and rotation frequency are described and it is shown how these equations should be modified to study the control problem. One is interested in the response of the system to the action of an external field, with the aim of unveiling problems which will have to be taken into account in the complete study of a feedback loop. A more detailed account of this work can be found in a paper by Lazzaro and Nave /16/.

2- THE THEORY OF RESISTIVE MODES

MHD stability

The operation of a tokamaks is characterised by two parameters, the inverse aspect ratio $\varepsilon = a/R_o$ and the safety factor $q(r) = (r/R)(B_\phi/B_\theta)$ which is a measure of the helicity of the field lines /22/. The geometry and notation is shown in figure 3.

The magnetic equilibrium configuration is obtained from the equations of magnetohydrodynamics /23/. The mhd treatment also provides the simplest model for the strongest instabilities observed in a tokamak /24/. In this model the plasma is described by the equations of a continuous fluid coupled to Maxwells equations through the currents. Two important mhd regimes are studied, the ideal case in the limit of infinite conductivity, and the resistive case.

201

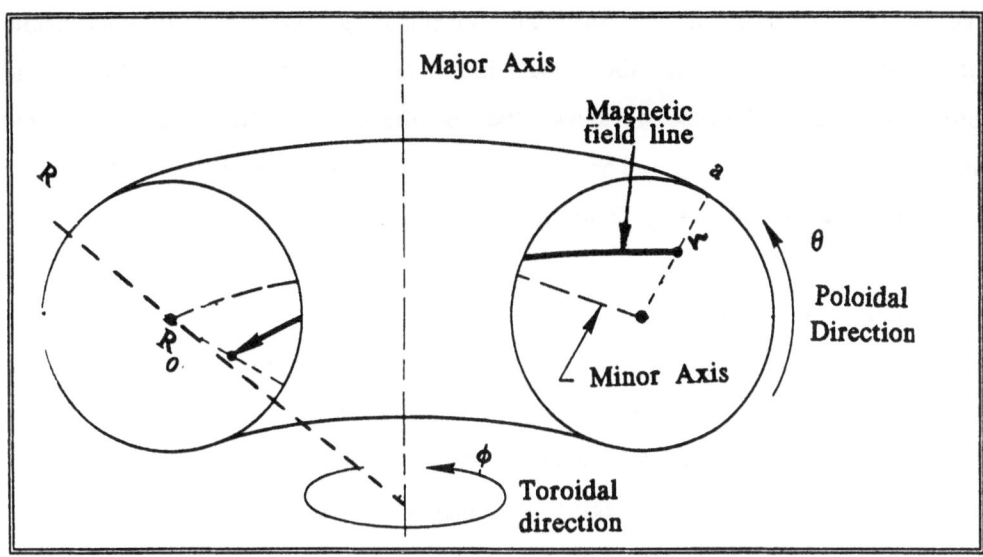

Figure 3. The geometry and notation.

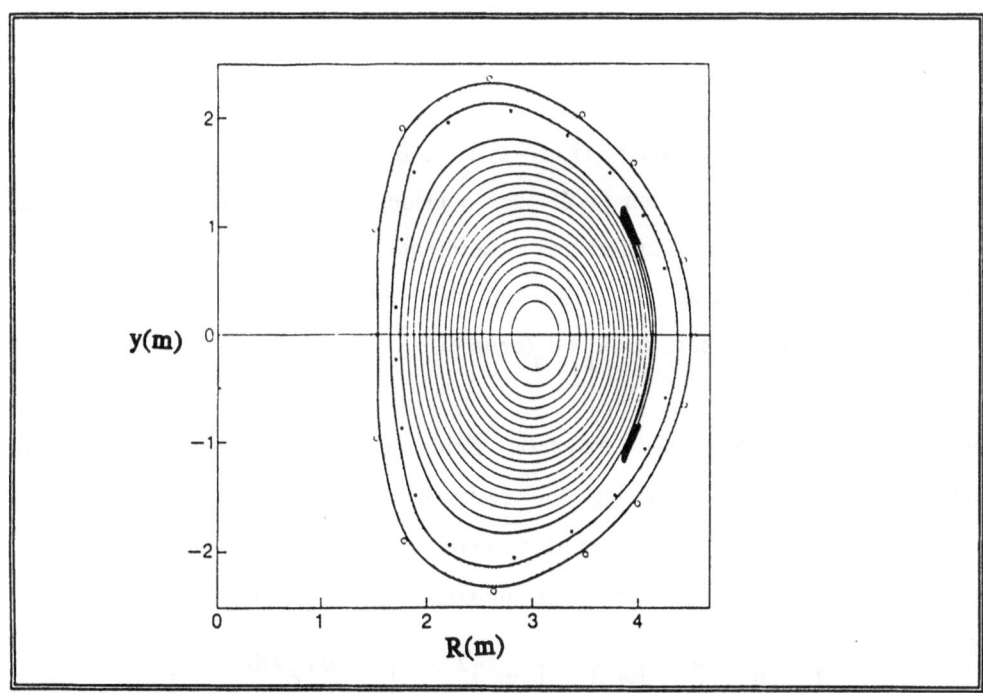

Figure 4. Cross-section of the JET plasma showing a typical equilibrium configuration.

Due to the symmetry in the toroidal coordinate ϕ, the magnetic equilibrium configuration, obtained in the ideal limit regime, consists of closed and nested surfaces. (Figure 4 show the equilibrium configuration calculated for a JET discharge /22/.) The twist of the magnetic field on each surface is characterised by the safety factor q(r).

This equilibrium configuration is subject to a variety of non-axisymmetric perturbations which can become unstable depending on the values of current and pressure gradients. Here, one is interested in the resistive modes and in particular in the 'tearing modes', which as the name suggests, 'break' the magnetic field lines, changing the magnetic topology with the formation of magnetic islands inside the plasma /4-8/.

The tearing mode equation

A large aspect ratio tokamak (i.e., $\varepsilon \ll 1$, and q(r) of the order of unity) with a circular cross-section will be considered. The toroidal coordinate ϕ is mapped into the cylindrical coordinate z and the magnetic field perturbation is expressed in terms of the helical flux perturbation ψ, as $\tilde{B} = Re(\nabla\psi \times h)$, where $h = e_z + e_\theta(nr/mR)$.

The equation governing the magnetic field perturbation due to a tearing mode is obtained by linearising the resistive mhd equations and by taking the space variation of the linearised perturbation as $\psi = \psi_{m,n}(r)e^{i(m\theta-n\phi)}$. This procedure leads after some manipulation to the tearing mode equation /24/:

$$\frac{1}{r}\frac{d}{dr}\left[r\frac{d\psi}{dr}\right] - \left[\frac{m}{r}\right]^2\psi - \left[\frac{dj_\phi/dr}{B_\theta(1-nq/m)}\right]\psi = 0$$

valid for $m \geq 2$. ψ is related to the transverse field perturbations by

$\tilde{B}_r = im\psi/r$ and $\tilde{B}_\theta = -\partial\psi/\partial r$. This equation is solved numerically for a given equilibrium current density profile, j_ϕ. The examples shown below are for parabolic $j_\phi = j_0(1-(r/a)^2)^\upsilon$ and mode numbers m=2,n=1.

The equation shows that tearing modes are destabilised by large values of the radial gradient of the equilibrium current j, and that the modes resonate with surfaces, s, where q has a rational value, $q=q_s=m/n$. The stability is determined by the jump of $\partial\tilde{B}_r/\partial r$ across the resonant layer /24/. Although there is an infinite spectrum of modes, the most dangerous ones, with larger growth rates, are the ones with low order mode numbers m and n.

The equation for the island growth

The resistive diffusion of the magnetic field with respect to the fluid allows in a non-linear stage the breaking and reconnection of the magnetic field lines with the consequent formation of magnetic islands /4-8/ (illustrated in figure 5).

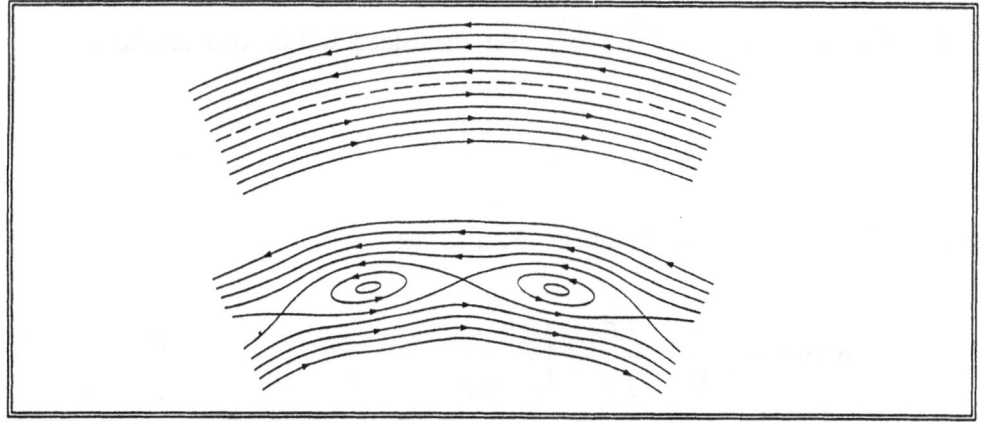

Figure 5. The tearing and rejoining of the magnetic field lines which occur during the non-linear phase of the instability as a consequence of finite resistivity /22/.

The width of the island, W, (see figure 6) is proportional to the square root of the radial field perturbation and is obtained by calculating the trajectory of a magnetic field line close to the resonant surface with the assumption that $\tilde{B}_r$ is constant over the island:

$$W = 4 \sqrt{\frac{rq\tilde{B}_r}{m\dfrac{dq}{dr} \cdot B_\theta}}\Bigg|_{r_s}$$

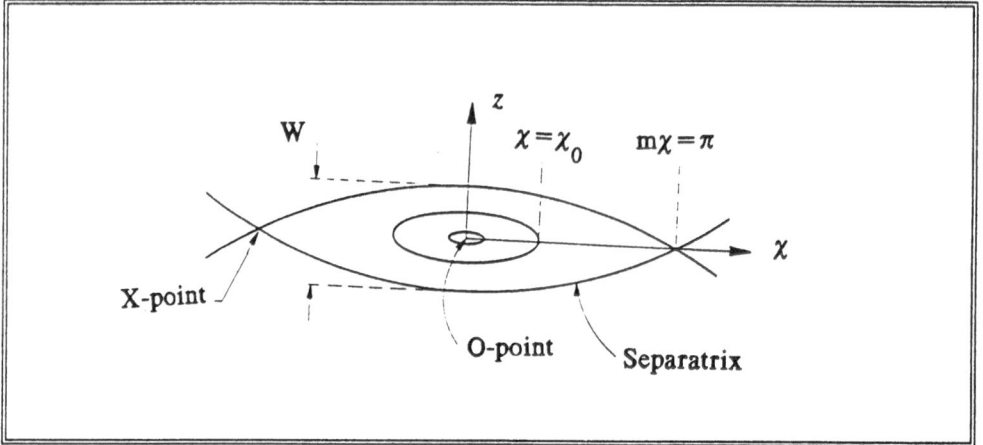

Figure 6. Geometry of a magnetic island in the plane (z,χ), where $z = r - r_s$ and $\chi = \theta - n\phi/m$ /22/.

The island growth is determined by integrating the field diffusion equation, $\partial \tilde{B}_r/\partial t = (\eta/\mu_0)\, \partial^2 \tilde{B}_r/\partial r^2$, over the island width. One obtains :

$$\frac{dW}{dt} = (\eta/2\mu_0)\, \Delta'(W) \tag{4}$$

where $\Delta'(W)$ is the stability parameter defined as:

$$\Delta'(W) = \frac{1}{\tilde{B}_r}\frac{\partial \tilde{B}_r}{\partial r}\Bigg|_{r_s - W/2}^{r_s + W/2} \tag{5}$$

$\tilde{B}_r$ is determined from the tearing mode equation (3).

Figure 7 shows $\Delta'(W)$ as a function of W for a linearly unstable m=2, n=1 mode /16/. Two boundary conditions are considered, Δ'_v for a perfect conducting wall close to the plasma edge and Δ'_∞ for a wall at the infinity. For a parabolic j-profile the dependence of Δ' on W is typically monotonically decreasing. The figure shows that the conducting wall has a beneficial effect on the stability because for any value of Δ' the corresponding island width is smaller when a wall is present.

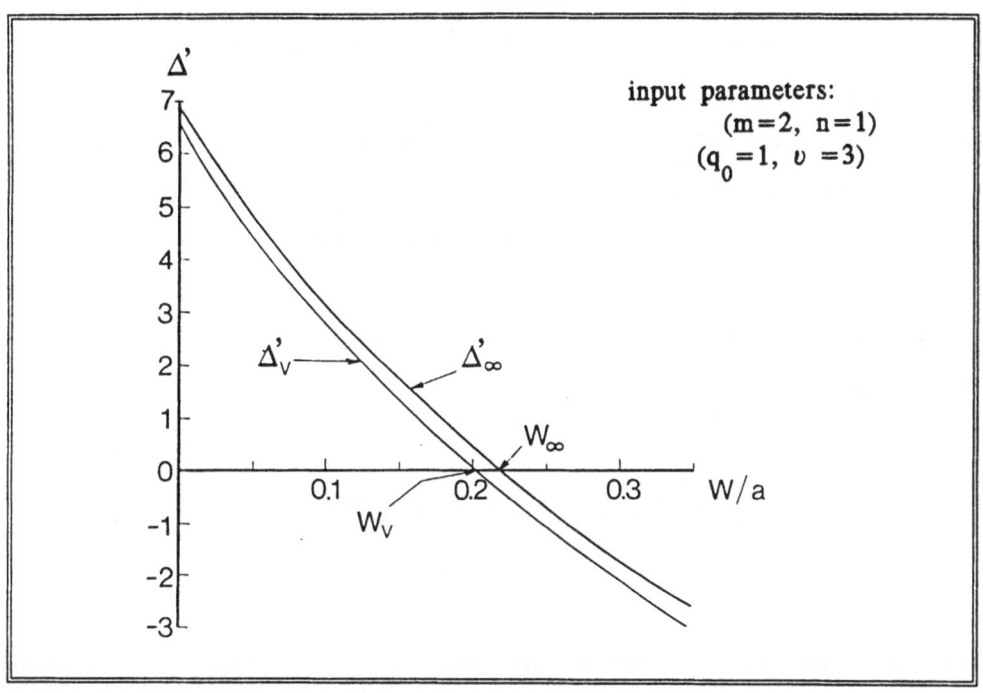

Figure7. The stability parameter Δ' versus the island width W, for a perfectly conducting vessel at radius 1.2a (Δ'_v) and no vessel(Δ'_∞).

The non-linear theory of tearing modes also predicts that the background axisymmetric profiles of pressure and current should relax locally to a force free configuration with vanishing gradients across the islands /7,8/.

The equation for the phase evolution

In the usual non-linear theory of tearing modes ψ is determined from equation (3) satisfying a real boundary condition. If the island is rotating, as is usually observed, the interaction with a resistive wall introduces a phase shift. The tearing mode equation now becomes complex and its solution gives the rate of change of the island velocity in addition to the island growth /25,26/.

For simplicity, we will assume that the plasma has a rigid rotation in the toroidal direction, and we express the time variation of the field perturbation as $\exp(-in\Phi_w(t))$,where $\Phi_w(t)=\int^t \omega(t')\,dt'$ is the phase and ω is the frequency. These oscillatory perturbations will induce eddy currents on the wall which in their turn will give rise to forces proportional to $\tilde{J} \times \tilde{B}$. Similar forces on the plasma result in a momentum exchange from the plasma to the wall, such that the evolution of the phase is then given by:

$$MR\ \frac{d^2\Phi}{dt^2}w = F_\phi = \int dV(\tilde{J} \times \tilde{B})_\phi \qquad (6)$$

where M is the mass of the plasma which rotates. F_ϕ can be calculated on the outer surface of the plasma resulting in

$$F_\phi = -(2n\pi^2/\mu_0)\ \mathrm{Im}(\psi^* d\psi/dr)\big|_{r=a}.$$

The explicit dependence of F_ϕ on ω and W will be shown below in section 4.

To illustrate the effect of the resistive wall on the mode rotation, figure 8 shows the solution of equations (4) and (6) for the case of a plasma with a parabolic j-profile /26/. At high frequencies the oscillatory fields seen by the wall as a result of the island motion make the wall behave like a good conductor and can keep the island size small. However, as the island grows the transfer of momentum between the plasma and the wall reduces the frequency. This allows penetration of the oscillating magnetic field into the wall. When the frequency falls to zero, known as the effect of mode-lock, the effect of stabilisation of the wall is lost completely and the growth rate of the island increases.

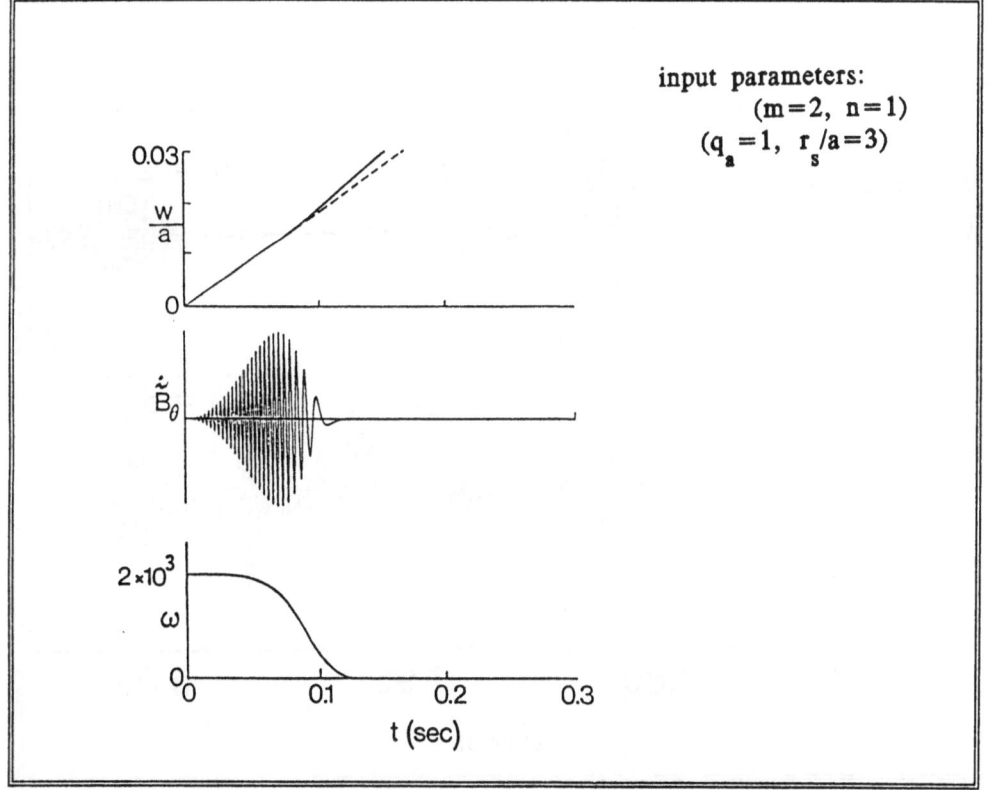

input parameters:
(m=2, n=1)
(q_a=1, r_s/a=3)

Figure 8. The time development of the island width, W, the magnetic perturbation d$\tilde{B}_\theta$/dt and the frequency ω.

3- THE EXPERIMENTAL EVIDENCE

The non-linear theory of tearing modes discussed above is in good agreement with observations. Several measurements provide experimental evidence for the existence of helical mhd resistive mode structure in tokamak discharges. With magnetic diagnostics placed on the tokamak vessel one can measure magnetic perturbations. These field perturbations, known as Mirnov oscillations, appear as bursts in the signals of $d\dot{\tilde{B}}_\theta/dt$. This is illustrated with the following results from JET. Figure 9 shows Mirnov oscillations observed before a disruption /3/. Typical frequencies are in the range of 1-10kHz . As the amplitude increases the frequency slows down. The amplitude, however, continues to increase even after the mode locks /27/. The magnetic perturbation signals are related to a bulk fluid rotation in the toroidal direction /28/. Mode analysis indicates low ratios of m/n /3/.

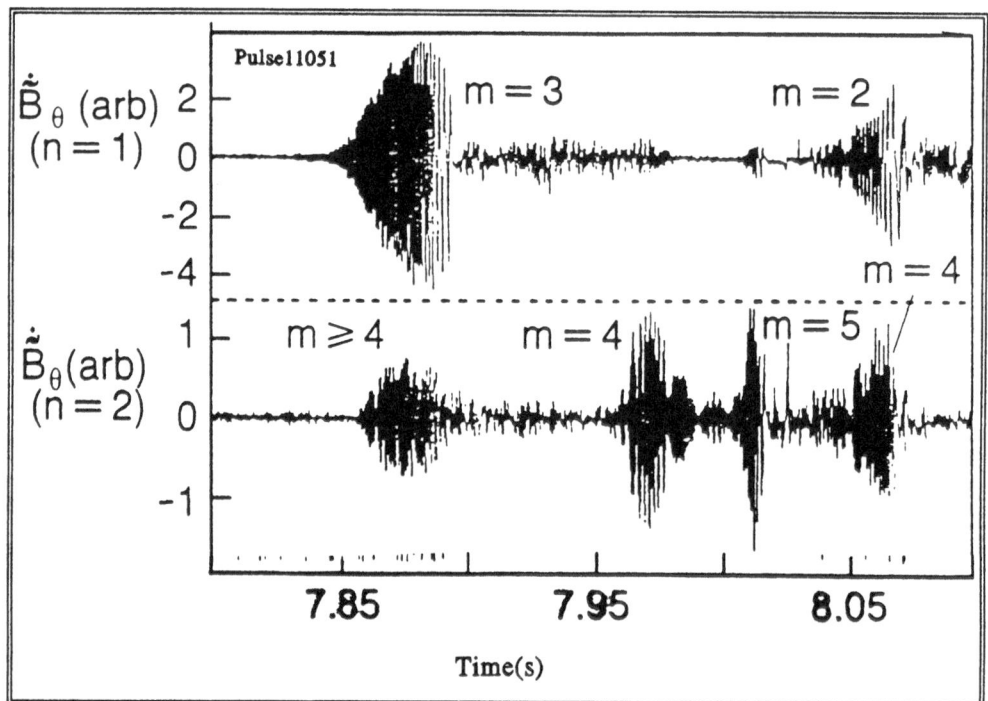

Figure 9. Oscillations observed in the signals of $d\dot{\tilde{B}}_\theta/dt$ before a disruption.

Soft X-ray emission and temperature signals show perturbations inside the plasma located at rational values of q. Figure 10 shows a flattening of the temperature profile close to the q=2 surface /21/. The flattening is not observed when the phase of the magnetic signal changed by 180°, showing the effect of the rotation. Soft X-ray tomography (figure 11) confirms the existence of island-like structures /29/.

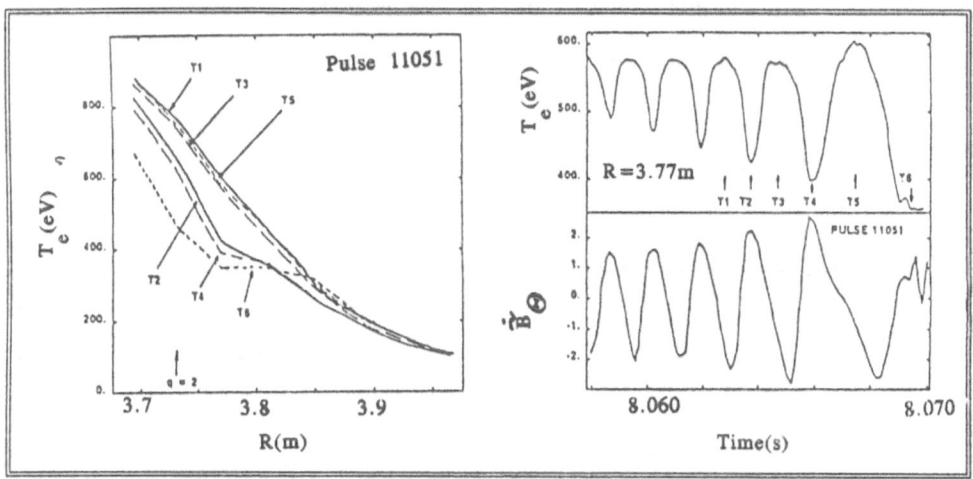

Figure   10.a)Electron temperature profiles during the m=2,n=1 oscillation,
10.b)Electron temperature near the q=2 surface and the magnetic
signal during the m=2,n=1 oscillation.

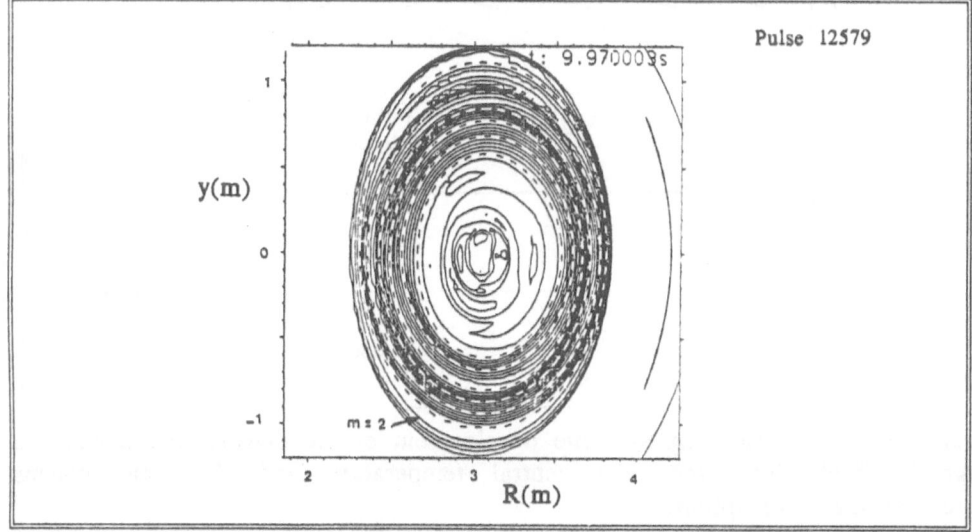

Figure 11. Soft-X-ray emission showing island-like structures.

The Mirnov instabilities are believed to play a critical role in tokamak disruptions. Figure 12 illustrates the sequence of events leading to a disruption /22/. Prior to the disruption changes in the plasma parameters lead to a more unstable configuration which allows the growth of instabilities. In the pre-cursor phase, typically lasting 10-100ms an $m=2, n=1$ instability is observed to grow. Then, usually after the frequency has come to zero, a rapid decay of the temperature is observed in a time scale of 1ms. Finally the plasma current decays to zero.

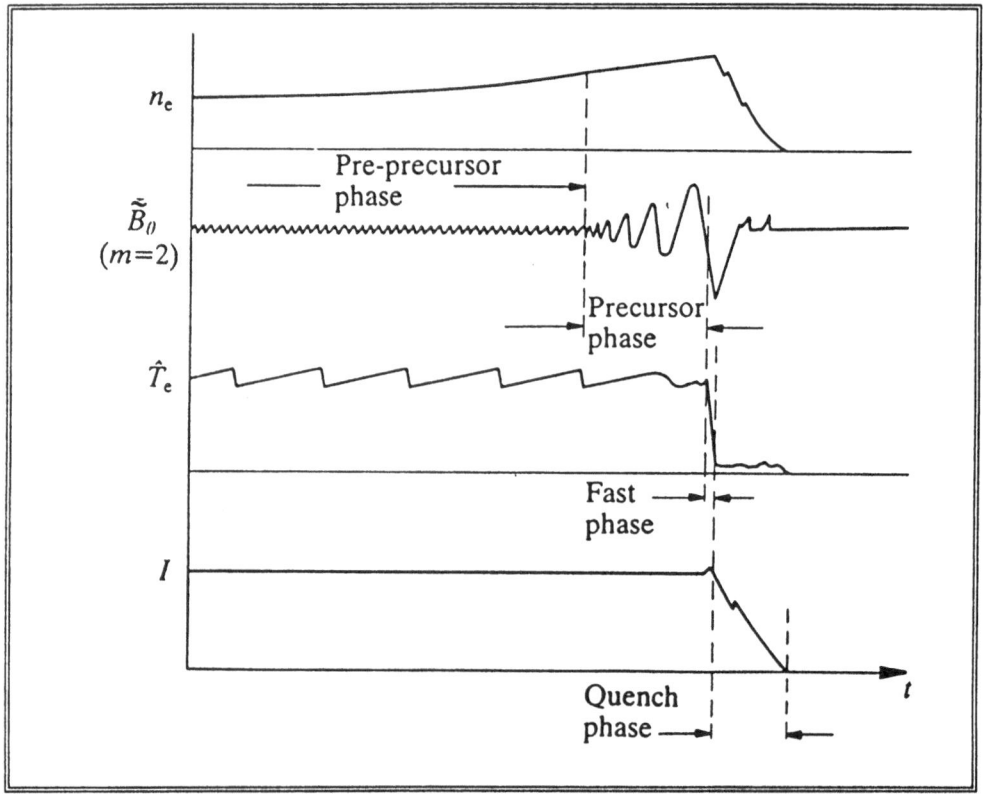

Figure 12. Graphs showing the time development of the density, the $m=2, n=1$ magnetic field fluctuation, the central temperature and the total plasma current during a disruption.

4 - THE CONTROL PROBLEM

The final phases of a disruption are very rapid and it is better to try to prevent the disruptions by controlling the events observed prior to the faster phases. The changes in the conditions before the precursor phase, however, are not always very obvious, so it makes sense to wait for the appearance of the Mirnov instabilities and then try to prevent their growth to disruptive proportions.

To reduce the amplitude of tearing modes, we must be able to reduce or dissipate the magnetic free energy associated with the stability parameter Δ'. We recall that Δ' is a function of dj/dr and of boundary conditions. Therefore two classes of control are possible:

a) try to affect locally either the current profile or plasma resistivity /30-34/;

b) try to change the boundary conditions. Here one is interested in this class of control when electrodynamic techniques are used /9-19/.

Although the metallic wall provides some stabilisation, for most cases of interest this is too small to prevent the growth of the instabilities. One could however envisage the enhancement of the wall effect by adding another conductor. By creating an external current with the same helicity as the mode to be stabilised an external field would be produced which could balance by negative feedback the field perturbation originating in the plasma. In these methods the tokamak vessel acts as a passive conductor, while the wiring carrying a helical oscillating current with variable amplitude act as an active conductor. The idea behind these methods is therefore to reduce the magnetic island size, as opposed to an alternative method where disruptions where controlled by applying a helical perturbing field to ergodize or break-up the $m=2$ magnetic island /35/.

The geometry of the problem is shown in figure 13. The plasma has a radius a. The position of the singular layer for the chosen m and n mode numbers is r_s. Outside the plasma there are two conductors, a thin wall at position d and inside the wall at radius b, $a < b \sim d$, there is a perfectly conducting time varying control surface current $I_A \exp(i\Phi_A)$ with a single helicity pitch m/n.

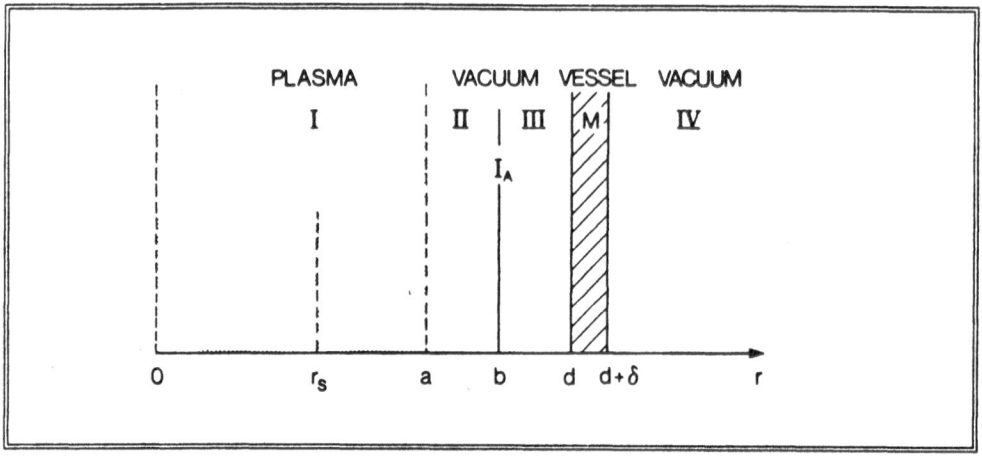

Figure 13. The various regions where ψ has to be determined.

The dynamical system describing the time evolution of the amplitude and the phase of the mode is given by equations (4) and (6).

ψ is calculated by solving the tearing mode equation in the plasma region satisfying a boundary condition which is a function of parameters characterising the external conductors

$$\left.\frac{\psi'}{\psi}\right|_{r=a} = f(a,\omega\tau_v,b,I_A,\Delta\Phi,d)$$

where $\tau_v = \mu_0\sigma\delta d/2$ is the wall time constant, σ is the wall conductivity and δ is the wall thickness (we assume that $\delta \ll d$),

$\Delta\Phi =n(\Phi_W-\Phi_A)$ is the difference between the phases of the mode and the applied current.

In order to identify the several effects embodied in the dynamic equations, it is useful to express ψ as a linear combination of two numerically well known eigenfunctions for the case of $I_A=0$ (ψ_v^0 for a perfectly conducting wall and ψ_∞^0 without a wall), in the form:

$$\psi_p = \psi_\infty^0 + \alpha\,(\psi_v^0+\psi_\infty^0)$$

Similarly, with the assumption that ψ_p is constant across the island, Δ' can be written as

$$\Delta'_p = \Delta'^{,0}_\infty + \alpha\,(\Delta'^{,0}_v+\Delta'^{,0}_\infty)$$

The boundary condition

The outer regions are described by the equations:

vacuum regions: $\nabla_\perp^2\psi = (\mu_0 I_A/b)\,\delta(r-b)\,\exp(i(\Delta\Phi+m\theta-n\phi))$

with $\psi = C_+r^m+C_-r^{-m}$

wall region: $\nabla_\perp^2\psi = in\dot\Phi_W\,\mu_0\sigma\psi$

with $\psi = C_+^M\,r^{p(r-d)} +C_-^M\,r^{-p(r-d)}$ (for $\delta/d\ll1$)

where $p^2=(m/r)^2-in\dot\Phi_W\mu_0\sigma$. The constants are determined from conditions of continuity for $\tilde{B}_r$ at interfaces and a jump condition for $\tilde{B}_\theta$ across current carrying layers. The full boundary-value problem consists of determining 8 constants by solving the system:

$$
\begin{bmatrix}
a^m & a^{-m} & 0 & 0 & 0 & 0 & 0 & (\psi_\infty^0 - \psi_v^0)_a \\
ma^{m-1} & -ma^{-m-1} & 0 & 0 & 0 & 0 & 0 & (\psi_\infty^{0\,\prime} - \psi_v^{0\,\prime})_a \\
b^m & b^{-m} & -b^m & -b^{-m} & 0 & 0 & 0 & 0 \\
mb^{m-1} & -mb^{-m-1} & -mb^{m-1} & -mb^{-m-1} & 0 & 0 & 0 & 0 \\
0 & 0 & d^m & d^{-m} & -1 & -1 & 0 & 0 \\
0 & 0 & md^{m-1} & -md^{-m-1} & -p & +p & 0 & 0 \\
0 & 0 & 0 & 0 & e^{p(d+\delta)} & e^{-p(d+\delta)} & (d+\delta)^{m-1} & 0 \\
0 & 0 & 0 & 0 & pe^{p(d+\delta)} & -pe^{-p(d+\delta)} & m(d+\delta)^{-m-1} & 0
\end{bmatrix}
\begin{bmatrix}
C_+^{I\,I} \\
C_- \\
C_+^{I\,II} \\
C_-^{I\,II} \\
C_+^M \\
C_-^M \\
C^{I\,V} \\
\alpha
\end{bmatrix}
=
\begin{bmatrix}
-\psi_v^0(a) \\
-\psi_v^{0\,\prime}(a) \\
0 \\
\mu_0 I_A e^{i\Delta\Phi} \\
0 \\
0 \\
0 \\
0
\end{bmatrix}
$$

From which one obtains /16/:

$$
F = \left(\frac{\psi_p'}{\psi_p} \right)_a = -\frac{m}{a}\,\frac{1+f(a/d)^{2m}}{1-f(a/d)^{2m}} - \frac{\mu_0 I_A e^{i\Delta\Phi}}{a\,\psi_p(a)}\left(\frac{a}{b}\right)^m \frac{1-f(b/d)^{2m}}{1-f(a/d)^{2m}} \qquad (7)
$$

$$
\underbrace{\hspace{4cm}}_{\text{Resistive wall}} \qquad \underbrace{\hspace{6cm}}_{\text{External current}}
$$

and
$$
\alpha = \alpha_0 (1 + \alpha_A), \qquad (8)
$$
where

$$
\alpha_0 = \left[1 - \frac{\psi_v^0}{\psi_\infty^0} \cdot \frac{F_v^0 - F_{vr}^0}{F_\infty^0 - F_{vr}^0} \right]^{-1}, \qquad \left.\right\}\ \begin{array}{l}\text{Resistive}\\ \text{Wall}\end{array}
$$

$$
\alpha_A = \frac{\mu_0 I_A e^{i\Delta\Phi}}{a}\left(\frac{a}{b}\right)^m \frac{1-f(b/d)^{2m}}{1-f(a/d)^{2m}} \cdot \frac{1}{\psi_\infty^0 (F_\infty^0 - F_{vr}^0)} \qquad \left.\right\}\ \begin{array}{l}\text{External}\\ \text{Current}\end{array}
$$

f is a complex function of the frequency ω:

$$f=f_R+f_I=((\omega^2\tau_v^2-im\omega\tau_v)/(\omega^2\tau_v^2+m^2))$$

The expressions for F_∞^0 (for *no* wall, i.e. $d/a \gg 1$), F_v^0 (for a perfectly conducting wall, i.e. $\omega\tau_v \gg 1$) and F_{vr}^0 (for a resistive wall) for the cases of $I_A = 0$ are obtained from (7) by taking the appropriate limits.

The dynamic system

The dynamic system describing the evolution of W and Φ_w in the presence of both a wall and an external current is obtained by substituting the expressions for F and α in (4) and (6). The system shown here was obtained in the approximation $(r_s/d)^{2m} \ll 1$ /16/.

$$\frac{dW}{dt} = \frac{\eta}{\mu_0}\left\{\Delta_\infty' - \left[f_R + \frac{I_A}{GW^2}\left|1-f\left(\frac{b}{d}\right)^{2m}\right|\cos(\Delta\Phi+\Theta_v)\right](\Delta_\infty'-\Delta_v')\right\} \qquad (9)$$

$$\frac{d^2\Phi_w}{dt^2} = C^2W^4\left(\frac{a}{d}\right)^{2m}\left[f_I + \frac{I_A}{GW^2}\left|1-f\left(\frac{b}{d}\right)^{2m}\right|\sin(\Delta\Phi+\Theta_v)\right] \qquad (10)$$

where

$$C^2 = (mn\pi^2/MR\mu_0 64)\left[B_\theta(q'/q)\right]_{r_s}^2 \psi_v^0\,\psi_\infty^0 \ ,$$

$$G = m\left(\frac{B_\theta q'}{q}\right)_{r_s}\frac{\psi_\infty^0}{8\mu_0}\frac{a^m b^m}{d^{2m}} \ ,$$

and

$$\Theta_v = \arctan \left\{ -f_I(b/d)^{2m} \; / \; [1-f_r(b/d)^{2m}] \right\} .$$

These equations show a superposition of effects from the plasma and from a system of active and passive conductors. These can be summarized as follows.

a) In the absence of both conductors
$dW/dt=(\eta/\mu_0)\Delta'_\infty$, describes the unstable growth of W up to a saturation size W_∞ (see illustration in figure 7).

b) Resistive wall effects :
-The factor $f_R(\Delta_v^{'0}+ \Delta_\infty^{'0})$ in eqn.(9) has a stabilising effect on the island width.

-The factor f_I in eqn.(10) is a frequency damping term. For $I_A=0$ eqn.(10) becomes
$$d\omega/dt=-C^2W^4(a/d)^{2m}(m\omega\tau_v/(\omega\tau_v^2+m^2))$$
whose solution was shown in figure 8.

c) Applied current effects:
-The currents induced on the resistive wall by the active conductor produce two effects:
   A reduction of the amplitude of the external current by the factor $|1-f(b/d)^{2m}|$.
   A phase shift Θ_v, leading to frequency runaway.

-The difference of phase between the mode and the applied current, $\Delta\Phi$, can lead to a phase 'flip' instability. The term $\sin\Delta\Phi$ in equation (10) leads to an exponentially growing solution for $\Delta\Phi$. When $\Delta\Phi$ flips to π the term $\cos\Delta\Phi$ in the equation (9) changes sign and the stabilising effect of I_A is lost.

Equations (9) and (10) show that the problem of amplitude stabilisation depends on the phase coherence between the mode and the applied current. A more detailed study of the response of the system to an external current can be found in the paper by Lazzaro and Nave /16/. Here will only be shown two examples that illustrate the effect of the phase shift caused by the wall, and the phase instability.

Firstly, figure 14 shows the response of the system to a step input controlling current $I_A \propto W^2$, with $\Delta\Phi = 0$. The island size growth is unstable up to the time when the controlling current is switched on. The behaviour of the mode rotation describes clearly the slowing down action of the resistive wall and, after I_A is switched on, the increase in rotation produced by the external field.

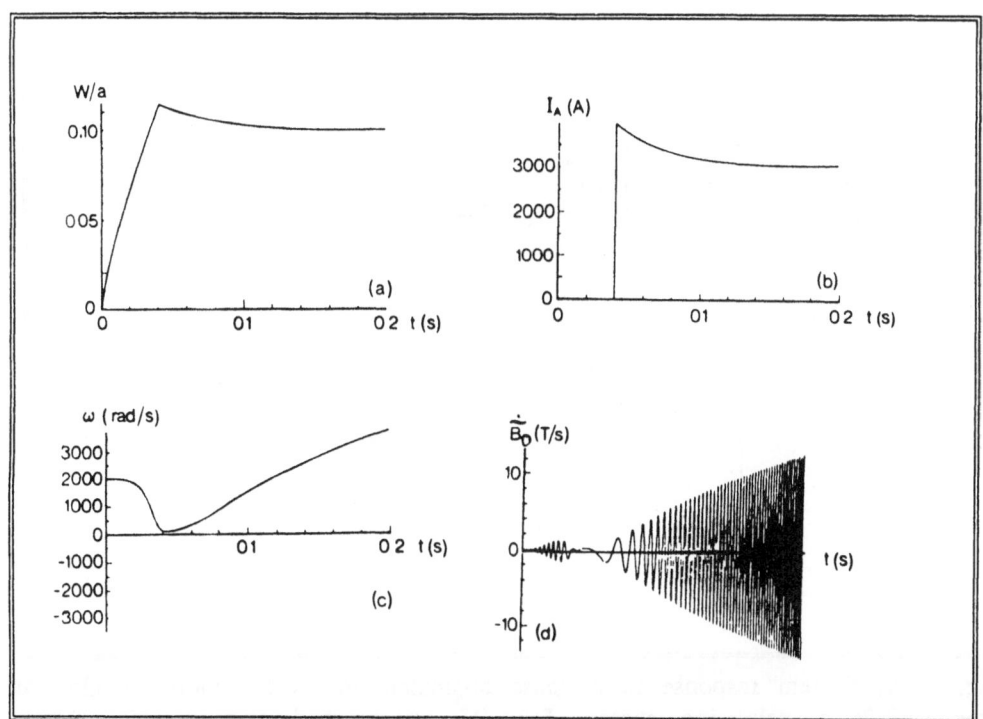

Figure 14. System response to a step control current $I_A \propto W^2$ switched on at $t = 0.04$s, for a perfect phase match $\Delta\phi = 0$.

In the second example, in figure 15, we can see the effect of the phase instability. We consider a mode that grows unstably but that initially does not rotate. Then we switch on an stabilising current $I_A \propto W^2$, and at a later time we introduce a small phase difference perturbation $\Delta\Phi = 0.5°$. The phase grows to π from the initial error over a time $t = \tau \ln(\pi/\Delta\Phi_0)$, where τ is proportional to the geometric average of the Alfven times of the equilibrium and the perturbed magnetic fields /16/. Associated with this phase instability the island width shows stable intervals when $\cos\Delta\Phi > 0$, however, overall the stabilisation effect of the external current is lost.

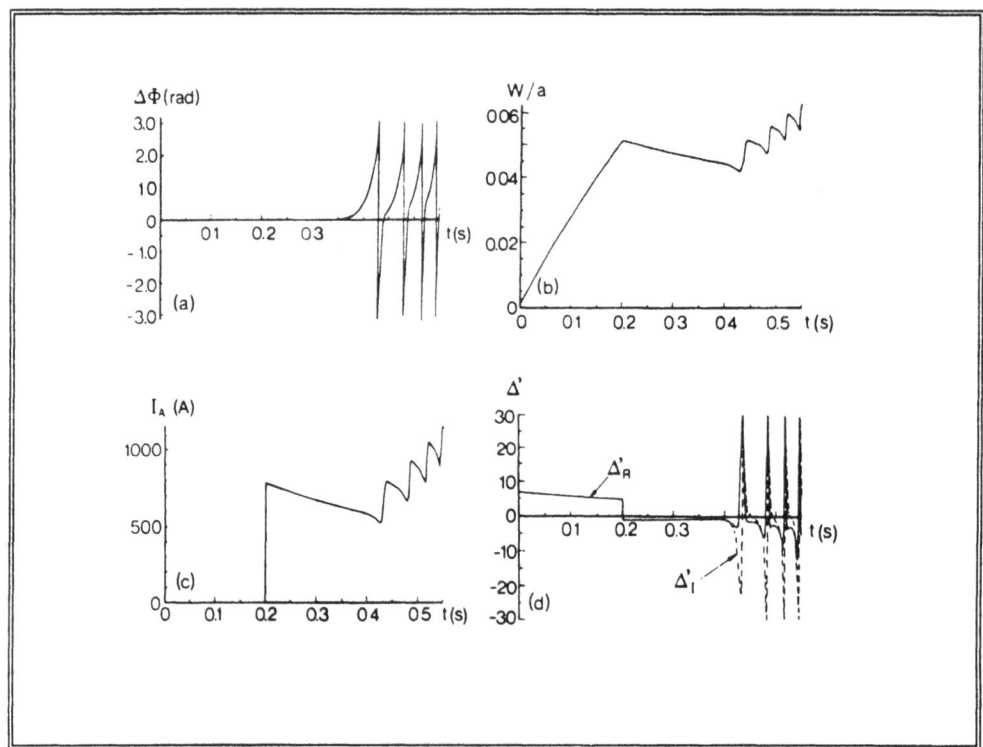

Figure 15. System response to a phase mismatch on a non-rotating mode. At time $t = 0.2$s a stabilising current $I_A \propto W^2$ is switched on, a perfect phase match $\Delta\Phi = 0$ is kept until $t = 0.3$s when a small phase difference $\Delta\Phi = 0.5°$ is introduced.

4-CONCLUSIONS

A model describing the time dependent non-linear evolution of the amplitude and phase of a tearing instability for a prescribed feedback and including the effect of a resistive vessel was developed by E.Lazzaro and M.F.F.Nave /16/. Here were shown the various assumptions used in that model which lead to a dynamic system consistent with the characteristics of the instabilities observed before a tokamak disruption.

The set of non-linear equations obtained show a superposition of effects from the plasma and from a system of active and passive conductors. These equations are appropriate to explain quantitatively previous experimental and theoretical results. In the equation for the magnetic island width the effect of the external conductors appear as a driving term which may be stabilising or destabilising depending on the phase coherence with the applied fields. The equation for the frequency is driven by a torque depending on the superposition of the currents induced in the passive wall and active current. The passive conductor terms describe a slowing down of the rotation and a partial stabilisation of the island width. Furthermore, the action of the active conductors is reduced if they are very close to the passive ones.

The equations show clearly that the problem of amplitude stabilisation is strictly related to the phase and frequency evolution and that one may expect regimes of both phase and frequency instabilities. A small uncontrolled error in the phase would lead to a rapid growth of that error on a time scale proportional to the geometric average of the Alfven times of the perturbed and equilibrium poloidal magnetic fields ($\simeq 3$ms for JET parameters). A phase shift due to the image currents produced on the resistive wall by the active conductors will increase the mode frequency, which may make the tracking of the phase more difficult.

Analysis of steady state conditions /16/ show that for small islands the amplitude feedback relation is proportional to the measured signal (i.e. $I_A \propto W^2$). (A fixed I_A for small islands would lead to an unstable situation). Having chosen the relation for the feedback of the amplitude, the stabilisation problem becomes reduced to a phase detection and control problem. With accurate phase information a successful feedback is possible with controlling currents of 10^{-3} the total plasma current.

However, apart from the many technical difficulties which may arise in the detection and feedback control of the instabilities, the theoretical study presented here may not yet tell the full story. Firstly, the regimes under which a chaotic behaviour may occur have not been studied. (Although, it has been clear from numerical simulations that such behaviour may arise.) Secondly, the system was derived for a single helicity mode (m,n). Toroidal effects not included in the model introduce mode coupling /36/.

The problem of control of tokamak disruptions is certainly as complex as many problems in economics. Apart from the phase instability, in a 1979 paper called 'The economic theory of the disruptive instability' /37/, Furth pointed out other adverse aspects which may arise and for which we still do not know the answear. In particular, he pointed out that the stabilisation of a single helicity mode may subsequently destabilise other modes with different helicities which could then grow to disruptive sizes.

The experimental viability of feedback stabilisation by means of an external helical current was demonstrated in the Russian tokamak TO-1 in 1978 /17/. Feedback stabilisation experiments are planned to be in operation in JET later in 1990 /21/. As it is not possible to install helical coils on a tokamak which is already cluttered with many devices for diagnostics and auxiliary heating systems, the helical current to stabilise the m=2,n=1 mode will be created by saddle coils. In order to minimise the time required to implement the JET feedback control, a similar system is being tested in the smaller tokamak DITE /20/. I finish with a photograph of the saddle coils and detectors being installed inside the wall of the DITE vacuum vessel (figure 16) and by wishing success to the DITE team.[1]

ACKNOWLEDGEMENTS

I am grateful to J.Wesson for permission to use figures 5,6 and 12 from his book /22/, to D.Robinson for permission to use the DITE photograph (figure16), to A.Vannuci and T.Stringer for comments on the text and T.Rowan for help with the typesetting. Finally, I would like to thank my friend and co-author E. Lazzaro for innumerable discussions, late working evenings and the latin sense of humour.

---

[1]

Initial results from DITE show reductions in the $m=2, n=1$ mode amplitude by factors up to 5, although care has to be taken to avoid phase instabilities /38/.

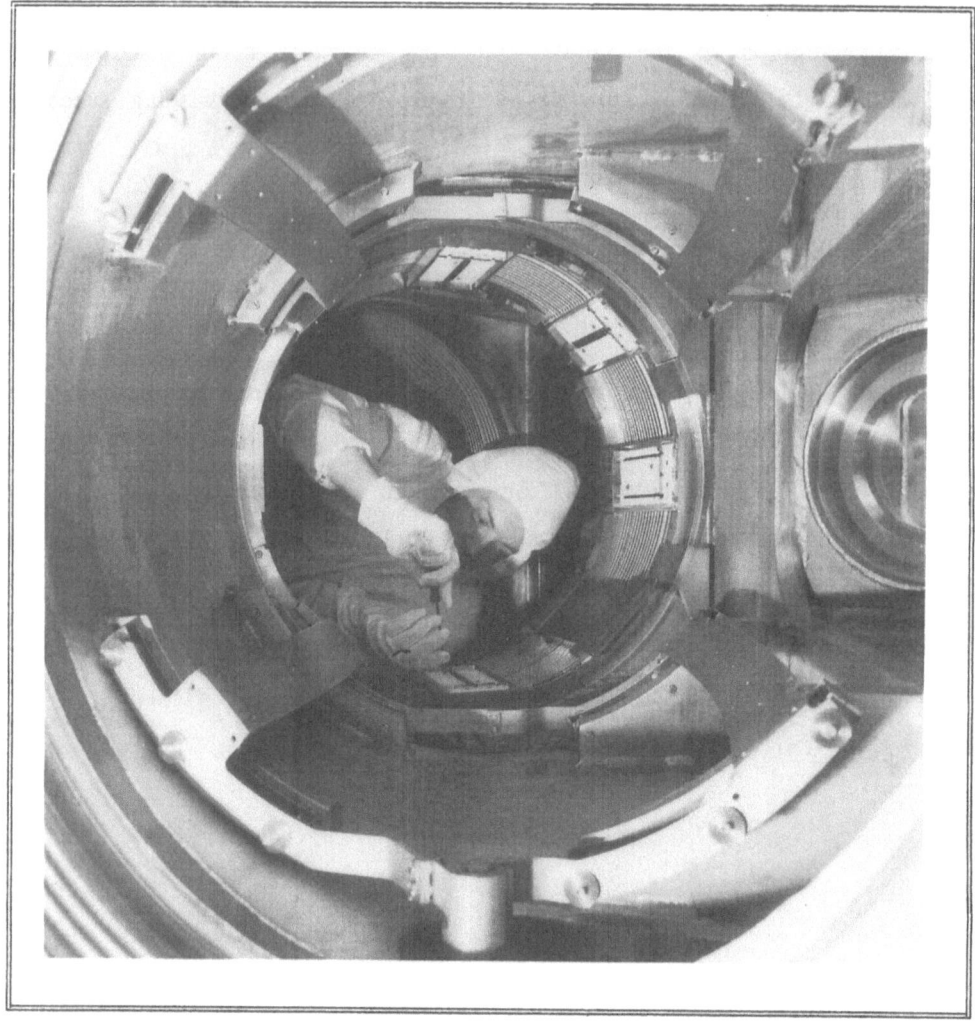

Figure 16. The detection system being installed inside the DITE vessel. The saddle coils can be seen in the background.

REFERENCES

/1/   P.H.Rebut and B.E.Keen, Fusion Technol.11,13 (1987)
/2/   B.B.Kadomtsev,Plasma Physics and Contr.Fusion **26**,217 (1987)
/3/   J.A.Wesson,R.D.Gill,M.Hugon et al., accepted by Nuclear.Fusion (1989)
/4/   P.H.Rutherford, Phys. Fluids **16**,1903 (1973)
/5/   R.White,D.A.Monticello,M.Rosenbluth,and   B.V.Waddell,   Phys.   Fluids **20**,800 (1978).

/6/ R.B.White, in Handbook of plasma physics, Eds.M.N.Rosenbluth and R.Z.Sagdeev,North-Holland Pub.Co., vol.1,chapter 3.5 (1983)

/7/ Bateman,G. and Morris,R.N.,Phys.Fluids 29,753(1986)

/8/ Battacharjee, A. et al., Phys.Fluids 29, 1 (1986)

/9/ K.Lackner and F.Karger, in Proc.of the 8th European Conference on Controlled Fusion and Plasma Physics (Czechoslovak Academy of Sciences,Prague),vol.1,p.5 (1977)

/10/ K.Yamazaki,K.Kawahata,Y.Abe et al.,in Proc. of the 11th International Conference on Plasma Physics and Controlled Nuclear Fusion Research (IAEA,Viena),vol.1,p.309 (1987)

/11/ Q.Zhao,J.Chen,J.Xie et al,in Proc. of the 10th Inernational Conference on Plasma Physics and Controlled Nuclear Fusion Research (IAEA,Viena),vol.1,p.345 (1985)

/12/ A.Vannucci,O.W.Bender,I.L.Caldas et al.,Nuovo Cimento 10,1193 (1988)

/13/ D.A.Monticello,R.B.White,and M.N.Rosenbluth, in Proc.of the 7th International Conference on Plasma Physics and Controlled Fusion Research (IAEA,Vienna),vol.1,p.605(1979)

/14/ V.V.Arsenin,L.I.Artemenkov,N.V.Ivanov et al.,in Proc.of the 7th International Conference on Plasma Physics and Controlled Fusion Research (IAEA,Vienna),vol.1,p.233 (1979)

/15/ E.Lazzaro and M.F.F.Nave,in Proc. of the 14th European Conference on Controlled Fusion and Plasma Physics (European Physical Society, Madrid),vol.3,p.1059 (1987)

/16/ E.Lazzaro and M.F.F.Nave,Phys.Fluids,31,1923 (1988)

/17/ V.V.Arsenin,L.I.Artemenov,N.V.I.Vanov,et al.,Proc.IAEA Conference on Controlled Fusion and Plasma Physics, Innsbruck (1978)

/18/ D.A.Monticello and R.B.White,Phys.Fluids 23,366(1980)

/19/ S.Von Goeler,G.Gernhart,W.Engelhart,et al. in Proc. of the IAEASymposium on Current Disruption in Toraidal Devices (Max-Planck-Institut für Plasmaphysik,Garching,1979),paper A5

/20/ T.Hender,J.Hugil,D.Robinson et al,'Disruption Control in the Tokamak', 12th IAEA Conf.on Plasma Physics and Contr. Nucl. Fus. Research, Nice (1988)

/21/ P.E.Stott,M.Hugon,P.Noll et al,'Stabilisation of disruptions',JET Report EUR/FU/87/JETSC32/6 (1987)

/22/ J.A.Wesson,'Tokamaks',Clarendon Press,Oxford (1987)

/23/ J.Blum,Gilbert,J.C.,J.L.Le Foll,and B.Thooris.,Proc. 8th Europhys.Conference on Computational Physics.,vol.10D,p.49,Eibsee 13-16 May 1986.

/24/ J.A.Wesson.,Nuclear Fusion 18,87 (1978)

/25/ P.H.Rutherford,in Proc. of the Course and Workshop on Basic Physical Processes of Toroidal Fusion Plasmas (Commission of the European Communities, Brussels),vol.2,p.531 (1985)

/26/ Nave,M.F.F.,Wesson,J.A.,Proc. 14th EPSConf. on Plasma Phys. and Cont.Fusion,Madrid,part 3,p.1103(1986)

/27/ J.Snipes,D.J.Campbell,P.S.Haynes et al.,Nuclear Fusion (1988)

/28/ D.Stork,A.Boileau,F.Bombarda et al.,in Proc. of the 14th European Conference on Controlled Fusion and Plasma Physics (European Physical Society,Madrid),vol.1,p.306 (1987)

/29/ Nave,M.F.F.,Lazzaro,E.,Gowers,C. et al.,Proc. 15th EPS Conf. on Plasma Phys. and Cont.Fusion,Dubrovnik,vol.1,p.441 (1988)

/30/ K.Toi,S.Itoh,K.Kadota et al.,Proc. of the IAEA Symposium on Current Disruption in Toroidal Devices (Max-Plank-Institut für Plasmaphysik,Garching),paper B8 (1979)

/31/ A.A.Borschegovskij,I.A.Popov and M.M.Stepanenko,in Proc. of the 12th European Conference on Plasma Ohysics and Controlled Fusion (European Physical Society,Budapest),p.307 (1985)

/32/ F.de Luca,A.Jachia and E.Lazzaro,Phys.Lett. A118,191 (1986)

/33/ R B.White et al.,Workshop on Magnetic Reconnection and Turbulence (Cargese,France) (1985)

/34/ V.Chan and G.Guest, Nuclear Fusion 22,272 (1982)

/35/ K.Yamazaki et al.,IAEA Conf. on Plasma Physics and Controlled Nuclear Fusion Research,Kyoto, 1986

/36/ D.Edery et al.,EUR-CEA-FC-1109 (1981)

/37/ H.P.Furth,Symposium on Current Disruption in Toraidal Devices (Max-Planck-Institut für Plasmaphysik,Garching,1979),paper B3

/38/ W.Morris,private communication (1989)

On a Cantor structure in a satellite scattering problem

Jean-Marc Petit
CNRS, Observatoire de Nice, BP 139, 06003 Nice, France
Michel Hénon
CNRS, Observatoire de Nice, BP 139, 06003 Nice, France

Abstract

The phenomenon of chaotic scattering is described in the context of satellite encounters. We consider a one-parameter family of orbits obtained by starting with two satellites on circular, coplanar and close orbits. We numerically find that this family exhibits a large number of discontinuities, probably an infinite number. This phenomenon seems to be due to the existence of homoclinic and heteroclinic points of unstable periodic orbits. We model the chaotic scattering by a simple billiard: a point particle bounces on two disks and in addition is subjected to a constant acceleration. This leads to a one-parameter family with chaotic scattering. With the help of symbolic dynamics, the structure of the family can be completely elucidated.

1. Introduction

In the last few years, many studies have been carried out on the chaos in bound classical hamiltonian systems and powerful methods have been developed and applied. In contrast, less work has been done on chaos in classical scattering systems. However, for nearly twenty years, there have been numerical observations of complicated - chaotic - behavior in continuous scattering problems: classical models for inelastic molecular scattering (Rankin and Miller 1971, Gottdiener 1975, Fitz and Brumer 1979, Schlier 1983, Noid et al. 1986), satellite encounters (Petit and Hénon 1986), vortex dynamics (Eckhardt and Aref 1989), potential scattering (Eckhardt and Jung 1986, Jung and Scholz 1987). But until recently, this phenomenon had not been studied for itself.

We found this kind of behavior in a simple physical problem: the encounter of two satellites on close circular orbits around a planet (namely, Saturn). In section 2, we describe in detail the physical problem and derive the equations of motion. One would notice that the equations of motion are non integrable and contain no true singularities. In section 3, we present more precisely the chaos that appears in our problem: the asymptotic behavior of the system is discontinuous with respect to the initial parameters. Then, in section 4, we give some hints on what creates this phenomenon. In view of the numerical difficulties which were encountered in exploring this problem, a simple "model" problem was developped (Hénon, 1988) which is just complex enough to exhibit all the features we are interested in. This model is described in section 5.

2. The physical problem

The physical problem we consider is a particular case of the three body problem. Two light bodies M_2 and M_3 describe initially coplanar and circular orbits, with slightly different radii, around a heavy central body M_1. Bodies M_2 and M_3 are initially far apart, so that their mutual attraction is negligible. However, the inner body has a slightly larger angular velocity and eventually catches up with the outer body; the distance from M_2 to M_3 becomes small and their mutual attraction is no longer negligible. We shall call this an *encounter*. This study can be applied to different problems in astronomy: planetary rings dynamics, motion of coorbital satellites (Janus and Epimetheus of Saturn) or planetary formation. For convenience, M_1 will be called the *planet* and M_2, M_3 will be called the *satellites*. The difference between the radii of the initial circular orbits will be called the *impact parameter*.

Analytic approximations of the solution are available in two cases:

(i) When the impact parameter is sufficiently large, the result of the encounter is only a slight deflection of M_2 and M_3 from their previous circular orbits. These deflections can then be obtained by a perturbation theory (Goldreich and Tremaine 1979, 1980).

(ii) When the impact parameter is very small, the interaction of M_2 and M_3 produces a "horseshoe" motion: M_2 and M_3 "repel" each other azimutally and never come in close proximity. This case can also be treated by a perturbation theory (Dermott and Murray 1981, Yoder et al. 1983).

Between these two asymptotic cases, however, no theory exists, and apparently only a numerical integration of the equations of motion can give the answer. In order to have an accurate numerical study, we first reduce the equations to a simpler form: the classical set of Hill's equations. Only a brief review of this reduction will be given here; details can be found in Hénon and Petit 1986. We assume that the mass of either satellite is small compared to the mass of the planet:

$$m_2 \ll m_1, \qquad m_3 \ll m_1, \tag{1}$$

where m_i is the mass of body M_i. We assume also that the distance between the two satellites is small compared to their distance to the planet. In a zero-order approximation, the two satellites can then be considered as a single body in orbit around the planet. This orbit will be called the *mean orbit*, and will be assumed to be circular. We call a_0 the radius of the mean orbit. (The precise definition of a_0 does not matter, as long as it is nearly equal to the radii of the satellite orbits). The angular velocity on the mean orbit is

$$\omega_0 = \sqrt{Gma_0^{-3}}, \tag{2}$$

where m is the total mass of the system:

$$m = m_1 + m_2 + m_3. \tag{3}$$

We define

$$\mu = \frac{m_2 + m_3}{m}. \tag{4}$$

Let X_i, Y_i be the coordinates of body i in an inertial system. We introduce dimensionless coordinates by

$$X_i' = \frac{X_i}{a_0}, \qquad Y_i' = \frac{Y_i}{a_0}, \qquad m_i' = \frac{m_i}{m}, \qquad t' = \omega_0 t, \tag{5}$$

and for simplicity we drop the primes in what follows. In the new variables, the radius of the orbit, the angular velocity, the mass of the system, and the gravitational constant are all equal to 1. We choose the origin of time so that the two satellites are in the vicinity of $X = 1$, $Y = 0$ at $t = 0$. We introduce new coordinates ξ, η, which will be called *Hill's coordinates:*

$$
\begin{aligned}
X_i - X_1 &= (1 + \mu^{1/3}\xi_i)\cos t - \mu^{1/3}\eta_i \sin t, \\
Y_i - Y_1 &= (1 + \mu^{1/3}\xi_i)\sin t - \mu^{1/3}\eta_i \cos t, \qquad (i = 2,3)
\end{aligned}
\tag{6}
$$

We go over to new coordinates ξ^*, η^*, ξ, η, describing, respectively, the position of the center of mass and the relative position of the two satellites:

$$\xi^* = \frac{m_2\xi_2 + m_3\xi_3}{m_2 + m_3}, \qquad \eta^* = \frac{m_2\eta_2 + m_3\eta_3}{m_2 + m_3}, \qquad \xi = \xi_3 - \xi_2, \qquad \eta = \eta_3 - \eta_2. \tag{7}$$

The equations for the motion of the center of mass are linear and easily integrated (Hénon and Petit 1986). The equations of relative motion are approximately

$$\ddot{\xi} = 2\dot{\eta} + 3\xi - \frac{\xi}{\rho^3}, \qquad \ddot{\eta} = -2\dot{\xi} - \frac{\eta}{\rho^3}, \qquad \rho = \sqrt{\xi^2 + \eta^2}, \tag{8}$$

which are *Hill's equations* (Hill 1978). The error in these equations is of order of $\mu^{1/3}$. They become exact in the limit of vanishing satellite masses. Taking this limit is equivalent to zoom on the two satellites and the main effect is to repel the planet to infinity and transform circular orbits in straight lines.

The most important points to notice on these equations are:

(i) There is no parameter left in the equations (the same equations are valid in every physical case).

(ii) As is easily shown, the initial conditions for relative motion of circular orbits are given by only one parameter: the impact parameter h. Therefore, the set of solutions is a one-parameter family, and it seems reasonable to try to study it. We can even reduce the study to positive values of h because of the symmetries of the equations of motion.

Hill's equations admit the integral

$$\Gamma = 3\eta^2 + \frac{2}{\rho} - \dot{\xi}^2 - \dot{\eta}^2 \tag{9}$$

which can be called the *Jacobi integral* by analogy with the restricted problem. We can write the Jacobi integral in terms of the initial conditions:

$$\Gamma = \frac{3}{4}h^2. \tag{10}$$

228

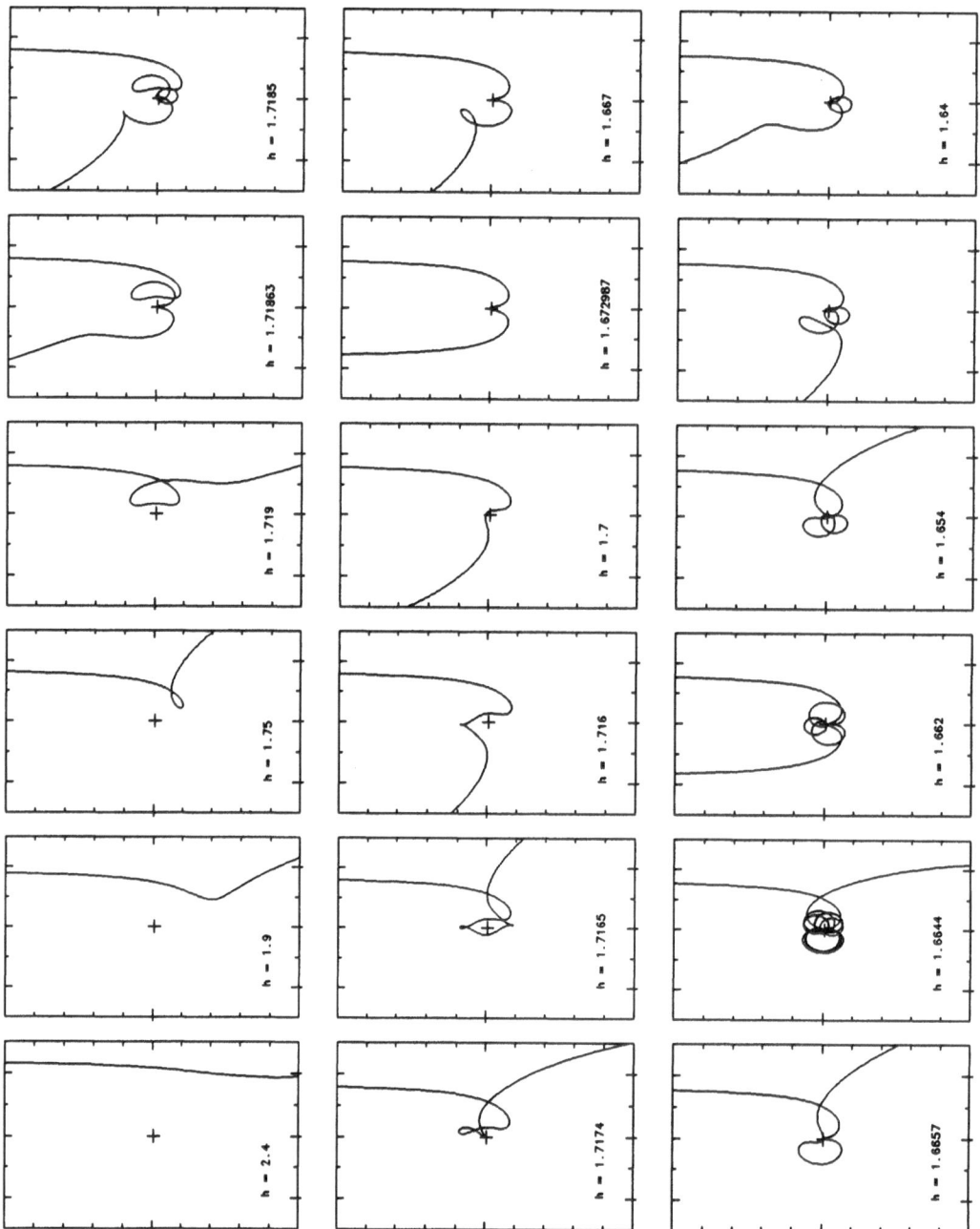

Figure 1. Family of encounter orbits. Each frame corresponds to one particular value of the reduced impact parameter h. The curve represents the relative motion of one satellite with respect to the other, in Hill's coordinate (ξ in abscissa, η in ordinate). The first approach is downwards from $\eta = +\infty$.

We have performed a detailed study of the entire familly $0 < h < \infty$. Figure 1, taken from a collection of several hundred pictures, represents the relative motion of the two satellites. For convenience, we shall think of the special case $m_2 \gg m_3$, and identify the origin of the (ξ, η) system with the satellite M_2; the curves then simply represent the motion of M_3. An interesting feature is that the third body always escapes either upward or downward, but never stays close for ever. This is in agreement with a general result by Marchal (1977) which shows that the set of "capture orbits" is of zero measure. For a more detailed explanation of the equation of motion and of the orbits, see Hénon and Petit (1986) and Petit and Hénon (1986).

3. Chaotic scattering

The family exhibits an interesting feature that we call "transitions". Roughly speaking, when h varies continuously, one can observe discontinuities. For given values of h, the orbit shape changes suddenly and an orbit that used to escape downward starts to escape upward, or the converse. This is the phenomenon that we want to develop now.

Consider an example. When h decreases from large values, the shape of the orbit changes continuously with h and the third body always escapes downward (first four plots of fig. 1). Suddenly, something happens and it escapes upward. The change (transition) occurs for $h_{max} = 1.718779940$. It can be thought of as a discontinuity of the shape of the orbit. But we need a more quantitative description of this discontinuity. If we look at the equations of motion, we find that there is no true singularity in it (the $1/\rho^2$ singularity can be removed by the Levy-Cevita regularization). So the position of the third body at a given time t is a continuous function of the impact parameter h. The orbits are absolutely not chaotic. But if we look at parameters describing the asymptotic motion as functions of h, we see very sharp variations at values of h corresponding to the changes of escape side. Especially, consider the final impact parameter h' (defined from the mean motion for $t \to \infty$). Using the Jacobi integral, it can be shown that $|h'| \geq h$. In the general case of asymptotically eccentric orbits, we call k the "reduced" (transformed into Hill's coordinates) eccentricity of the relative motion. One can show that

$$\Gamma = \frac{3}{4}h^2 - k^2. \tag{11}$$

When we start with circular orbits and finish with eccentric orbits, the equality

$$\Gamma = \frac{3}{4}h^2 = \frac{3}{4}h'^2 - k'^2 \tag{12}$$

holds. Therefore, $h'^2 \geq h^2$, which leads to the previous inequality. A downward escape corresponds to $h' > 0$ and an upward escape to $h' < 0$. Therefore a change of escape side leads to a change of sign for h' and a discontinuity of step at least $2h$. This is what we mean by discontinuity. The set of discontinuity values of h being very complex, we shall speak of "chaotic" behavior of the familly.

We will now describe rapidly the set of discontinuities. Consider an orbit defined by an arbitrary value h_0. Typically, the following happens: when decreasing h from h_0, the orbit changes continuously down to h_1 where there is a discontinuity in the sense defined above. We call this a "transition value". Numerically, it is not difficult (even if time consuming) to localize this value with any accuracy. Similarly, if we increase h

from h_0, we reach a second transition value h_2. The interval between h_1 and h_2 is called a "continuity interval". There are two particular cases: a continuity interval ranges from h_{max} to ∞, an other one ranges from 0 to $h_{min} = 1.336117188$ (figure 2). Suppose we have localized an interval of continuity. We do it again, starting from another value h_0 out of the range $[h_1, h_2]$. We find another interval of continuity and so on. One could expect to find all the intervals to be contiguous. This would give an exhaustive description of the orbit family.

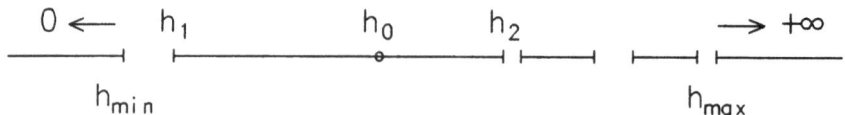

Figure 2. A Schematic representation of the largest continuity intervals.

But life is not that simple. Experiment shows that intervals are never contiguous. If one takes a point in an unexplored interval, one will find a new continuity interval which doesn't touch a previous interval neither on the left nor on the right. This gives birth to two new unexplored intervals. This goes on and on to infinity. One result is that there is no possible exhaustive description of the family. This must remind the reader of the classical definition of the Cantor set. The difference here is that the intervals are not regularly ordered. The actual structure can be seen on the function $h' = f(h)$ (figure 3). We can easily see the two external intervals $[0, h_{min}]$ and $[h_{max}, +\infty]$ and the three largest inner continuity intervals.

4. Some hints

We shall now try to explain how the discontinuities occur. In order to reconcile the continuity of the orbits with the discontinuity of the asymptotic behavior (h'), the family must go through an orbit with infinite capture time. This is achieved by having an orbit asymptotic to a periodic orbit. For example, figure 4 represents the transition orbit we find when decreasing h from large values: $h = h_{max}$. The orbit tends to a bean shaped periodic orbit. This limiting orbit is easily identified: it belongs to the one-parameter family a of periodic orbits, emanating from the Lagrangian point L_2 (Hénon 1969, Fig. 2). It is an unstable periodic orbit, which is necessary in a Hamiltonian system since it admits an asymptotic orbit.

It will be helpful to introduce at this point a *surface of section* defined for instance by $\eta = 0$ and $\dot\eta > 0$: for each crossing of an orbit with the ξ axis in the positive direction (η increasing), we plot a point with the coordinates ξ, $\dot\xi$ (figure 5). An orbit is then represented by a sequence of points. For a given value of Γ, a point in the surface of section defines completely the corresponding orbit: ξ, $\dot\xi$, η are immediately known and $\dot\eta$ can be computed from (9). In particular the next intersection point can be found. This defines a mapping of the surface of section onto itself, known as a *Poincaré map*.

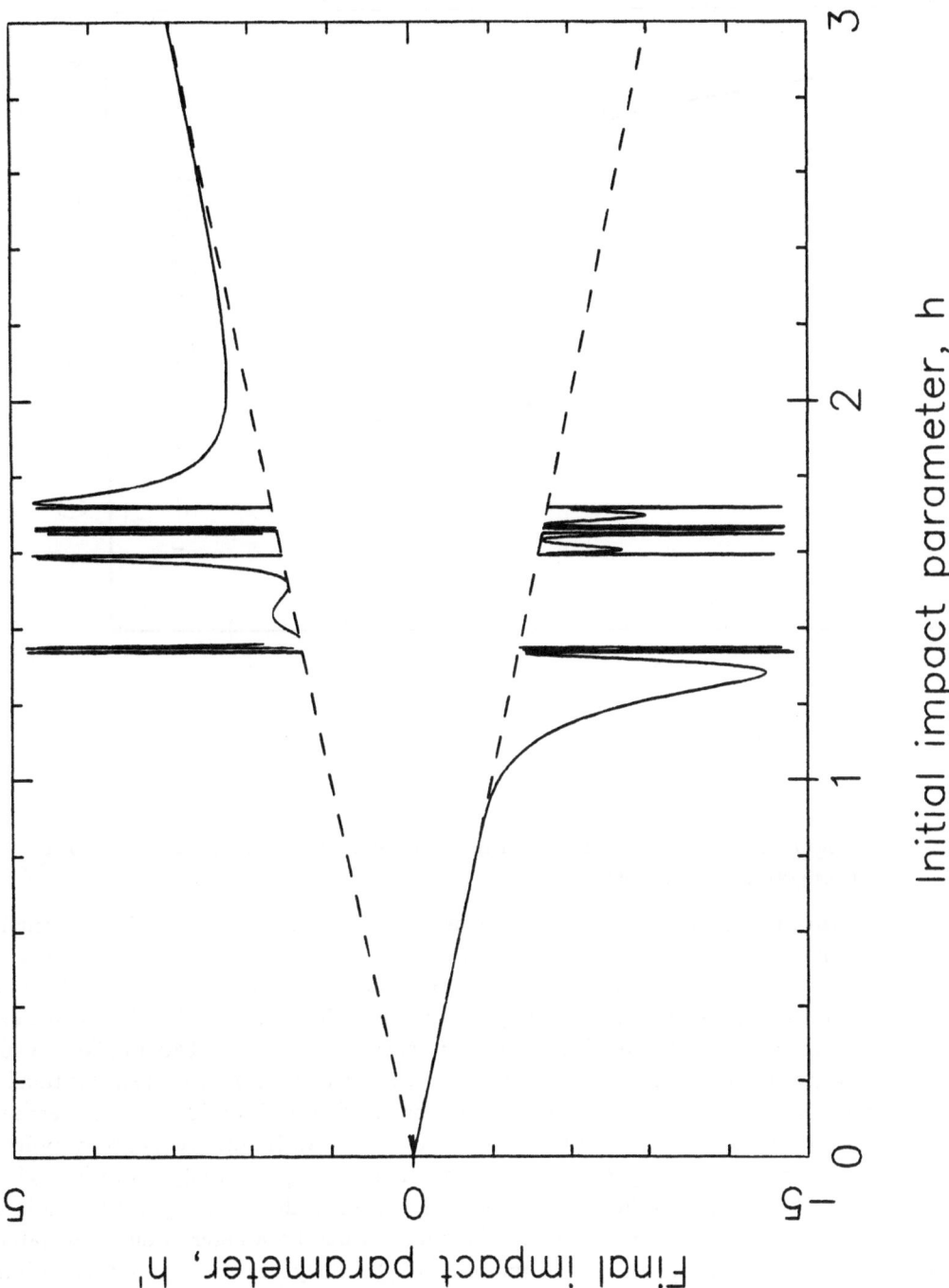

Figure 3. Final impact parameter h' as a function of the initial impact parameter h. The region between the two dashed lines is forbidden.

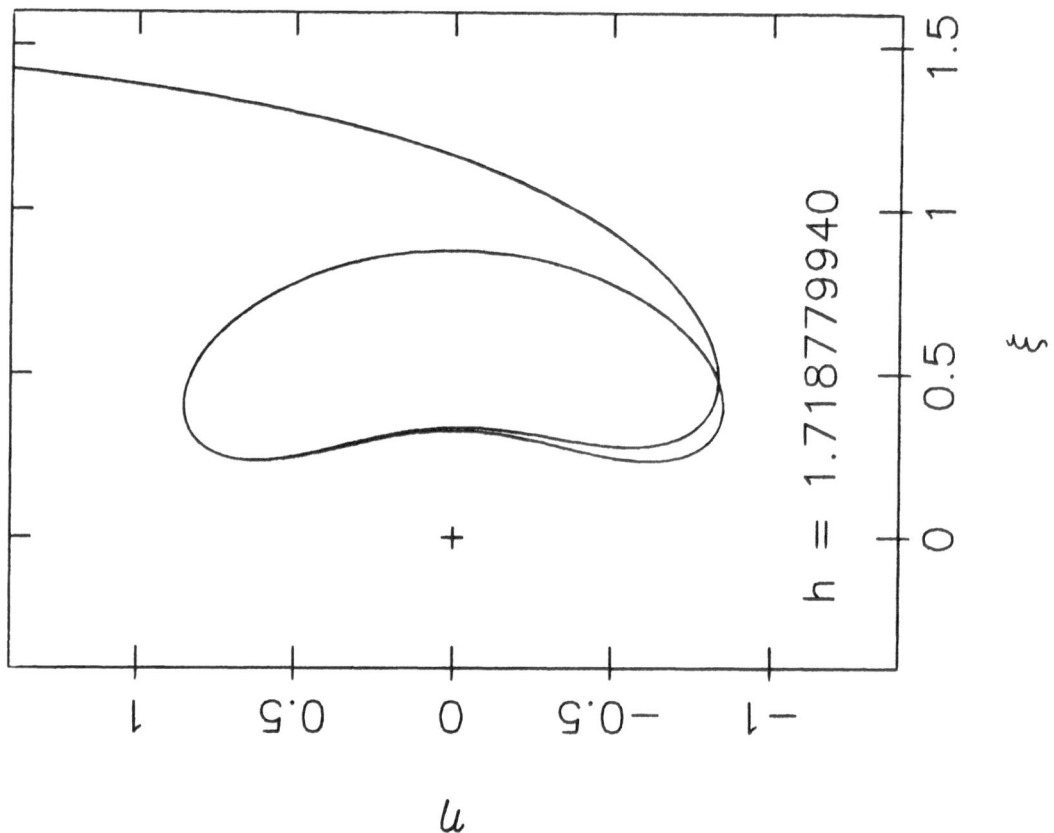

Figure 4. An orbit of the Satellite Encounter family which is asymptotic to an unstable periodic orbit.

Note that for the orbits we are concerned with, we get generally a finite (small) number of points in the surface of section: three points for the orbit with $h = 1.71863$ for instance. An orbit can also have no point at all in that surface ($h > 2.4$). The periodic orbit is represented by a fixed point P (figure 5). The stability index of that orbit is of order 320, corresponding to two real eigenvalues $\lambda_1 = 1/640$ and $\lambda_2 = 640$. The eigenvalue smaller than 1 in modulus (λ_1) is associated with a one-parameter family of *incoming orbits* tending towards the periodic orbit. The orbit of figure 4 is a member of this family. An orbit of this family is represented by an infinite sequence of points on the *stable invariant manifold* W_s of P and converges exponentially towards P (Y_0, Y_1, Y_2, ...). Since the periodic orbit is unstable, there are also *outgoing orbits*, tending towards the periodic orbit for $t \to -\infty$. They form a one-parameter family associated with the eigenvalue larger than 1 in modulus (λ_2). They are represented by points $(\ldots, Z_{-2}, Z_{-1}, Z_0, \ldots)$ located on the *unstable invariant manifold* W_u of P and which diverge exponentially from P.

233

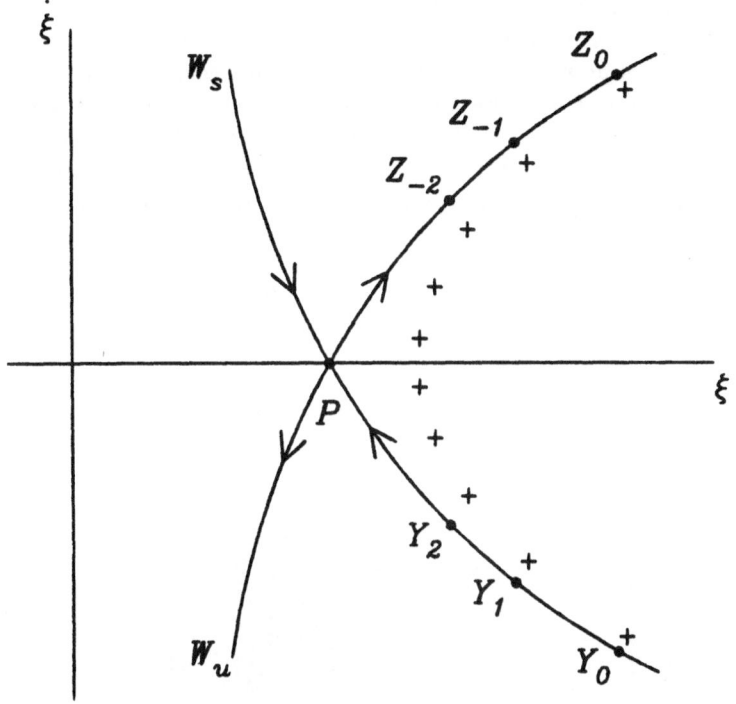

Figure 5. Sketch of the surface of section. The value of λ_1 has been artificially increased to show the structure more clearly.

Consider now an orbit of our family with h slightly different from h_{max}, say larger. The points in the surface of section are slightly beside W_s (crosses on the picture). They stay close to W_s until they reach the vicinity of P, then they go away along W_u. An important point is that λ_2 is positive. So the points go along only one branch of W_u. Here, it is the upper right branch. The corresponding orbits are quite regular. Particularly, they all escape downward and vary continuously when h increases (figure 6a). This accounts for the continuity interval for $h > h_{max}$.

For $h = h_{max}$, the point crosses W_s and for $h < h_{max}$, the points escape along the left branch of W_u. The two branches of W_u are in two different parts of phase space. This explains the transitions. The orbits for $h < h_{max}$ are shown on figure 6b. Things are much more complicated than before. Sometimes orbits escape upward, some time downward. So there is no continuity interval on the left of h_{max}. This explains the complex structure of the continuity intervals. For $h < h_{max}$, instead of escaping directly, the orbit will first go in the vicinity of an other unstable periodic orbit. This orbit will itself give birth to a transition phenomenon, that we shall call a *second order transition*. In this way, one can construct a hierarchical structure of transitions of higher and higher order. Suppose we have an orbit going close to one periodic orbit then close to a second one. By changing h, we can push the points in the surface of section closer to the first fixed point. Particularly, one can manage to have the same pattern along W_u and one or more additional points in the vicinity of P. This corresponds to orbits with the same escape but with one or more additional turns around the first periodic orbit (figure 7). In the first plot, the orbit follows the periodic orbit during half a turn, in the second during one and a half and in the third during two and a half (even if this is not visible one the figure). This gives rise to a geometrical progression of ratio λ_1 in the values of h.

(a)   (b)

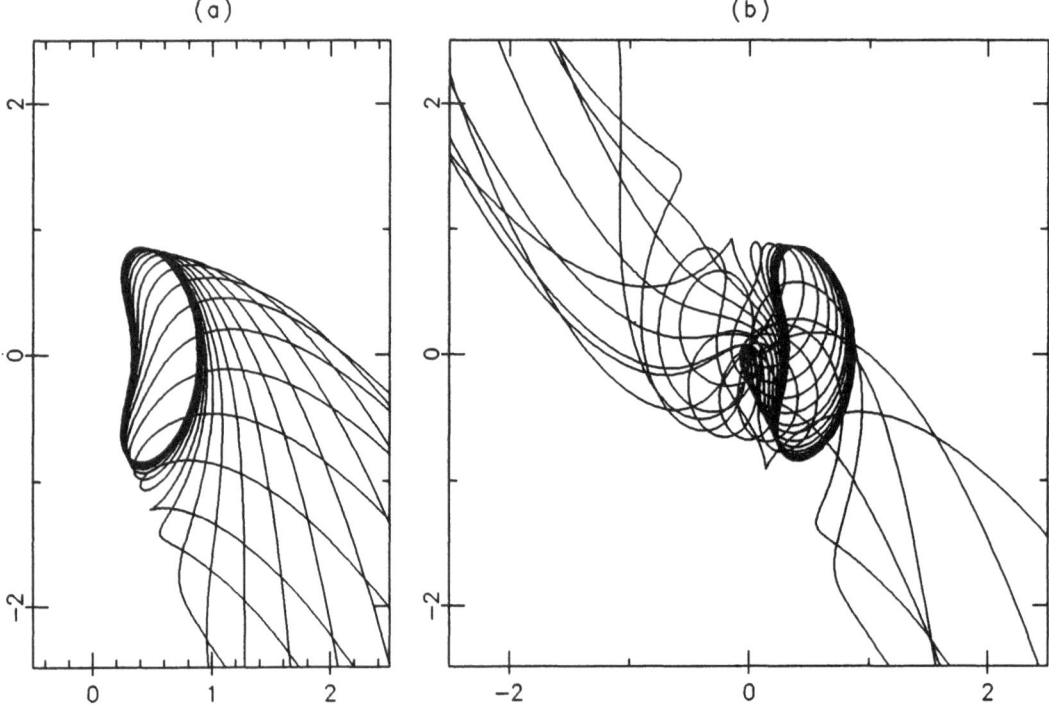

Figure 6. (a) Outgoing orbits for h above the critical value h_{max}. (b) Outgoing orbits for h below h_{max}.

From all our numerical integrations, it seems that only two family of periodic orbits are involved: family a mentioned above and the symmetrical family b also described in Hénon 1969.

The necessary ingredients for this kind of behavior is the existence of periodic orbits and heteroclinic or homoclinic points (intersection points of invariant manifolds of two different or one single periodic orbit). But it is very difficult to go any further with this problem due to the large value of the eigenvalue (~ 640).

Other authors have observed similar behavior in scattering problems. Recently, Jung and Scholz (1988) have studied the scattering of a charged particle by a magnetic dipole. Due to the smaller value of the eigenvalue, they have been able to compute the stable and unstable manifolds of the periodic orbit with great details (solid line in Fig. 6 of their paper). It happens in their case that they need only one periodic orbit due to the presence of a homoclinic point. The dots in that figure represent the first intersection of the surface of section when varying the initial parameter. They also transported the invariant manifold into the space of initial parameters (Fig. 11 of the same paper). If now we draw a line across the figure, corresponding to the variation of one of the initial parameters, we shall see immediately the existence of a complex structure. For comparison, we did the same thing in the satellite problem (figure 8) but the structure is far too thin to be seen.

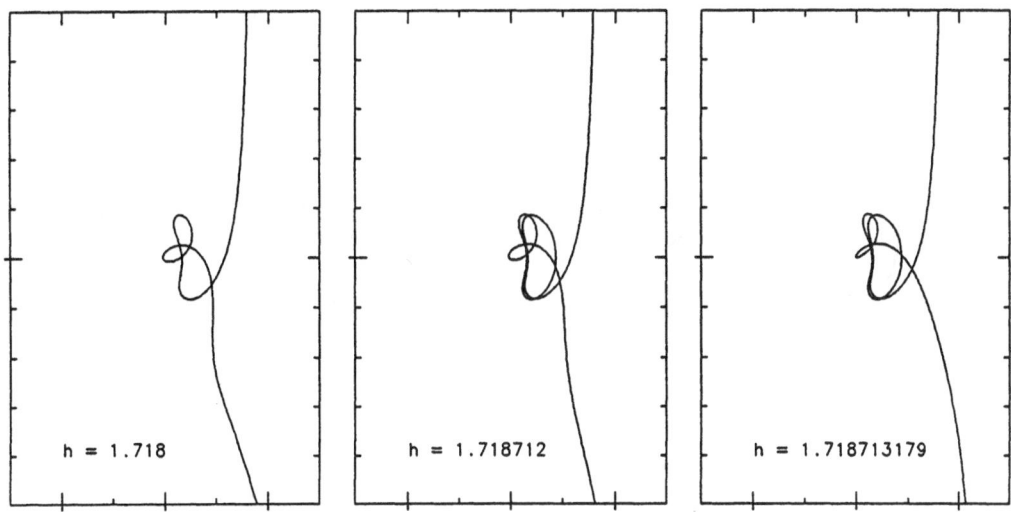

Figure 7. Three orbits with essentially the same outgoing but different behavior during the close encounter. The orbit goes along the unstable periodic orbit for half a turn on the left, for one and a half in the middle and for two and a half on the right plot.

5. The inclined billiard

According to Moser (1973), in the vicinity of a homoclinic point, it is theorically possible to define a symbolic dynamics which is Bernoulli. This gives a better description of the dynamics of the system. But in our problem, we haven't been able to define it so far. So a model problem was designed which is complex enough to exhibit all the features we are interested in, and simple enough so that all the calculations can be done analytically. This model is the *inclined billiard* (Hénon, 1988). It is defined as follows: a particle moves in the (X, Y) plane and bounces elastically on two fixed disks with radius r and with their centers in $(-1, -r)$ and $(1, -r)$ respectively. In addition, it is subjected to a constant acceleration g which pulls it in the negative Y direction. To make the computation affordable, one considers the limit where r is large and approximates the circles (disks) by parabolas. The "disks" extend then from $-\infty$ to ∞ in the X direction and the number of rebounds of the particle on them is now always infinite. We suppose that initially the particle is started at (h, Y_0) where Y_0 is large. So we can neglect the thickness of the profile of the disks. Only the slopes are of consequence.

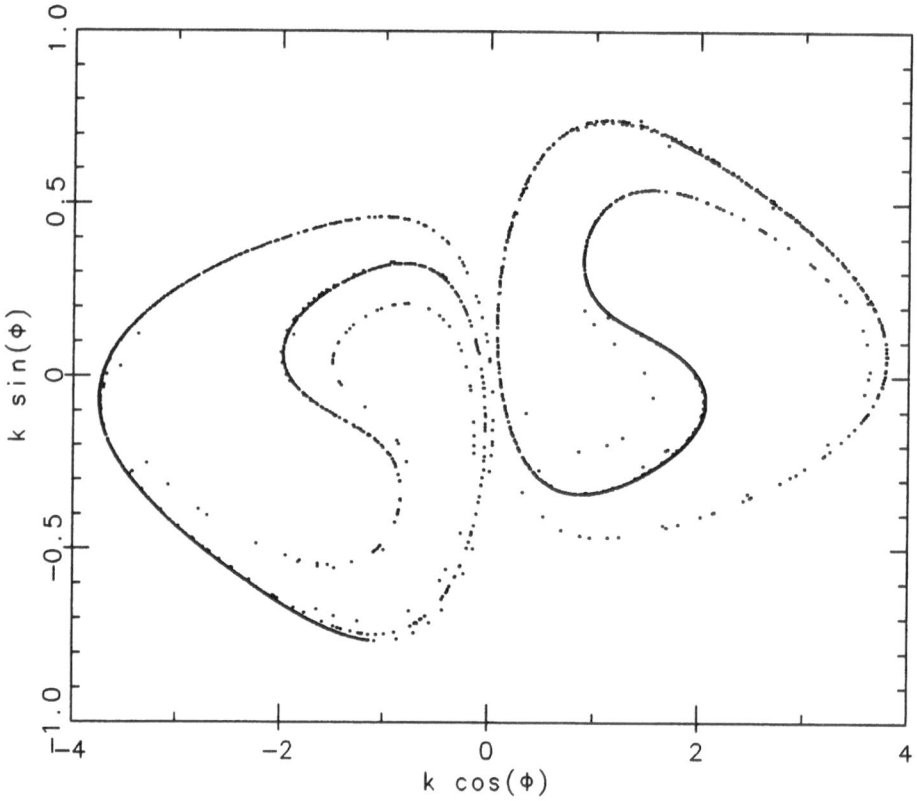

Figure 8. Image of the invariant manifolds in the space of initial parameters.

We consider the surface of section defined by the collisions. The coordinates in this surface are X, the horizontal abscissa of the collision and W, the tangential velocity of the particle. After a tedious calculation, one can finally obtain the mapping:

$$u_{j+1/2} = W_j + \frac{\sqrt{2E}}{r} f(X_j),$$

$$X_{j+1} = X_j + \frac{2\sqrt{2E}}{g} U_{j+1/2}, \qquad (13)$$

$$W_{j+1} = U_{j+1/2} + \frac{\sqrt{2E}}{r} f(X_{j+1}),$$

with

$$f(X) = \begin{cases} X+1 & \text{for } X < 0, \\ X-1 & \text{for } X \geq 0, \end{cases}$$

where $U_{j+1/2}$ is the horizontal velocity between the collisions and E the energy of the particle. Arbitrarily, we have decided that the intersection of the two disks belongs to the right disk. We introduce a new parameter Φ, related to the energy by

$$\cosh \Phi = 1 + \frac{4E}{gr}, \tag{14}$$

and the new variables u and w

$$U = u\sqrt{\frac{g}{2r}\left(2 + \frac{4E}{gr}\right)}, \qquad W = w\sqrt{\frac{g}{2r}\left(2 + \frac{4E}{gr}\right)}, \tag{15}$$

in order to rewrite the mapping in a dimensionless form:

$$X_{j+1} = X_j \cosh \Phi + w_j \sinh \Phi - s_j(\cosh \Phi - 1),$$
$$w_{j+1} = X_j \sinh \Phi + w_j \cosh \Phi - (s_j \cosh \Phi + s_{j+1}) \tanh \frac{\Phi}{2}, \tag{16}$$
$$s_j = \mathrm{sign} X_j.$$

The parameter Φ cannot be eliminated since it is related to the eigenvalue of the fixed point. It is easy to show that there are five kinds of asymptotic regimes:

1. right-escaping orbit: $X_j \to +\infty$, $w_j \to +\infty$.
2. right-asymptotic orbit: $X_j \to +1$, $w_j \to 0$.
3. left-escaping orbit: $X_j \to -\infty$, $w_j \to -\infty$.
4. left-asymptotic orbit: $X_j \to -1$, $w_j \to 0$.
5. oscillating orbit: X_j and w_j are bounded but have no limit.

We will now define a symbolic dynamics to represent the essence of the dynamics of the billiard. To each orbit we associate a semi-infinite sequence $d_1, d_2, \cdots$ of 0 and 1. The orbit is described by the sequence of points in the surface of section and each point is represented by 0 if it is a collision on the left disk and 1 if it is a collision on the right disk, rather than by its coordinates. Then we define a number A by its binary representation:

$$A = 0.d_1 d_2 d_3 \cdots = \sum_{j=1}^{\infty} 2^{-j} d_j.$$

Clearly, $0 \le A \le 1$. A given sequence defines one value of A, but there might be *two* sequences with the same value of A:
- If A is of the form $k.2^{-n}$, where k and n are natural numbers, A is called a round number and has two representations: $0.d_1 \cdots d_{n-1}0111 \cdots$ and $0.d_1 \cdots d_{n-1}1000 \cdots$.
- In the other case, A is a non-round number and the sequence is oscillating.

There is a simple correspondence between the types of orbits, the D sequence and A.

238

orbit	D sequence	A
right-escaping	1-ending	round
right-asymptotic	1-ending	round
left-escaping	0-ending	round
left-asymptotic	0-ending	round
oscillating	oscillating	non-round

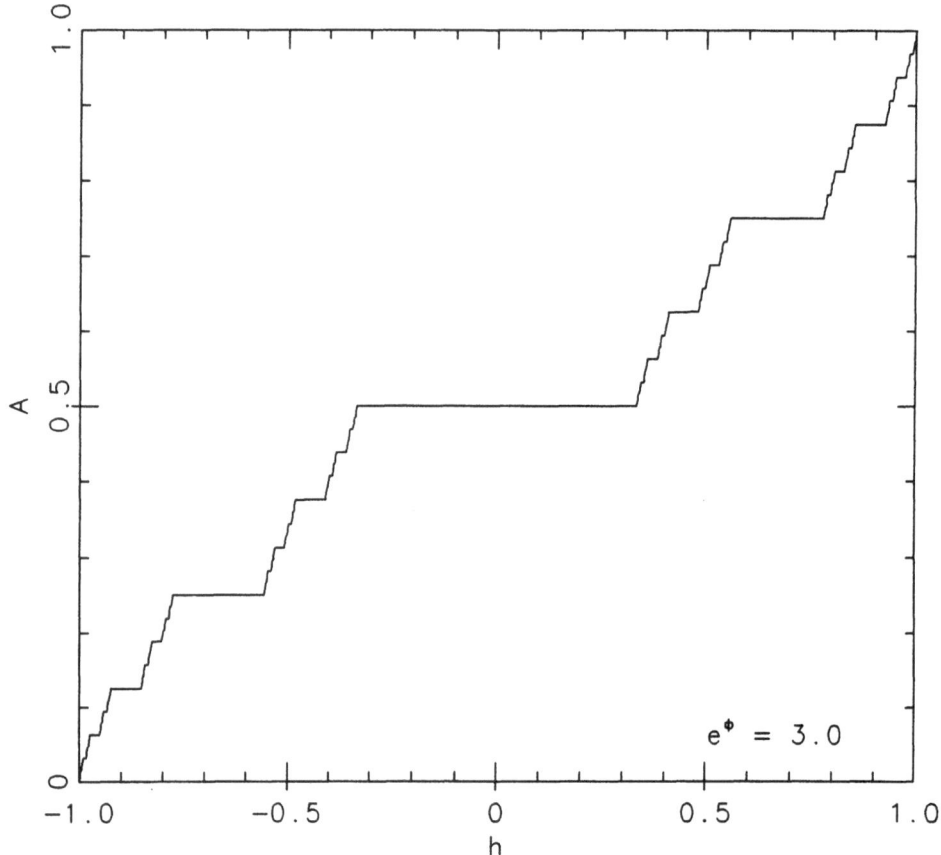

Figure 9. The function $A(h)$ for $e^{\Phi} = 3$.

In a continuity interval, the orbit changes continuously, so A is constant. This suggests to look at the function $A(h)$. Figure 9 shows the numerical result for $\lambda = e^{\Phi} = 3$. The reader will have recognized a *Devil's staircase*. It is possible to explain completely this figure, provided that $e^{\Phi} \geq 3$. One can show the following:

- A is a non-decreasing function of h.
- A is a continuous function of h.

- If A is non-round, it corresponds to a unique value of h defined by:

$$h = (e^{\Phi} - 1) \sum_{j=1}^{+\infty} e^{-j\Phi} s_j. \tag{17}$$

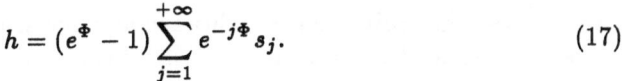

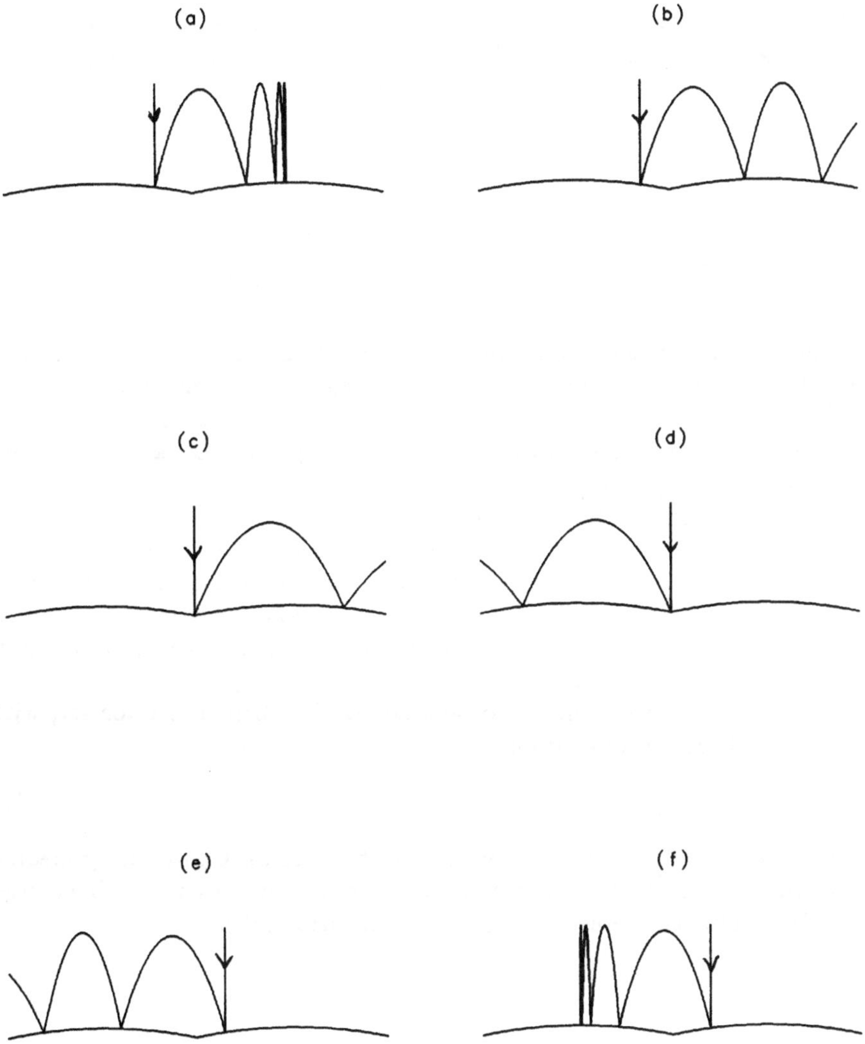

Figure 10. Structure of the h-orbits corresponding to a round value of A (here $A = \frac{1}{2}$). (a) right-asymptotic orbit; (b), (c) right-escaping orbits; (d), (e) left-escaping orbits; (f) left-asymptotic orbit.

- If A is round, things are more interesting. A corresponds to a horizontal step, that is to a continuity interval $h_- \leq h \leq h_+$. All the steps have the same structure.

Consider the central step on figure 9 (the largest). The leftmost point corresponds to an asymptotic orbit (a) (figure 10) on the right: $A = 0.0111\cdots = 1/2$. When h increases, the orbit escapes rightward and the sequence doesn't change. This is true for interval (b). After orbit (c), there is a discontinuity: the slope for the first collision changes suddenly and we go to orbit (d). Now, $A = 0.1000\cdots$, that is still $1/2$. Here, there is a small miracle: the orbit changes completely through (c) but not A. The discontinuity at $X = 0$ disappears completely in A. Then there is the interval (e) of left escaping orbits and finally the left asymptotic orbit (f). One can easily compute the values of h_- and h_+ from A:

$$
\begin{aligned}
h_- &= (e^\Phi - 1) \sum_{j=1}^{n-1} e^{-j\Phi} s_j - (e^\Phi - 2) e^{-n\Phi}, \\
h_+ &= (e^\Phi - 1) \sum_{j=1}^{n-1} e^{-j\Phi} s_j + (e^\Phi - 2) e^{-n\Phi}.
\end{aligned}
\tag{18}
$$

We have thus fully solved the "inverse problem" and obtained a complete classification of h-orbits. For non-round A, there is a single h given by (17), corresponding to an oscillating orbit. For round A, there is a finite closed interval of h values $h_- \le h \le h_+$ (or an infinite interval in the special cases $A = 0$ and $A = 1$). This explains the *Devil's staircase* appearance of the figures.

There are two additional results that can be proved:

- The curve $A(h)$ has exact self-similarity. The curve as a whole extends from $h = -1$ to $h = +1$ and from $A = 0$ to $A = 1$. In the lower left corner is an exact replica of the whole picture, reduced by a factor e^Φ horizontally and 2 vertically, extending from $h = -1$ to $h = -1 + 2e^{-\Phi}$ and from $A = 0$ to $A = 1/2$. There is an identical replica in the upper right corner.
- The set of values of h corresponding to bounded orbits forms a Cantor set, with measure 0 and with fractal dimension

$$
\ln(2)/\Phi. \tag{19}
$$

In the borderline case $e^\Phi = 3$, we obtain exactly the classical Cantor set (repeated exclusion of the middle third). The asymptotic orbits form an enumerable subset of the bounded orbits; this subset also has the dimension (19).

References

Dermott S. F. and Murray C. D. 1981: Icarus 48, 1

Eckhardt B. and Aref H. 1989: Phil. Trans. R. Soc. Lond. A 326, 655

Eckhardt B. and Jung C. 1986: J. Phys. A 19, L829

Fitz D. E. and Brumer P. 1979: J. Chem. Phys. 70, 5527

Goldreich P. and Tremaine S. 1979: Nature 277, 97

Goldreich P. and Tremaine S. 1980: Astrophys. J. 241, 425

Gottdiener L. 1975: Molecular Physics 29, 1585

Hénon M. 1969: Astron. Astrophys. 1, 223

Hénon M. 1988: Physica D 33, 132

Hénon M. and Petit J-M. 1986: Celes. Mech. 38, 67

Jung C. and Scholz H-J. 1987: J. Phys. A 20, 3607

Jung C. and Scholz H-J. 1988: J. Phys. A 21, 2301

Marchal C. 1977: J. Differ. Equations 23, 387

Noid D. W., Gray S. K. and Rice S. A. 1986: J. Chem. Phys. 84, 2649

Petit J-M. and Hénon M. 1986: Icarus 66, 536

Rankin C. C. and Miller W. H. 1971: J. Chem. Phys. 55, 3150

Schlier C. G. 1983: Chemical Physics 77, 267

Yoder C. F., Colombo G., Synnott S. P. and Yoder K. A. 1983: Icarus 53, 431

CONTROLLING CHAOS THROUGH PARAMETRIC EXCITATIONS

Marco Pettini

Osservatorio Astrofisico di Arcetri
Largo E.Fermi 5, 50125 Firenze
and I.N.F.N. Sezione di Firenze, Italy

Abstract.
As in many physical and non physical systems chaos can have harmful consequences, the possibility is discussed of reducing or suppressing it without radically modifying the system.

An heuristic reasoning is proposed, then it is shown on a Duffing-Holmes oscillator, that a resonant effect can kill chaos when parametric perturbations are introduced with suitable frequencies.

Introduction.

Chaos is rather ubiquitous in nonlinear dynamical systems and has been observed in many real physical and non-physical systems.

Chaos is not noise, even though for a long time it has been observed as such in mechanical or electronic devices [1]. Chaos is a very complicated order, arising also from low dimensional deterministic dynamics. Sometimes it can be useful; this is for instance the case of the ergodic divertor in tokamaks, where a stochastic layer of magnetic field is produced at the plasma edge to improve the confinement. In many other cases chaos can have harmful consequences: plenty of engineering devices could be mentioned; we address the interested reader to ref.[1] which provides an interesting survey in this field.

Among physical systems where chaos is harmful, we want to mention magnetic confinement devices for controlled thermonuclear fusion, where the intrinsic chaoticity of particle dynamics is responsible for an enhanced diffusion across the confining magnetic field; this *anomalous* transport is much larger than the loss rate predicted by collisional transport theory (see, e.g., ref.[2]). The destruction of regular magnetic surfaces, due to chaotic instability, is another unpleasant effect in these systems [3].

Also particle accelerators of betatron type are afflicted by chaotic instabilities, these can be caused by beam-beam interactions in storage-ring colliders [4].

In some cases one can a-priori suggest how a machine should be designed in order to avoid the onset of chaos: an example has been given for stellarators [5] for which the dangerous parameter ranges have been investigated.

More generally, if a given physical or non-physical system is satisfactorily described by some nonlinear dynamical model, then by studying - analytically or numerically - its parameter space, it is possible to know how chaos could be avoided.

But, let us consider those situations where one cannot make a system operate in a safe domain of parameter space. In other words, assume that chaos is unavoidable for the operating conditions of your system. For example, this is the case of anomalous transport in tokamaks. Then the only thing you can dream of is to *perturb* your system in a skilful way to reduce or even suppress chaos. This idea, obviously, is not new, though to the best of the author's knowledge the only attempt, explicitly aiming at the above stated goal, dates back to an old preprint [6] (in russian) by Izrailev and

Chirikov. These authors studied how a perturbation of an area preserving map can change dramatically the phase space structure, hence the diffusion properties of the model; the drawback is in the choice of the perturbation, which is critical, and on its amplitude, which is not small.

In a more recent paper [7], a white noise, added to a map modeling the Belusov-Zhabotinsky reaction, has been proved useful to reduce or suppress chaos. The explanation is related to the peculiar structure of the invariant density $\rho(x)$ of the map, which is strongly peaked in the region of $\mid \partial f / \partial x \mid$ that gives the largest contribution to the Lyapunov characteristic exponent. The introduction of additive noise smears out this peak of $\rho(x)$ thus reducing chaos.

In another work [8], the problem of nonlinear filtering has been investigated, and it has been shown how a chaotic excitation of a nonlinear system can produce a periodic response.

In what follows we suggest another possibility of reducing or suppressing chaos which is based on parametric excitations: one wonders whether a suitable time dependent variation of a parameter can produce the desired effect. Moreover, one looks for some "resonant" effect, so that a *small* relative variation of a parameter could be effective, provided that some "resonance" condition is satisfied. The advantage of such a possibility is that the hardware of a given chaotic system should be only slightly modified. At variance, the addition of new couplings in the system could be hardly feasible without deep modifications.

A question that naturally arises is whether, due to some general theorem, the above sketched program could be unfeasible. A major obstacle could be represented by structural stability of chaos. Following the definition á la Andronov-Pontriaguin [9] of stability, we define as structurally stable the flow of a dynamical system if it is homeomorphic to the flow of a perturbed version of the system, the homeomorphism being close to the identity. Anosov flows or Smale diffeomorphisms are structurally stable: perturbing an Anosov flow another Anosov flow is obtained [10] and the same happens with Smale's limiting sets [11]. Anyway, the overwhelming majority of chaotic systems are likely to be structurally unstable; for dissipative flows the denomination "non-hyperbolic strange attractors" is now frequently used to remark an important difference of chaotic dynamics without Axiom-A attractors.

A weaker definition of structural stability has been recently proposed [12] just to circumvent the severe conditions imposed by the request of topological equivalence between perturbed and unperturbed flows. This definition, which embraces a wider class of dissipative dynamical systems, is based on limiting properties of an ϵ-smoothing of the Bowen-Ruelle measure of an attractor; it is defined through a Fokker-Planck equation with an ϵ-diffusion and it results less constraining than the Andronov-Pontriaguin definition of structural stability.

In conclusion, as most chaotic systems are not structurally stable (in the strong sense) there is no general argument that can rule out a-priori our program.

Let us now discuss an heuristic argument which has led to consider parametric excitations. The idea arises from the following observations:

a) parametric perturbation can modify the stability properties of fixed points of linear (or linearized) systems [13];

b) Jacobi equation for geodesic variations is a linear equation whose stable and unstable solutions correspond to regular and chaotic motions.

The first item means that the elliptic fixed point $(\dot{x}(0), x(0)) = (0, 0)$ of the linearized pendulum equation

$$\ddot{x} + {\omega_0}^2 x = 0 \tag{1}$$

can be made unstable substituting ${\omega_0}^2 \rightarrow {\omega_0}^2(1 + \epsilon f(t))$, where $f(t) = f(t + T)$. This is a parametrically excited oscillation.

Near the hyperbolic fixed point $(\dot{x}(0), x(0)) = (0, -\pi)$ the same equation reads

$$\ddot{x} - {\omega_0}^2 x = 0 \tag{2}$$

and the same substitution can make stable the unstable position $(0, -\pi)$ provided that the pivot of the reversed pendulum is in sufficiently rapid oscillation [13].

The second item is used *only heuristically* as follows. At least for newtonian systems, Lagrange equations of motion describe the geodesics of a Riemannian manifold (the configuration space) equipped with the Jacobi metric [14] $g_{ij}(\mathbf{x}) = 2[E - U(\mathbf{x})]\delta_{ij}$, where E is the total energy of the system and $U(\mathbf{x})$ is the potential energy; then the Jacobi equation for the second variation of the action functional describes the local stability of geodesics with respect to a reference geodesic $\gamma : \{x^i = x^i(t)\}$; when expressed in local coordinates it reads [15]

$$\nabla_{\dot{x}}^2 \xi^i + \dot{x}^j \dot{x}^k \xi^l R^i_{jkl} = 0 \tag{3}$$

where $\nabla_{\dot{x}}$ is the covariant derivative, R^i_{jkl} is the curvature tensor associated to g_{ij}, t is the natural parameter along the geodesic and ξ^i is the Jacobi field of geodesic variation.

For two-dimensional manifolds of constant curvature eq.(3) becomes

$$\frac{d^2 \xi_\perp}{dt^2} + K \xi_\perp = 0 \tag{4}$$

where $\xi_\perp$ is the perpendicular component of the Jacobi field ξ and K is the gaussian curvature of the manifold.

From eq.(4) it is clear that on a sphere S^2 the geodesics are stable because $K > 0$. At variance, on a Lobatchevsky plane M, defined by the metric $ds^2 = (dx^2 + dy^2)/y^2$, the geodesics are unstable because $K = -1 < 0$ everywhere, and the geodesic flow defined on the unitary tangent bundle $T_1 M$ is an Anosov flow.

Loosely speaking we have recovered, at a different level, equations (1) and (2) to describe regular and chaotic dynamics. Letting $K \rightarrow K(1 + \epsilon f(t))$, as with eq.(1), one can make exponentially unstable nearby geodesics on a manifold of positive (unperturbed) curvature. In fact, consider the integrable nonlinear system $\ddot{x} + \sin x = 0$, the solutions $x(t)$ are regular and stable. But when a parametric perturbation is added:

$$\ddot{x} + (1 + \epsilon \cos \omega t) \sin x = 0 \tag{5}$$

chaos shows up. Therefore we can hope that parametric perturbation of eq.(4), when $K < 0$, might act to stabilize the exponentially unstable (chaotic) trajectories, in analogy with eq.(2). Within this analogy the sign of K should periodically change in time: this should be a strong modification of the system and not merely a perturbation.

Anyway, in general chaotic flows are not topologically equivalent to geodesic flows on manifolds of *constant* negative curvature, if this were the case one should have structural stability (after the Lobatchevsky-Hadamard theorem [10]), thus ergodicity, mixing, etc., but this is not the generic situation. Finally, notice that there is not a trivial relationship between perturbation of K and parametric perturbations of the equations of motion.

The above discussed conjecture is tested on a particular dynamical system in the following paragraph.

A paradigmatic system

We report here some results, recently obtained [16], for the so called Duffing - Holmes oscillator. This model, defined by the equation

$$\ddot{x} - x + \beta x^3 = -\delta \dot{x} + \gamma \cos \omega t \tag{6}$$

is one of the simplest nonlinear dissipative ODE undergoing a chaotic transition. With some approximations of Galerkin type [11], it can be derived from a PDE describing the dynamics of a buckled beam; in a different context, it can also be used to describe plasma oscillations [17].

Equation (6) can be trivially rewritten as

$$\begin{pmatrix} \dot{x} \\ \dot{y} \end{pmatrix} = \begin{pmatrix} y \\ x - \beta x^3 \end{pmatrix} + \epsilon \begin{pmatrix} 0 \\ -\frac{\delta}{\epsilon} y + \frac{\gamma}{\epsilon} \cos \omega t \end{pmatrix} \tag{7}$$

which is in the form

$$\dot{\mathbf{x}} = \mathbf{f}_0(\mathbf{x}) + \epsilon \mathbf{f}_1(\mathbf{x}, t). \tag{8}$$

The unperturbed part $\dot{\mathbf{x}} = \mathbf{f}_0(\mathbf{x})$ can be derived from the Hamiltonian

$$H = \frac{1}{2} y^2 - \frac{1}{2} x^2 + \frac{1}{4} \beta x^4 \tag{9}$$

and is integrable. Its phase space has only one hyperbolic fixed point from which an "eight-shaped" separatrix originates. The motion on this separatrix is given by

$$x^{(0)}(t) = \sqrt{\frac{2}{\beta}} \operatorname{sech} t$$

$$\dot{x}^{(0)}(t) = -\sqrt{\frac{2}{\beta}} \operatorname{sech} t \tanh t. \tag{10}$$

The separatrix, parametrically defined by eq.(10), is also called homoclinic loop and results from the superposition of the so called stable and unstable manifolds, W^s and W^u, respectively tangent at the origin to the stable and unstable eigenspaces E^s and E^u of the hyperbolic point. W^s and W^u are defined as those trajectories which converge asymptotically to the hyperbolic fixed point: W^s for $t \to +\infty$ and W^u for $t \to -\infty$. When the system $\dot{\mathbf{x}} = \mathbf{f}_0(\mathbf{x})$ is perturbed only by a dissipative term, the two manifolds $W^{s,u}$ never meet and the solutions are still regular. If a forcing term is also added (i.e. an energy supply is added to balance friction losses) then $W^{s,u}$ may have an homoclinic intersection and hence an infinity of subsequent intersections [4]. We briefly recall how

Melnikov's method works to determine the condition of homoclinic intersection of W^s and W^u and so of the onset of chaos. Let $\Gamma^{(0)}(t) = (\dot{x}^{(0)}(t), x^{(0)}(t))^T$ be the unperturbed motion on the homoclinic loop, write

$$W^{s,u}(t,t_0) \simeq \Gamma^{(0)}(t-t_0) + \epsilon W^{s,u}{}_1(t,t_0) \tag{11}$$

to describe how $W^{s,u}$ are perturbed up to first order in ϵ (due to $\mathbf{f}_1$ in eq.(8)) starting from $\Gamma^{(0)}$; t_0 is an arbitrary reference time and $W^{s,u} \equiv (\dot{x}^{s,u}, x^{s,u})^T$ are column vectors. One gets

$$\frac{d}{dt}W^{s,u}{}_1 = \mathbf{J}(\Gamma^{(0)}(t-t_0))W^{s,u}{}_1 + \epsilon\mathbf{f}_1(\Gamma^{(0)}(t-t_0),t) \tag{12}$$

where $\mathbf{J}$ is the Jacobian matrix of $\mathbf{f}_0$ computed at $\Gamma^{(0)}(t-t_0)$.

Then the Melnikov distance is defined as

$$\Delta(t,t_0) = \mathbf{n} \cdot (W_1^s(t,t_0) - W_1^u(t,t_0)) \tag{13}$$

where $\mathbf{n}$ is the normal to $\Gamma^{(0)}(t-t_0)$.

After some algebra one finally finds the Melnikov function

$$\Delta(t_0) = -\int_{-\infty}^{\infty} dt\,(\mathbf{f}_0 \wedge \mathbf{f}_1)_{\Gamma^{(0)}(t-t_0)} \tag{14}$$

which in principle can be explicitly computed; if $\Delta(t_0)$ changes sign for some t_0, then an infinity of homoclinic intersections between W^u and W^s will take place and chaos will set in. This is the only *general* predictive method to study the condition for the onset of chaos in ODE.

Notice that for Hamiltonian systems there are always homoclinic intersections when a non integrable perturbation $\epsilon\mathbf{f}_1(\mathbf{x},t)$ is added to an integrable system; in this case the Melnikov function [18] is

$$M(t_0) = -\int_{-\infty}^{\infty} dt\,\{H_0,H_1\}_{\Gamma^{(0)}(t-t_0)}, \tag{15}$$

where curly brackets are Poisson brackets, of the unperturbed Hamiltonian H_0 with the perturbation Hamiltonian H_1, computed along the unperturbed separatrix $\Gamma^{(0)}$; $M(t_0)$ is useful to evaluate the thickness of the stochastic layer. The analytical computation of $\Delta(t_0)$ for eq.(7) is standard and yields

$$\Delta(t_0) = 2\pi\sqrt{\frac{2}{\beta}}\gamma\omega\,\text{sech}(\frac{\pi\omega}{2})\sin\omega t_0 + \frac{4\delta}{3\beta}. \tag{16}$$

Unfortunately there are not so many models for which explicit computation of $\Delta(t_0)$ can be performed. Therefore we chose Duffing - Holmes model because it is not difficult to compute $\Delta(t_0)$ when a parametric perturbation is introduced.

Let us modify eq.(7) to

$$\begin{pmatrix} \dot{x} \\ \dot{y} \end{pmatrix} = \begin{pmatrix} y \\ x - \beta(1+\eta\cos\Omega t)x^3 \end{pmatrix} + \epsilon\begin{pmatrix} 0 \\ -\frac{\delta}{\epsilon}y + \frac{\gamma}{\epsilon}\cos\omega t \end{pmatrix}, \tag{17}$$

if $\eta \ll 1$ we are allowing a periodic modulation of small amplitude of the parameter β.

Accounting for the modulating term $\beta\eta \cos\Omega t\, x^3$ in the perturbation function $\mathbf{f}_1$, the new Melnikov function $\Delta_p(t_0)$ is given by

$$\Delta_p(t_0) = \Delta(t_0) - \frac{\eta}{\beta}\int_{-\infty}^{\infty} dt\ \dot{x}^{(0)}(t-t_0)[x^{(0)}(t-t_0)]^3 \cos\Omega t \qquad (18)$$

using (10), after simple but tedious computations, one finds

$$\Delta_p(t_0) = 2\pi\sqrt{\frac{2}{\beta}}\gamma\omega\,\text{sech}(\frac{\pi\omega}{2})\sin\omega t_0 + \frac{4\delta}{3\beta} + \frac{\pi\eta}{6\beta}(\Omega^4 - 6\Omega^2 + 1)\text{cosech}(\frac{\pi\Omega}{2})\sin\Omega t_0. \quad (19)$$

Let us now consider a set of parameters for which $\Delta(t_0)$ changes sign, thus predicting the existence of chaos; then we add the parametric perturbation as in eq.(17) and we compute, from the numerical tabulation of eq.(19), the time needed between two successive homoclinic intersections of W^u and W^s, denote it τ_M.

In fig.1, τ_M^{-1} is reported vs the parametric perturbation frequency Ω for the following set of parameters: $\gamma = 0.088$, $\delta = 0.154$, $\omega = 1.1$, $\beta = 4$, $\eta = 0.1$.

A "resonance line" is found which is centered at $\Omega \equiv \omega$; for this value of Ω, τ_M becomes infinite; this means that W^u and W^s never intersect, hence chaos should disappear.

Similar results are obtained for different sets of parameters. If η is smaller than some critical value η_c, then τ_M remains finite and homoclinic intersections are not suppressed. By increasing η above η_c, a *line broadening* is observed. Finally, outside the interval of Ω where the quartic polynomial in eq.(19) is negative, there is no way to avoid intersections of W^u and W^s.

Accurate numerical experiments can be performed to make a comparison with these predictions. Equation (17) is integrated using a Hammings modified predictor-corrector of fourth order, time integration steps $\Delta t = 0.001 - 0.003$ and integration times of $t = 10000 - 20000$ after a transient of $t = 500$. Simulations are performed on a Cray x-mp computer.

By means of a standard technique [19], the maximal Lyapunov characteristic exponent λ is computed to detect chaos and measure its strenght.

For the same set of parameters of fig.1, except for η which is 0.03, λ is computed for different values of the parametric perturbation frequency Ω. The results are shown in fig.2 and are rather striking. The resonance line $\tau_M^{-1}(\Omega)$, $\Omega \sim \Omega_R{}^0 = 1.1$ in fig.1 corresponds to the first order resonance of $\lambda(\Omega)$ in fig.2. Moreover, other resonances show up: at $\Omega = 2.2$ and $\Omega = 3.3$, second and third harmonics of $\Omega_R{}^0$ respectively, at $\Omega = 0.55$ and $\Omega = 2.75$, i.e. at the first subharmonic of $\Omega_R{}^0$ and at the third harmonic of the subharmonic. Both these last resonances correspond to a non negligible weakening of chaos but not to its suppression ($\lambda = 0.$) as for the principal resonances.

We have thus provided an example showing that chaos can be reduced or eliminated in a dissipative system by means of parametric perturbations. A 3% modulation of a parameter is able to make regular the chaotic dynamics when the modulation frequency is resonant with the forcing frequency.

Let us just spend few words about Hamiltonian systems. We have already mentioned that in this case chaos is always present, therefore one can "only" hope either to stabilize some broken KAM tori, or to modify the diffusion properties of the model. Very preliminary results, obtained with the Hamiltonian version of the Duffing-Holmes oscillator and with autonomous systems, show that sometimes several small and sticky islands can be produced by parametric perturbations; the major consequence is a slowing down of diffusion in phase space due to intermittent trapping near islands. An increase of intermittency is revealed by a worse convergence of maximal Lyapunov exponent caused by enhanced fluctuations of local divergence rate of nearby trajectories. In general, stronger perturbations are necessary to produce measurable effects.

Concluding remarks.

We have briefly shown that the general claims presented in the Introduction do not belong to the realm of science fiction: at least in one case some analytical prediction can be made (using Melnikov's theory) that indicate the possibility of suppressing chaos by parametric perturbations.

Numerical simulations confirmed these predictions revealing at the same time a richer phenomenology, which is not surprising because Melnikov's theory is based on an approximate method.

The final message of the present work is essentially twofold: *i)* the general problem of stabilization of chaotic instability deserves a lot of further work, *ii)* it might be worth trying to go farther along the line here proposed.

Many problems are open and need a deeper understanding to assess the actual interest of what is presented here. Among the open questions we mention that:

i) we do not know to what extent our result is generic, i.e. if it can be reproduced for any non Axiom-A chaotic system;

ii) Melnikov's theory is general but applies to weakly perturbed systems and it is of practical utility only in a few simple cases;

iii) in order to understand *if* and *how* chaos can be suppressed, one should be able to understand it and describe it better than now;

iv) one could argue that the results presented here are not of practical utility because of the so called "stability dogma" [11]: if you observe chaos in real systems it must be structurally stable (in the strong sense) and the same should hold for the theoretical models used. Theoretical models will never take into account all the interactions, perturbations, noise etc. which are present in real systems. Therefore, it should be very interesting to try to kill chaos in a real chaotic system reasonably modeled by a Duffing - Holmes equation of motion.

As a final remark, we mention that, after the Birkhoff - Smale homoclinic theorem [11], the existence of homoclinic intersections for the Duffing-Holmes oscillator ensures the existence of a hyperbolic invariant set Λ. Other considerations [11] rule out, for the same model, the possibility for Λ to be an attractor. There are several reasons to believe that noise, which is always present both in real systems and in numerical models, plays an important role together with Λ to stabilize chaotic transients or, at least, to make them very long with respect to practical observational times.

We hope that this contribution will stimulate some research work on this subject.

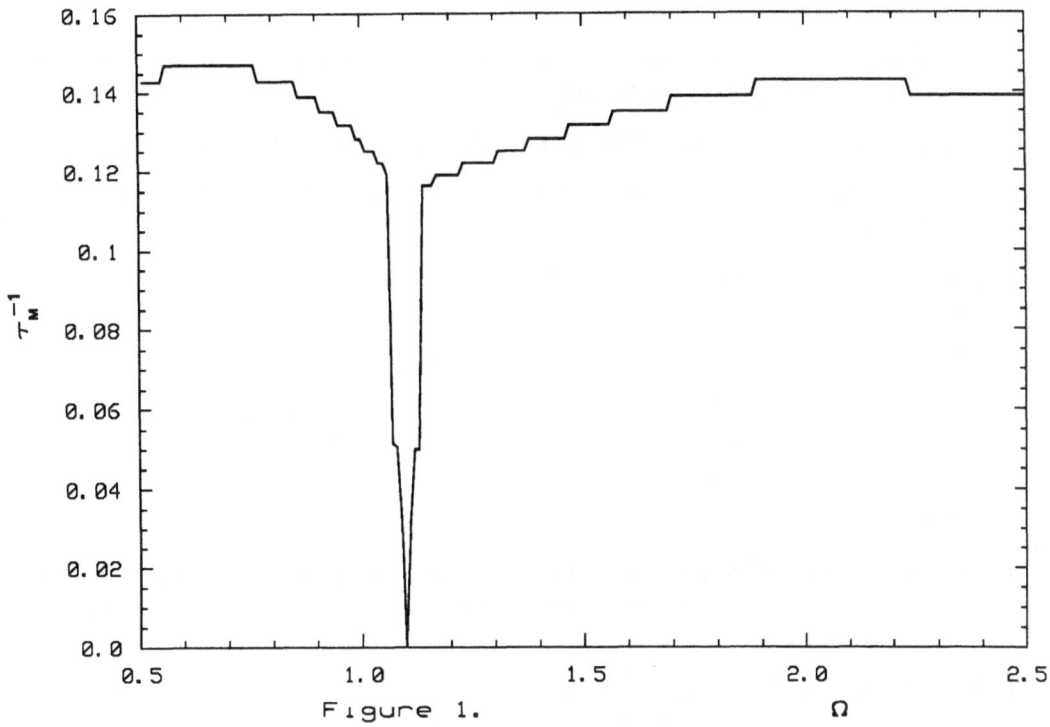

Figure 1.

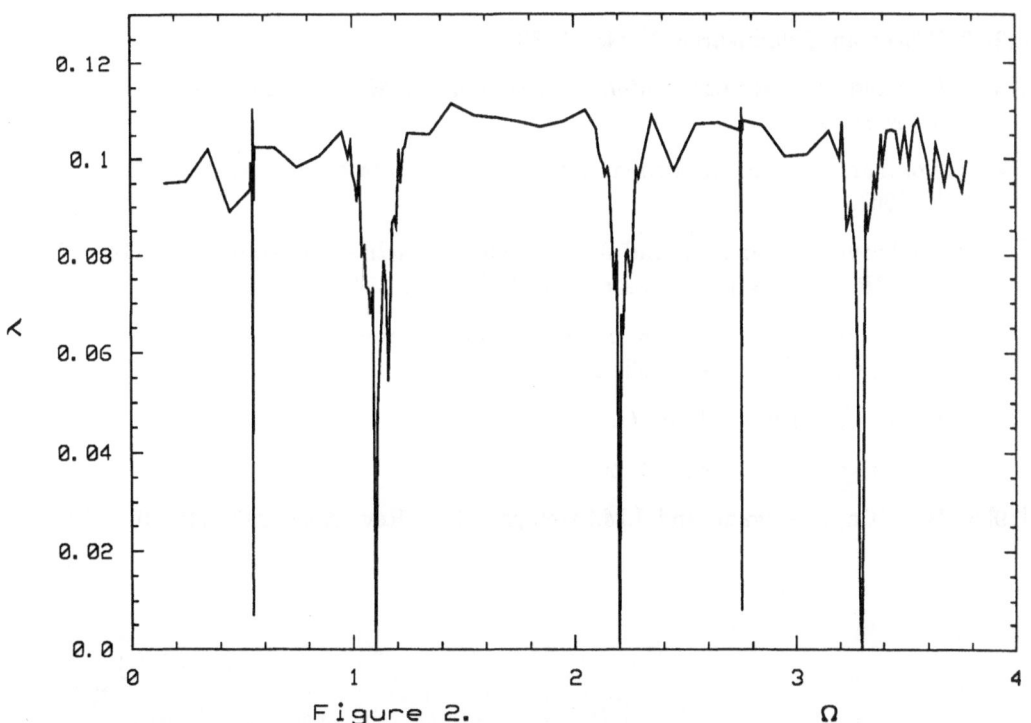

Figure 2.

References

[1] F.C.Moon, *Chaotic Vibrations. An Introduction for Applied Scientists and Engineers*, John Wiley & Sons, N.Y. 1987.

[2] M.Pettini *et al.*, Phys. Rev. **A38**, 344 (1988), and references quoted therein.

[3] M.N.Rosenbluth, R.Z.Sagdeev, J.B.Taylor, and G.M.Zaslavsky, Nucl. Fusion **6**, 297 (1966).

[4] M.Rasetti, *Modern Methods in Equilibrium Statistical Mechanics*, World Scientific, Singapore 1987, p.73.

[5] J.D.Hanson, and J.R.Cary, Phys. Fluids **27**, 767 (1984).

[6] F.M.Izrailev, and B.V.Chirikov, *Numerical experiments on stabilization of stochastic instability with the use of computer in interactive regime* (in russian), I.Ya.F. preprint 74-13, Novosibirsk 1974.

[7] K.Matsumoto, and I.Tsuda, J. Stat. Phys. **31**, 87 (1983).

[8] I.Purica, *Mathematical models for the correlation of equipment quality with their behavior in operation*, Doctoral Thesis, National Committee for Nuclear Energy - IBNE, Bucarest 1987.

[9] V.I.Arnold, *Chapitres Supplémentaires de la Théorie des Equations Différentielles Ordinaires*, Editions MIR, Moscow 1980 (French translation).

[10] V.I.Arnold, and A.Avez, *Ergodic Problem of Classical Mechanics*, W.A.Benjamin Inc.,N.Y. 1968.

[11] J.Guckenheimer, and P.Holmes, *Nonlinear Oscillations, Dynamical Systems, and Bifurcations of Vector Fields*, Springer-Verlag, N.Y. 1983.

[12] E.C.Zeeman, Nonlinearity **1**, 115 (1988).

[13] V.I.Arnold, *Les Méthodes Mathématiques de la Méchanique Classique*, Editions MIR, Moscow 1976.

[14] R.Abraham, and J.E.Marsden, *Foundations of Mechanics*, W.A.Benjamin Inc., N.Y. 1967.

[15] B.Doubrovine, S.Novikov, and A.Fomenko, *Géometrie Contemporaine*, Vol. I, Editions MIR, Moscow 1979 (French translation), p. 354.

[16] R.Lima, and M.Pettini, *Suppression of chaos by resonant parametric perturbations*, preprint, CPT-CNRS Marseille 1989.

[17] R.A.Mahaffey, Phys. Fluids **19**, 1387 (1976).

[18] B.V.Chirikov, Phys. Rep. **52**, 263 (1979).

[19] G.Benettin, L.Galgani, and J.M.Strelcyn, Phys. Rev. **A14**, 2338 (1979).

DYNAMICAL SYSTEMS METHODS IN ACCELERATOR PHYSICS: THE DYNAMIC APERTURE PROBLEM

Rui Alves Pires[1] and Rui Dilão[2]

1) CERN, PS Division, CH–1211 Geneva 23, Switzerland
2) CFMC, Av. Prof. Gama Pinto 2, 1699 Lisboa Codex, Portugal

1. Introduction

The aim of this paper is to show how qualitative methods of the theory of non–linear Dynamical Systems are used to design and optimize the layout of magnets in a circular synchrotron particle accelerator. The dynamic aperture problem consists in the optimization of the limits of stability in the plane transversal to the machine design orbit, in order to avoid losses in the bunch of particles at injection. Here we derive a formula for the linear dimension of the stability region in the transversal plane to the accelerator, as a function of the lengths of the straight sections that surround dipoles with a parasitic sextupolar field.

The general linear theory of the alternating–gradient synchrotron accelerator has been developed by Courant and Snyder in 1958, [CS]. The tools used by these authors are essentially based on Floquet theory of linear differential equations and the resonance analysis of externally forced linear differential equations. These techniques are still important for the design of the present days machines. However, in the future, with the use of superconducting magnets, non–linear fields will be strong enough to affect the stability limits of the transversal oscillations of the bunch of particles travelling inside the vacuum chamber of the accelerator.

An important point that distinguishes externally forced linear systems from parametrically excited systems is that, in the latter amplitude growth at instability is exponential in "time", whereas in the former amplitude growth is only linear. Therefore, in particle accelerators, where the resonances are excited parametrically, an accurate knowledge of the mechanisms of resonance excitation is required. In the Courant and Snyder classical paper , [CS], this distinction is not stressed and all the theories of resonance in accelerators have been developed in the framework of small denominators, characteristic of externally forced linear systems.

In references [D] and [APD] we have developed a formalism to deal with non–linear fields strongly localized along the vacuum chamber of the accelerator. This description produces physical criteria for the optimization of the layout of magnets along the machine and for the choice of the working point of the accelerator in a suitable parameter space.

We derive here a formula for the length of a drift space that follows a superconducting dipole magnet with a sextupolar magnetic field, in order to optimize the limit of stability for oscillations of the beam of particles in the direction transversal to the design orbit of the accelerator. This is obtained through the Poincaré map associated to the equations of motion derived in [D] and by simple topological considerations about the conservative Hénon map. The main result of this paper is summarized in Theorem 3.

2. Equations of motion and the dynamics of sextupoles

The simplest structure for a synchrotron machine is a sequence of dipole magnets, quadrupoles, and drift spaces arranged in a closed design orbit, Fig. 1. The problem we want to deal with concerns the stability of oscillations of a bunch of particles in the direction transversal to the machine design orbit and in this case the longitudinal and transversal motions can be decoupled, [CS]. In the transversal plane the equations of motion of a charged test particle of mass m and logitudinal momentum p_0 are,

$$\ddot{x} + K_x(s)x = \frac{m}{p_0^2} Q_x(x, z, s) \tag{1a}$$

$$\ddot{z} + K_z(s)z = \frac{m}{p_0^2} Q_z(x, z, s), \tag{1b}$$

where a dot over a letter represents differentiation with respect to s, $K_x(s)$ and $K_z(s)$ are periodic step functions with period L (the length of the machine closed design orbit), describing the distribution and strength of dipoles and quadrupoles. $Q_x(x, z, s)$ and $Q_z(x, z, s)$ are the non–linear force components in the x and z directions and are periodic in the variable s. For example, in the case of dipoles and quadrupoles $Q_x = 0$. For symmetric sextupoles, $Q_x = p_0^2 k_2(x^2 - z^2)/2m$, where k_2 measures the sextupolar strength.

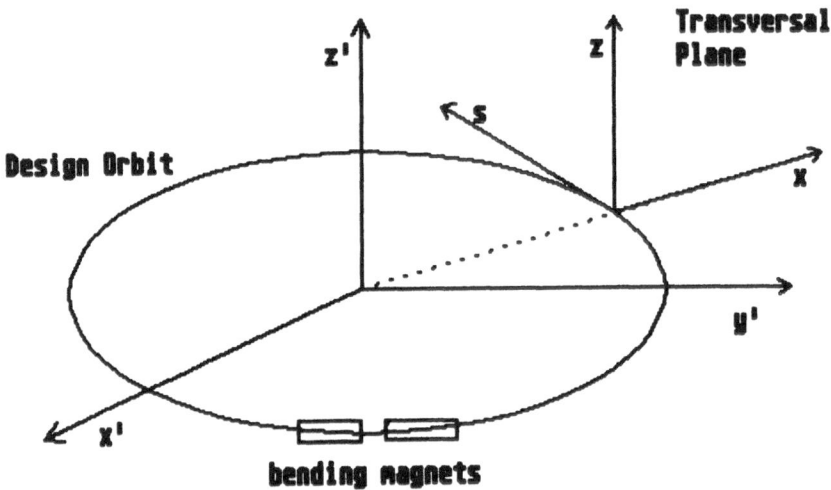

FIG. 1

In general dipoles bend the particles in the (x', y') plane and if we consider that the non–linear forces are symmetric in relation to the (x', y') plane we can drop the dependence in z in equation (1a) and neglect equation (1b), (see [D] for details). Hence, for this simple case and with $Q_x = 0$, we have the simpler equation

$$\ddot{x} + K(s)x = 0$$

whose solutions are, by Floquet theorem,

$$x(s) = a\sqrt{\beta(s)}\cos(\nu_0\phi(s) + \delta)$$

where a and δ are integration constants, $\beta(s)$ is the Courant and Snyder beta function and,

$$\nu_0 = \frac{1}{2\pi}\oint\frac{dt}{\beta(t)}, \quad \phi(s) = \frac{1}{\nu_0}\int_0^s\frac{dt}{\beta(t)}. \tag{2}$$

The parameter ν_0 is the tune of the betatronic motion and $\phi(s)$ is the phase advance. For $s = L$, the length of the closed orbit, $\phi = 2\pi$.

It is easy to prove that for machines with dipoles and quadrupoles it is always possible to find a function $K(s)$ such that $\beta(s)$ is strictly positive and periodic. In such conditions, choosing the new reduced variable $u = x/\sqrt{\beta(s)}$ and ϕ for independent variable, equation (1a) reads, (dropping the z dependence),

$$\frac{d^2u}{d\phi^2} + \nu_0^2 u = \nu_0^2\frac{m}{p_0^2}\beta^{3/2}(\phi)Q_x(u, \phi). \tag{3}$$

If we make $Q_x = 0$ in the above equation, it describes the motion of a test particle along the structure of dipoles, quadrupoles and drift spaces (straight sections) of the accelerator. If $Q_x \neq 0$ we are incorporating the effect of symmetric non–linear fields. For example, in the case of symmetric sextupoles, $Q_x(u, \phi) = p_0^2 k_2(\phi)\beta(\phi)u^2/2m$, where $k_2(\phi)$ gives there distribution and strengths along the design orbit.

In general, in synchrotron accelerators the tune ν_0 of the betatronic motion is greater than one and the right hand side of equation (3) has period 2π in ϕ. So, we are in presence of "superharmonic" parametric excitation and the techniques of perturbation theory do not apply. This motivates the introduction of some physical simplifications in the study of equation (3). In reference [D] we have proved the following:

Theorem 1. *Let us suppose that there are $N \geq 1$ strongly localized sextupolar fields (thin lens approximation) at phases ϕ_m, $m = 1, ..., N$, along the machine design orbit. Thus, the right hand side of equation (3) reads:*

$$\nu_0^2\frac{m}{p_0^2}\beta^{3/2}(\phi)Q_x(u, \phi) = \nu_0^2\beta^{5/2}(\phi)k_2(\phi)u^2/2 =$$

$$= \nu_0^2\sum_{k=-\infty}^{\infty}\sum_{m=1}^{N}\beta^{5/2}(\phi_m)\Delta\phi_m\frac{k_2(\phi_m)}{2}\delta(\phi - \phi_m - 2\pi k)u^2$$

where

$$\Delta\phi_m = \frac{1}{\nu_0}\int_{s_m-L_m/2}^{s_m+L_m/2}\frac{dt}{\beta(t)} \simeq \frac{L_m}{\nu_0\beta(\phi_m)},$$

and s_m is the mean position of the localized sextupolar interaction. L_m is the interaction length. Under these conditions we have:

a) Let $u_1 = u(2\pi)$, $v_1 = \dot{u}(2\pi)/v_0$, $u_0 = u(0)$ and $v_0 = \dot{u}(0)/v_0$. Then, the Poincaré map or one–turn map of equation (3) is

$$\begin{pmatrix} u_1 \\ v_1 \end{pmatrix} = \begin{pmatrix} \cos(2\pi v_0) & \sin(2\pi v_0) \\ -\sin(2\pi v_0) & \cos(2\pi v_0) \end{pmatrix} \begin{pmatrix} u_0 - \sum_{m=1}^{N} \alpha_m \sin(v_0 \phi_m) u^2(\phi_m) \\ v_0 + \sum_{m=1}^{N} \alpha_m \cos(v_0 \phi_m) u^2(\phi_m) \end{pmatrix}$$

where

$$u(\phi_m) = \cos(v_0 \phi_m) u_0 + \sin(v_0 \phi_m) v_0 + \sum_{p=1}^{m-1} \sin(v_0(\phi_m - \phi_p)) \alpha_p u^2(\phi_p)$$

$$\alpha_m = \frac{\beta^{3/2}(\phi_m) k_2(\phi_m) L_m}{2}.$$

Moreover, the Jacobian determinant of the above map is constant and equals one.

b) If in addition for all $m = 1, ..., N$ there exist integers $k(m) = 1, 2, ..$, such that,

$$v_0 \phi_m = \int_0^{s_m} \frac{dt}{\beta(t)} = k\pi \ ,$$

the map in a) reduces to

$$\begin{pmatrix} u_1 \\ v_1 \end{pmatrix} = \begin{pmatrix} \cos(2\pi v_0) & \sin(2\pi v_0) \\ -\sin(2\pi v_0) & \cos(2\pi v_0) \end{pmatrix} \begin{pmatrix} u_0 \\ v_0 + u_0^2 \sum_{m=1}^{N} (-1)^{k(m)} \alpha_m \end{pmatrix}.$$

In the case of one sextupole , $N = 1$, the Poincaré map in Theorem 1 a) reduces to

$$T : \begin{pmatrix} u_1 \\ v_1 \end{pmatrix} = \begin{pmatrix} \cos(2\pi v_0) & \sin(2\pi v_0) \\ -\sin(2\pi v_0) & \cos(2\pi v_0) \end{pmatrix} \begin{pmatrix} u_0 - \alpha_1 \sin(v_0 \phi_1) u^2(\phi_1) \\ v_0 + \alpha_1 \cos(v_0 \phi_1) u^2(\phi_1) \end{pmatrix} \tag{4}$$

where

$$u(\phi_1) = u_0 \cos(v_0 \phi_1) + v_0 \sin(v_0 \phi_1),$$

which is a quadratic map of the plane. In the case of $N > 1$ sextupoles the Poincaré map has polynomial order 2^N and the coefficients of the non–linear terms grow very fast, [D]. However we show below that even with a simple map like (4) we can in fact obtain some optimizing results for the design of a particle accelerator.

We now recall the following Theorem:

Theorem 2 ([E],[H]). Let

$$H : \begin{pmatrix} u_1 \\ v_1 \end{pmatrix} = \begin{pmatrix} \cos(2\pi v_0) & \sin(2\pi v_0) \\ -\sin(2\pi v_0) & \cos(2\pi v_0) \end{pmatrix} \begin{pmatrix} u_0 \\ v_0 \end{pmatrix} + \begin{pmatrix} f(u_0, v_0) \\ g(u_0, v_0) \end{pmatrix}$$

where f and g are homogeneous polynomials of degree 2. If the Jacobian of H is non-zero and everywhere constant, then there exists a linear change of coordinates such that

H is linearly conjugated to the map

$$H' : \begin{pmatrix} x_1 \\ y_1 \end{pmatrix} = \begin{pmatrix} \cos(2\pi\nu_0) & \sin(2\pi\nu_0) \\ -\sin(2\pi\nu_0) & \cos(2\pi\nu_0) \end{pmatrix} \begin{pmatrix} x_0 \\ y_0 + x_0^2 \end{pmatrix} \tag{5}$$

In view of Theorem 2 we conclude that maps (4) and (5) are linearly conjugated. More precisely, there exists a linear function $h(x, y) = (u, v)$, with non–zero Jacobian determinant, such that $H' = h^{-1} \circ T \circ h$ and so the qualitative structure of the orbits in phase space of map (4) can only change if the parameter ν_0 is varied. The parameter α_1 is simply a scale factor and ϕ_1 rotates orbits in phase space (see the transformations of coordinates defined below in (7)).

An important problem in the design of the accelerator lattice concerns the length of straight sections between regions of strong non–linear fields([D]). This can be easily seen by Theorem 1. Suppose we make all the $\phi_2, ..., \phi_N$ decrease simultaneously to ϕ_1 or take the values $\phi_m = k(m)\pi/\nu_0$, with $m = 1, ..., N$. In both limits the dynamics of the test particle is described by maps linearly conjugated to the Hénon map. Between these two limits, where very high order polynomial terms come into play, we can have a drastic change in the dynamics of the system, [D]. The strongest change in the phases ϕ_m can be otained by changing the lengths of straight sections because in those situations $\beta(s) \sim s^2$. Moreover, between dipoles producing a strong non–linear parasitic sextupolar field, straight section lengths are the simplest parameters we can change to control the non–linear coupling between them, mantaining acceptable limits of stability.

Most of the accelerator lattice magnets are dipoles producing a strong sextupolar parasitic field. We want to obtain with the limits of stability around the design orbit, that is, the region of Lyapunov stability of the point $(0, 0)$ in the transversal phase space, Fig. 1. We consider a beam line as shown in Fig.2, consisting of a sequence of smaller lines *dipole + drift space* with a thin sextupolar field in the middle of the dipole.

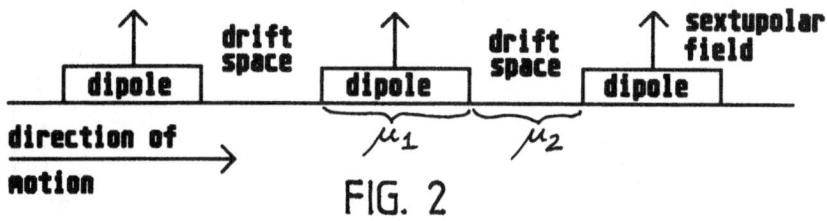

FIG. 2

Let μ_1 and μ_2 be the phase advances as indicated in figure 2, $2\pi\mu = \int dt/\beta(t)$, (2). With this notation the transfer map that relates the normalized position and velocity of a test particle between the begining and the end of the accelerator small line is, by (4),

$$\begin{pmatrix} u_1 \\ v_1 \end{pmatrix} = \begin{pmatrix} \cos(2\pi\mu) & \sin(2\pi\mu) \\ -\sin(2\pi\mu) & \cos(2\pi\mu) \end{pmatrix} \begin{pmatrix} u_0 - \alpha_1 \sin(\pi\mu_1)u^2(\mu_1/2) \\ v_0 + \alpha_1 \cos(\pi\mu_1)u^2(\mu_1/2) \end{pmatrix} \tag{6}$$

where
$$u(\mu_1/2) = u_0 \cos(\pi\mu_1) + v_0 \sin(\pi\mu_1),$$

and $\mu = \mu_1 + \mu_2$ is the total phase advance in the beam line. By Theorem 2 this map is linearly conjugated to the Hénon map.

To obtain information about the region of stability around the point $(0,0)$ in phase space we use the results known for map (5). To be more precise we take the two sets of new variables,
$$\begin{aligned}(w, z) &= (\alpha_1 u, \alpha_1 v) \\ (x, y) &= (w \cos(\pi\mu_1) + z \sin(\pi\mu_1), -w \sin(\pi\mu_1) + z \cos(\pi\mu_1))\end{aligned} \tag{7}$$

and a straightforward calculation shows that (6) is linearly conjugated to the Hénon map (5), $(\nu_0 = \mu)$.

The Hénon map has two period one fixed points, $x_0 = (0,0)$ and $x_1 = (2\tan(\pi\nu_0), -2\tan^2(\pi\nu_0))$. The point x_1 is of saddle type for $\nu_0 \neq k$ and $\nu_0 \neq k + 1/2$, where k is a positive integer. For $\nu_0 = k$ and $\nu_0 = k + 1/2$, map (5) is readily integrable. It is known numerically that for all $\nu_0 \in (k, k+1/2)$ the stable and unstable manifold of x_1 intersect transversally and so there exist a trapping region around the origin $(0,0)$ where orbits remain limited. Far from the origin the iterated points go to infinity.

For the Hénon map the trapping region around the origin can be easily computed numerically. Take the straight line with equation $y = -x \tan(\pi\nu_0)$ and test, for each ν_0 and for a sufficiently large number of iterates, if the orbits of the points $(x, -x \tan(\pi\nu_0))$ remain bounded. The maximum value that the variable x can reach with bounded iterates gives an estimate of the dimensions D of the stable region around the origin $(0,0)$. In figure 3 we represent this maximum value as a function of the parameter ν_0.

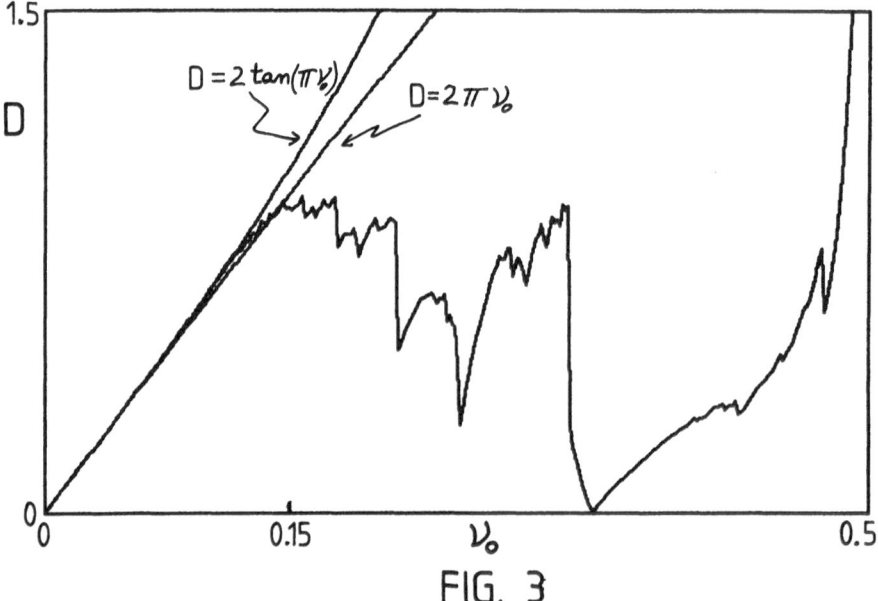

FIG. 3

Figure 3 clearly shows how the linear dimension D of the stability region around the origin changes drastically. For $0 \le \nu_0 \le 0.15$ the linear dimension of the stability region along the x direction goes as (see fig. 3),

$$D = 2\tan(\pi\nu_0) = 2\pi\nu_0 + 2\pi^3\nu_0^3/3 + \cdots \qquad (8)$$

which is the x coordinate of the saddle point x_1. In fact, if $\nu_0 \le 0.15$, the stable (W^s) and unstable (W^u) manifolds of x_1 are almost coincident. For $\nu_0 > 0.15$ the homoclinic intersections develop too fast in the formation of horseshoes. This can be seen in figure 4 where we present the phase space for map (5) for $\nu_0 = 0.15$ and $\nu_0 = 0.2$, as well as pieces of the stable (W^s) and unstable (W^u) manifolds of x_1.

For real accelerator lattices the phase advance μ between consecutive bending dipoles is typically in the interval $[0.02, 0.1]$ as is the case of the design lattice of the Large Hadron Collider (CERN) for protons at 450 GeV. Therefore, we can use the linear approximation for D to derive the dependence of the dimension of the stability region as a function of the accelerator parameters.

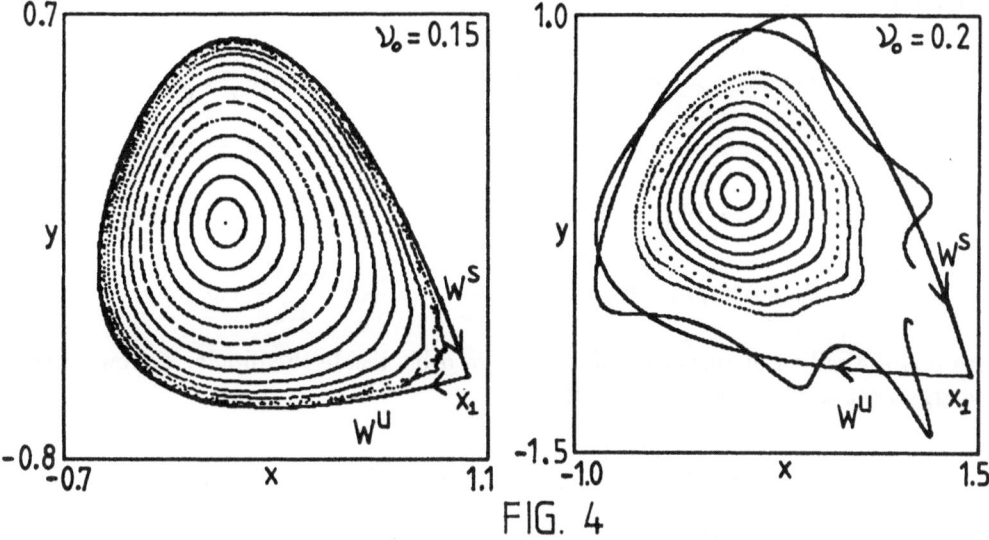

FIG. 4

As we have seen in Theorem 2 the only parameter driving the bifurcation structure of map (6) is the tune, or phase advance, ν_0. So,we now derive an explicit formula for the phase advance μ in the beam line of figure 2. In a drift space the beta function as a function of the length s is, [S],

$$\beta(s) = \beta_0 + \dot{\beta}_0 s + \frac{4 + \dot{\beta}_0^2}{4\beta_0}s^2 \qquad (9)$$

where β_0 and $\dot{\beta}_0$ are the values of $\beta(s)$ and $\dot{\beta}(s)$ at $s = 0$. Hence, the phase advance μ as a function of s becomes,

$$\mu = \frac{1}{2\pi} \int_0^s \frac{dt}{\beta(t)} = \frac{1}{2\pi} \left(\arctan(\frac{\dot{\beta}_0}{2} + \frac{4 + \dot{\beta}_0^2}{4\beta_0} s) - \arctan(\frac{\dot{\beta}_0}{2}) \right).$$

We now calculate the dynamic aperture (the linear dimension of the stability region in phase space) as a function of the straight section length for the machine line in figure 2. Denote by s_1 the length of the drift space. As, the total phase advance μ is given by $\mu = \mu_1 + \mu_2$ we have,

$$\mu = \mu_1 + \frac{1}{2\pi} \left(\arctan(\frac{\dot{\beta}_{02}}{2} + \frac{4 + \dot{\beta}_{02}^2}{4\beta_{02}} s_1) - \arctan(\frac{\dot{\beta}_{02}}{2}) \right) \tag{10}$$

where β_{02} refer to the beta function at the end of the dipole and μ_1 is the phase advance in the dipole.

As we have seen previously in (8), for $\mu \leq 0.15$ the linear dimension of the stability region is $D = 2\pi\mu$, so, the dynamic aperture of the beam line, in the Hénon map coordinates is, by (10),

$$D = D_{dipole} + \arctan(\frac{\dot{\beta}_{02}}{2} + \frac{4 + \dot{\beta}_{02}^2}{4\beta_{02}} s_1) - \arctan(\frac{\dot{\beta}_{02}}{2}) \tag{11}$$

where D_{dipole} is the dynamic aperture corresponding to the phase advance inside the dipole.

In the dipole we have, [St],

$$\beta(s) = \beta_{01} \cos^2(s/\rho) + \dot{\beta}_{01} \rho \sin(s/\rho) \cos(s/\rho) + \rho^2 \frac{4 + \dot{\beta}_{01}^2}{4\beta_{01}} \sin^2(s/\rho)$$

where ρ is the radius of curvature of the design orbit inside the dipole. Following the same argument as above we conclude that,

$$D_{dipole} = \arctan(\frac{\dot{\beta}_{01}}{2} + \frac{4 + \dot{\beta}_{01}^2}{4\beta_{01}} \rho \tan(L_1/\rho)) - \arctan(\frac{\dot{\beta}_{01}}{2}) \tag{12}$$

where L_1 is the length of the curved orbit inside the dipole and β_{01} refers to the beta function at the begining of the dipole.

In non–normalized coordinates , $x = \sqrt{\beta_{01}} u$ and by (7), the dynamic aperture D^* of the machine line in figure (2) is,

$$D^* = \frac{\sqrt{\beta_{01}}}{\alpha_1} D \simeq$$

$$\frac{\sqrt{\beta_{01}}}{\alpha_1} \left(\frac{L_1}{\beta_{01}} + \frac{s_1}{\beta_{02}} \right) = \frac{2}{\beta^{3/2}(\mu_1/2)\sqrt{\beta_{01}} k_2} + \frac{\sqrt{\beta_{01}}}{\beta_{02} \alpha_1} s_1$$

where $\alpha_1 = \beta^{3/2} k_2 L_1 / 2$ is the sextupolar strength defined in Theorem 1 and L_1 is the dipole length; this last expression has been obtained under the hypothesis of small lengths in (11) and (12).

Therefore, we conclude that the increase of straight sections that surround dipoles originates an increase in the acceptance stability of the machine. This conclusion has been derived under the assumption that the phase advance is kept in the interval $[0, 0.15]$. The same qualitative conclusion still holds for μ in the neighbourhood of 0.5 (see figure 3), although this is in general too restrictive if we want to keep the machine as small as possible. Another possibility to increase the dynamic aperture of the beam line is to decrease the sextupolar constant α_1 (see Theorem 1) or to decrease the values of the beta function.

The conclusions of this paper can be summarized in the following theorem.

Theorem 3. *Let B be the machine line*

$$B = dipole\ with\ sextupolar\ field + drift\ space$$

and μ the phase advance inside B. Then, in the framework of the thin lens approximation (Theorem 1) and under the hypothesis that $\mu \leq 0.15$ in B, the dynamic aperture of the machine line B is,

$$D^* = \frac{\sqrt{\beta_{01}}}{\alpha_1} \Big(\arctan(\frac{\dot{\beta}_{02}}{2} + \frac{4 + \dot{\beta}_{02}^2}{4\beta_{02}} s_1) + \arctan(\frac{\dot{\beta}_{01}}{2} + \frac{4 + \dot{\beta}_{01}^2}{4\beta_{01}} \rho \tan(L_1/\rho)) -$$

$$\arctan(\frac{\dot{\beta}_{01}}{2}) - \arctan(\frac{\dot{\beta}_{02}}{2}) \Big) \simeq \frac{2}{\beta^{3/2}(\mu_1/2)\sqrt{\beta_{01}}k_2} + \frac{\sqrt{\beta_{01}}}{\beta_{02}\alpha_1} s_1$$

where, β_{01}, $\dot{\beta}_{01}$, β_{02} and $\dot{\beta}_{02}$, are the values of the beta function and its derivative at the begining and end of the dipole. $\beta(\mu_1/2)$ is the value of the beta function at the middle of the dipole. α_1 [resp. k_2] is the sextupolar strength as defined in Theorem 1a), L_1 is the length of the dipole, s_1 is the length of the drift space and ρ is the radius of curvature of the particle orbit inside the dipole.

References

[APD] R. Alves Pires and R. Dilão, Non–linear phenomena in circular accelerators II: beam–beam interaction and ionic impurities, in preparation.

[CS] E. D. Courant and H. S. Snyder, Theory of the alternating gradient–synchrotron, *Ann. of Phys.* **3** (1958) 1–48.

[D] R. Dilão, Non–linear phenomena in circular accelerators I: a model with a non–linear diffeomorphism of the plane, CERN SPS/88–47.

[E] W. Engel, Ganze Cremona–Transformationen von Primzahlgrad in der Ebene, *Math. Ann.* **136** (1958) 319–325.

[H] M. Hénon, Numerical study of quadratic area–preserving mappings, *Quart. Appl. Math.* **27** (1969) 291–312.

[S] P. Schmüser, Basic course on accelerator optics, CERN Yellow Report 87–10, 1987.

[St] K. Steffen, Basic course on accelerator optics, CERN Yellow Report 85–19, 1985.

DETERMINISTIC CHAOS VERSUS RANDOM NOISE: FINITE CORRELATION DIMENSION FOR COLORED NOISES WITH POWER-LAW POWER SPECTRA

A. Provenzale and A. R. Osborne

Istituto di Cosmo-Geofisica del C.N.R.

Corso Fiume 4, Torino 10133, Italy

ABSTRACT

We show that simple "colored" random noises characterized by power-law power spectra generate a finite and predictable value of the correlation dimension. This result is a counter-example to the traditional expectation that stochastic processes lead to a non convergence of the correlation dimension in computed or measured time series. These results also indicate that the observation of a finite dimension from the analysis of one or a few time series is not sufficient to infer the presence of a strange attractor in the system dynamics.

INTRODUCTION

In this paper we review and extrapolate some recent results on the problem of disentangling between random noise and low-dimensional deterministic chaos [29, 30, 37]. An interesting issue in the study of the irregular and apparently random behavior of physical systems (such as turbulent fluid flows) is in fact to determine whether the dynamics is governed by some stochastic processes (associated with the presence of a large number of active degrees of freedom) or if it is dominated by the action of a few excited modes exhibiting chaotic behavior, see for example Eckmann and Ruelle [9] for an introduction to these topics. If a small number of excited modes dominate the system, then an approach based on the concept of low-dimensional deterministic chaos may be appropriate. In this case the system evolution may be modelled in terms of a few ordinary differential equation.

In recent years, the attempt of addressing the above issue in a phenomenological context has benefited from the development of a variety of new time series analysis techniques based on dynamical systems theory [2, 4, 6, 9, 10, 14, 16, 17, 18, 22, 33, 35, 40, 43, 44]. In the majority of cases these methods have been applied to the study of forced, dissipative systems in which the low dimensional chaotic dynamics is in general associated with the presence of a strange attractor in phase space. Experimental evidence of low dimensional chaos has been obtained in several detailed analyses of carefully controlled laboratory systems (see for example [1, 5, 7, 8, 19, 24, 40] for results in fluid dynamics).

Among the various methods available, the calculation of the fractal dimension of the attractor which underlies the system evolution in phase space has probably received the widest attention. Traditionally, systems whose dynamics are governed by stochastic processes are thought to be associated with an infinite value for the fractal dimension in phase space. This is because random noises are generally expected to fill very large dimensional regions of the available phase space (i.e. they are associated with a very large number of excited degrees of freedom). By contrast, finding a finite, non integer value of the dimension is usually considered to be a strong indication of the presence of low-dimensional deterministic chaos. The precise value of the fractal dimension of the attractor has in addition an important physical significance. For forced, dissipative systems in fact the number of variables needed to describe the dynamics has been shown to be strictly related to the attractor dimension (see Mane' [27] and Takens [41]). If the system evolution is dominated by a strange attractor with fractal dimension D then an upper limit to the number of variables required to describe the dynamics may be fixed at n=2D+1. This in turn implies that *at most* 2D+1 ordinary differential equations are needed to rigorously describe the system evolution, if the appropriate collective variables can be defined. This is in sharp contrast to the behavior of systems dominated by a very large number of excited modes which are better described by a stochastic approach.

For the above implications, and thanks to the relative simplicity of the methods for computing approximations to the fractal dimension, a number of investigations on the behavior of *uncontrolled* natural or laboratory systems have also been pursued [11, 12, 20, 21, 28, 29, 38, 42]. In several cases apparent evidence of low dimensional chaos has been found. In some cases, however, the supposed presence of a low dimensional strange attractor was based only on the detection of a finite fractal dimension from the analysis of one or a few time series. The finite fractal dimension found from the data was then considered to be representative of the dimension of the underlying attractor.

It is important to note, however, that although the relationship between the fractal dimension and the number of excited modes is rigorous for systems which *are known* to be dominated by deterministic chaos, in the analysis of experimental data one in general does not know *a priori* if a low dimensional attractor exists. Nevertheless, the observation of a finite fractal dimension from a measured signal is often considered as *evidence* of low dimensional chaos (and hence as a statement about the system dynamics). Thus a common conclusion is that by estimating the fractal dimension of the attractor one can easily distinguish between random noise and low dimensional chaos. In the present paper we provide quantitative evidence that this expectation may be misleading. *We discuss a simple class of stochastic processes with power-law spectra which give a finite (and predictable) value for the fractal dimension.* This in turn implies that *detecting a finite and non integer value for the fractal dimension is not sufficient to indicate the presence of a strange attractor.*

In the following we use the method developed by Grassberger and Procaccia [16], which is a fast and reliable technique for computing the attractor dimension. However, the results discussed here are completely independent of the particular method employed as they hold in general for all techniques for computing the fractal dimension.

CORRELATION FUNCTION AND CORRELATION DIMENSION

Here we briefly recall the method proposed by Grassberger and Procaccia [16] for computing the correlation dimension of strange attractors. Given a measured scalar time series $X(t_i)$ the first step is an embedding procedure to reconstruct a pseudo phase space for the system considered. The reconstructed space may be obtained by a time embedding procedure introduced by Takens [41] (see also Packard et al [33]), in this case a vector time series in $\mathbf{R}^N$ is defined as

$$X(t_i) = \{X(t_i,),X(t_i + \tau),...,X(t_i + (N-1)\tau)\} \tag{1}$$

Here τ is an appropriate time delay multiple of the sampling time Δt and N is the dimension of the vector $X(t_i)$. See for example Atten et al [1], Eckmann and Ruelle [9] and Fraser and Swinney [13] for discussions on the best choice of the delay τ. The time embedding method is rigorous (i.e. it furnishes a correct reconstruction of the phase space) for time series with infinite length (i.e. with an infinite number of points) and with a finite variance [41].

Given the vector time series $X(t_i)$, one defines the correlation function $C_N(\varepsilon)$ as

$$C_N(\varepsilon) = \frac{1}{M^2-M} \sum_{i \neq j}^{M} H\{\varepsilon - \|\mathbf{X}(t_i) - \mathbf{X}(t_j)\|\} \qquad (2)$$

where H is the Heaveside step function, M is the number of points in the vector time series $\mathbf{X}(t_i)$ and the vertical bars indicate the norm of the vector. If an attractor for the system exists then

$$C_N(\varepsilon) \underset{\varepsilon \to 0}{\sim} \varepsilon^{\nu_N} \qquad (3)$$

and

$$\nu_N \underset{N \to \infty}{\sim} \nu \qquad (4)$$

where ν is the correlation dimension of the attractor. For further details on this method see the original paper by Grassberger and Procaccia [16].

FINITE CORRELATION DIMENSION FOR COLORED RANDOM NOISES

The method proposed by Grassberger and Procaccia has been developed for determining the dimension of the attractor, given that an attractor exists. In the study of experimental data, however, this route has been somewhat reversed: If, given a measured time series from a system with apparently random behavior, a time embedding procedure and the subsequent calculation of the correlation function lead to determining a finite and non-integer value for ν through equations (3) and (4), then the system is considered to be dominated by low-dimensional deterministic dynamics. As mentioned above, systems dominated by stochastic processes are by contrast expected to provide a very different output. For random systems the exponent ν_N is supposed not to saturate at any finite value ν but it is thought to increase without bound. The widely adopted example of white noise supports this view. In this Section, however, we show that this expectation is not satisfied for every type of random noise.

To quantitatively exploit this observation we start with the usual Fourier representation of a discrete scalar signal

$$X(t_i) = \sum_{k=1}^{M/2} \{P(\omega_k) \, \Delta\omega_k\}^{1/2} \cos(\omega_k t_i + \phi_k) \; ; \; i=1,M \qquad (5)$$

where $\omega_k = 2\pi/N\Delta t$, Δt is the sampling interval and M is the number of points in the time series. $P(\omega_k)$ is the power spectrum of the signal and the ϕ_k's are the Fourier phases. A simple and widely adopted method to generate a random signal using equation (5) is to consider a fixed power spectrum (which gives the energy distribution among the various Fourier modes) and random phases.

Let us now consider signals of the form (5) whose power spectrum has a power law dependence

$$P(\omega_k) = C \, \omega_k^{-\alpha} \qquad (6)$$

and the phases are randomly, uniformly distributed on the interval $(0,2\pi)$. The choice of a power-law spectrum is physically significant since many experimental measurements from widely different systems have approximate power-law spectra. For example 3-D turbulence, 2-D and geostrophic turbulence [23, 32, 34, 39], internal waves in the ocean [15], passively advected scalars [3], drifter trajectories in large scale flows [29, 31, 36] are well-known examples in fluid flows. In what follows each time series obtained by inverting the spectrum (6) for a fixed value of α (i.e. obtained using equation (5)) is a particular realization of a member of this one parameter (α) family of "colored" random noises. Each choice of the set of random phases corresponds to a different realization of the same stochastic process. The constant C in the spectrum is fixed by the requirement that the time series have unit variance.

To proceed with the analysis we have selected a number of values of the spectral exponent α and for each of these we have generated a time series, say $x(t_i;\alpha)$, where $i=1,M$. The latter notation indicates that each process is viewed as a function of the spectral exponent α. The number of points in each realization $x(t_i;\alpha)$ is M=8192. The classical time embedding procedure is then used: For each selected value of the spectral exponent α we consider forty different embedding spaces with increasing dimension N, $1 \leq N \leq 40$. An N-dimensional curve $x(t_i;\alpha)$ is obtained by taking N time delayed values of the signal $x(t_i;\alpha)$ for the selected value of α. We stress that the use of a time embedding procedure is not critical since other embedding methods (such as the use of independent realizations of the random process, as done in [30]) lead to the same results. The goal is now to search for convergence in the calculated value of the correlation dimension and, in the case of convergence, to see whether a relation may be found between the spectral exponent α and the value of the correlation dimension.

The first example is shown in Figure 1 for a value of the spectral exponent $\alpha=1.50$. We first compute the time series of 8192 points and we then compute the correlation functions $C_N(\varepsilon)$ for $N = 1,2,..,40$ for an appropriate value of the time delay. If scaling holds for every N then one can define the exponents v_N through equation (2.3) and study their behavior for increasing N. Figure 1a reports the time series $x(t_i;\alpha)$ and Figure 1b reports the forty correlation functions $C_N(\varepsilon)$ for $N = 1,.., 40$ when a value of the time delay $\tau = 100$ is chosen. A scaling behavior is evident for all N's, i.e. each $C_N(\varepsilon)$ is a straight line on the log-log plot. Consequently, the values of the various v_N can be determined. Figure 1c shows v_N versus N. A clear saturation of v_N to a finite value v is observed, even if the time series $x(t_i;\alpha)$ is the output of a stochastic process. This traditionally unexpected result implies that a finite value for the correlation dimension v may be found even for non deterministic, random signals. The error bars on the individual values of v_N are purely statistical and represent the 95% confidence limits of the least-squares-fit of log $C_N(\varepsilon)$ versus log ε.

Figure 2a shows the time series $x(t_i,\alpha)$ for the case $\alpha=2.00$. Figure 2b reports $C_N(\varepsilon)$, N = 1,.., 15 for $\tau = 100$. Again scaling is evident. The values of v_N can be easily defined for each N, and, as shown in figure 2c, they saturate at a finite value $v \cong 2.0$ when N is increased. Finally, in Figure 3a we show the time series $x(t_i;\alpha)$ for $\alpha=2.75$ and in Figure 3b we show the correlation functions $C_N(\varepsilon)$ for this case when a time delay $\tau=100$ is used. Again there is an evident scaling range in the correlation functions, and the plot of v_N versus N reported in figure 3c shows the saturation of v_N to the small value $v \cong 1.23$ for the correlation dimension.

From the above figures one can see that the scaling region is limited at very small values of $C_N(\varepsilon)$ by a "knee" in the correlation function. This behavior is well known in the analysis of computer generated or experimental data (see for example Eckmann and Ruelle [9]) and is due either to the presence of experimental or numerical errors or, more importantly, to the lack of statistics at small scales induced by the finite length of the signals analyzed. This however presents no difficulty in the analysis as long as the scaling region is sufficiently large.

266

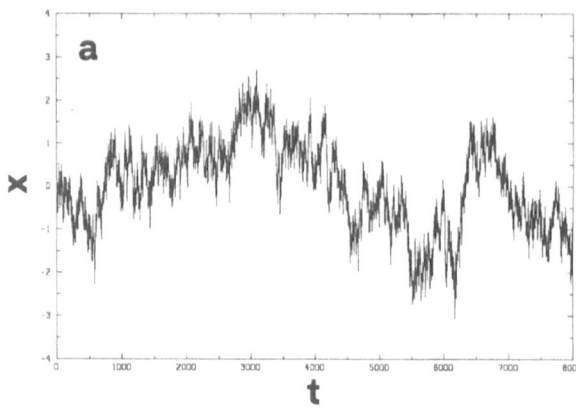

Figure 1. (a) The time series $x(t_i;\alpha)$ for $\alpha=1.50$; (b) The forty correlation functions $C_N(\varepsilon)$ for this value of α and $\tau=100$; and (c) the correlation dimension ν_N versus the embedding dimension N.

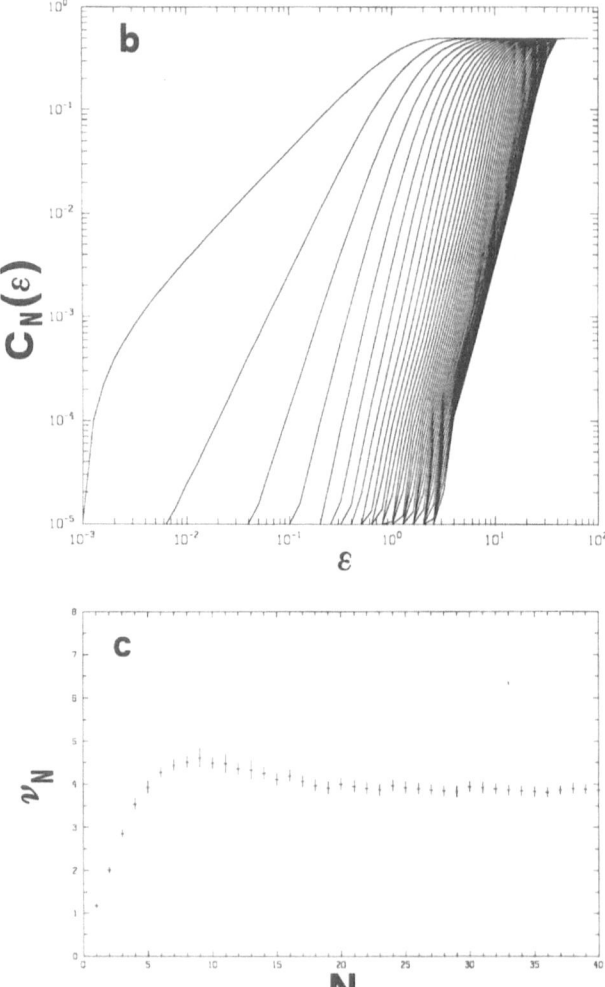

Figure 2. (a) The time series x(t_i;α) for α=2.00; (b) The forty correlation functions C_N(ε) for this value of α and τ=100; and (c) the correlation dimension v_N versus the embedding dimension N.

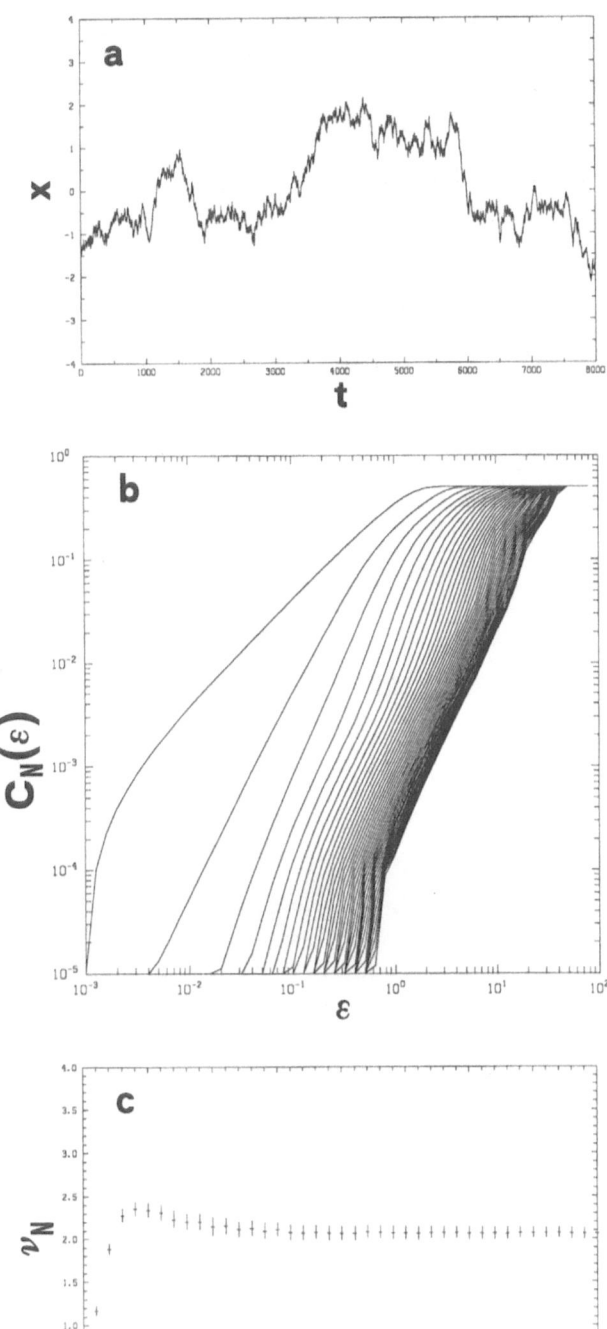

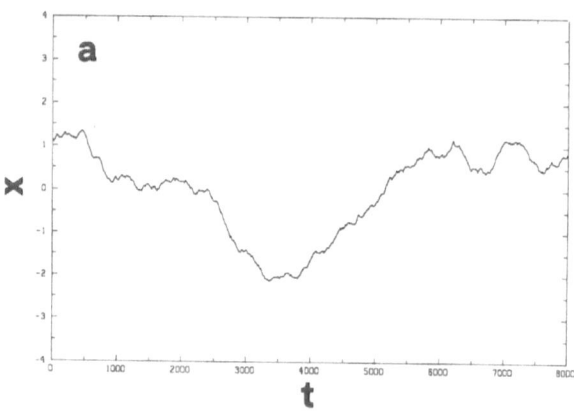

Figure 3. (a) The time series $x(t_i;\alpha)$ for $\alpha=2.75$; (b) The forty correlation functions $C_N(\varepsilon)$ for this value of α and $\tau=100$; and (c) the correlation dimension ν_N versus the embedding dimension N.

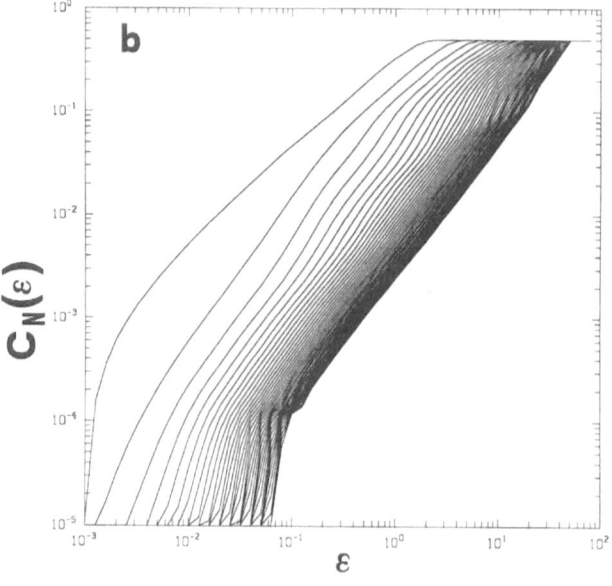

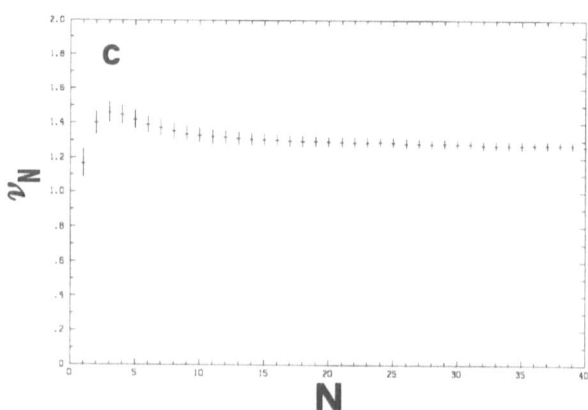

Repeating the above analysis for a number of values of the spectral exponent α allows one to search for a quantitative relation between the exponent α and the correlation dimension v. Figure 4 shows the value of the correlation dimension versus the exponent α. For each value of α the dimension v has been determined as the average of the individual values v_N for $N > N_0$, where N_0 is the phase space dimension beyond which no systematic increase of v_N is observed. The error bars in Figure 4 represent the standard deviation of the average dimension for $N > N_0$. It is evident how the correlation dimension is a well defined monotonically decreasing function $v(\alpha)$ of the spectral exponent α for this class of random processes. The solid curve in Figure 4 represents a theoretical expression which is discussed below. From Figure 4 one realizes that the expectation of a non saturating dimension is really satisfied in the case of white noise, but it is *not* true in general for random noises with power-law power spectra. Consequently, the traditional belief that a finite correlation dimension is indicative of the presence of low-dimensional chaos may be deeply misleading.

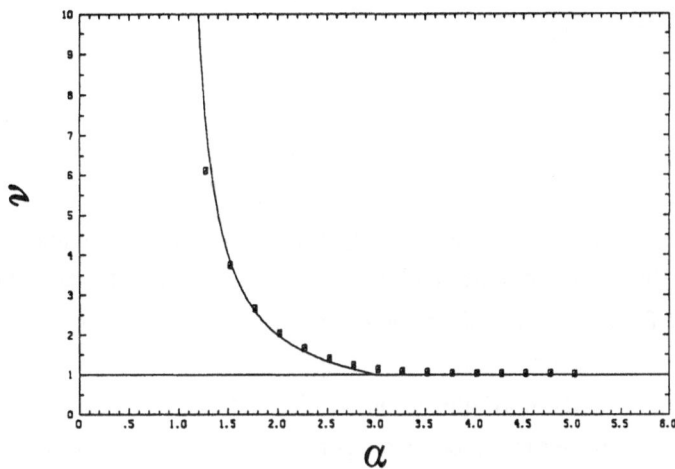

Figure 4. *The correlation dimension v versus the spectral exponent α. The correlation dimension is a well defined, monotonically decreasing function $v(\alpha)$ of the spectral exponent α for these random noises.*

Before discussing the interpretation of these results, we briefly recall that changes in the set of random phases, or changes in the number of points of the time series do not modify the above results [30]. Also, changing the value of the time delay τ does not affect the form of the relationship between ν and α [30, 37].

Now the obvious question arises as to whether the values of the correlation dimension determined above have any physical or mathematical meaning. It is clear that they are not related to any deterministic dynamics. The correct answer is that these values of ν are generated by the fractal nature of the time series $x(t_i;\alpha)$. A brief definition of a fractal is "a mathematical object whose topological dimension is strictly smaller than its Hausdorff dimension" [25, 26]. In dynamical systems theory a strange attractor (of a system of differential equations) with fractal dimension D>1 is an object which is differentiable along the direction of motion and which can be "fractal" in some direction perpendicular to the direction of motion. Fractal curves however exist which are everywhere continuous and non differentiable. Examples of these curves are the various types of fractal "random paths" such as classic Brownian motion. These paths are realizations of a stochastic process and they possess a finite Hausdorff dimension which may be larger than their topological dimension.

A "fractal" analysis of the phase-space trajectories $x(t_i,\alpha)$ discussed above reveals that these are self-similar fractal curves. In a previous paper [30] we have analyzed the curves $x(t_i;\alpha)$ by using several methods for computing the dimension and the scaling properties of fractal curves (such as the calculation of the scaling exponent H and calculation of the divider dimension D_L, see [30] for definitions), and we demonstrated the fractal nature of $x(t_i,\alpha)$. The interpretation of the finite correlation dimensions found for the random processes studied above is thus clear. The curves generated by inverting power-law spectra and random phases are random fractal paths, their Hausdorff dimension is finite, and thus their correlation dimension is finite as well. The Grassberger and Procaccia algorithm is independent of the ordering of the points in the signal and it is not able to test the differentiability of the curve under study. *Thus this method cannot distinguish between fractal attractors and fractal random curves if the two have the same dimension.*

As already mentioned above, in Figure 4 a theoretical curve is drawn together with the results of numerical analysis. There is in fact a close theoretical relationship between fractal, self-affine signals and their power spectra. Self-affinity implies that there is no preferred length scale in the dynamics. By an easy extension of the classic argument which relates the structure function of a signal with a power law spectrum to the spectral exponent, one finds that a self-affine signal has a power law spectrum $P(\omega) = C\omega^{-\alpha}$ with $\alpha = 2H + 1$ (see Osborne and Provenzale [30] and Panchev [34]). The fractal dimension of a curve is thus related to the slope of the power spectrum of each of its components (if they are independent and have the same scaling exponent H) by the equation

$$D = \frac{2}{\alpha - 1} \tag{7}$$

The constraint $0<H<1$ and the requirements on the convergence of the integral which defines the structure function makes equation (7) valid only for $1<\alpha<3$. For $\alpha > 3$ the Hausdorff dimension of the curve is equal to its topological dimension and the curve ceases to be fractal. For $0\leq\alpha\leq1$ the scaling exponent is zero and the fractal dimension has an unbounded (infinite) value, i.e. the signal in this case really generates a non saturating fractal dimension. Equation (7) has been used to generate the solid curve shown in Figure 4.

Looking at the above Figures one realizes that there is some difference between the theoretical curve given by equation (7) and the numerical dimension estimates, particularly for $\alpha\cong1$ and $\alpha\cong3$. The numerical results are systematically below the theoretical curve for $\alpha\cong1$ and above it for $\alpha\cong3$. Careful study of these effects led us to the conclusion that they may be entirely understood in terms of the low- and high-frequency cutoffs to the power spectrum implied by the use of time series with a finite number of points [30]. The low- and high-frequency cutoffs in the spectrum correspond to the finite values respectively of the length T of the time series and of the sampling interval Δt. When the total length T is increased and the sampling time Δt is decreased the deviations tend (although very slowly) to disappear and the numerical values of the fractal dimension become closer to the theoretical curve. A generalized expression for $\nu(\alpha)$ may anyhow be obtained also for finite values of T and Δt; an excellent agreement is observed between the numerical estimates of the dimension and this generalized relation (see [30]).

Finally, an interesting point concerns the reason for the failure of the phase space reconstruction procedure based on the time embedding method. In this regard we note that the time embedding furnishes a rigorous reconstruction of phase space only in the case of time series with infinite length and with finite variance. The class of stochastic processes considered here, however, are random noises with stationary increments (see Panchev [34]) and are in some sense similar to diffusion processes. Hence, one of the requirements needed by the rigorous demonstration of the time embedding method is not satisfied; in this case this procedure does not reconstruct the system phase space.

SUMMARY AND CONCLUSIONS

We have reviewed some recent results on the study of stochastic processes with power-law power spectra and random, uniformly distributed Fourier phases. We have shown that *these colored random noises generate a finite and predictable value of the correlation dimension.* The results of this research provide a counter-example to the general belief that random signals give a non saturating value for the correlation dimension. The finite dimensions have been shown to be generated by the fractal properties of the signals. These "colored" random noises in fact are self-affine signals which generate fractal curves whose dimension is completely determined by the power spectral exponent α. Thus, for these random processes there is a one-to-one correspondence between the power-law dependence of the spectrum and the fractal behavior. We note that this correspondence is not necessarily observed for all systems with power-law spectra since the randomness of the Fourier phases is the other important ingredient.

In our opinion the results discussed here have some implications on experimental studies of deterministic chaos. An usual expectation here is in fact that, if a finite and small fractal dimension is determined, then the system is thought to be dominated by a low dimensional strange attractor. *The present analysis shows on the contrary that observing a finite correlation dimension from experimental data does not necessarily imply the presence of deterministic chaos, as there are stochastic processes which give a finite value of the fractal dimension that is clearly not related to any deterministic dynamics.* Care must thus be exercised in concluding that a strange attractor exists from the sole calculation of the fractal dimension from an experimental (or numerical) system whose exact dynamics is not sufficiently known. In particular finite fractal dimensions from systems with power-law spectra must be interpreted with caution since they may be generated by the fractal (non deterministic) properties of stochastic processes with this kind of spectra. We recall that also the K_2 entropy (another well-known "test" for chaoticity, see Grassberger and Procaccia [17]) has a finite and predictable value for the random processes considered here [37]. Hence, also this procedure apparently does not allow for distinguishing between deterministic chaos and random noise. Consequently, more refined tests must be applied to experimental data in order to univoquely determine the presence of low dimensional chaos.

REFERENCES

[1] P. Atten, J. G. Caputo, B. Malraison and Y. Gagne, "Determination of attractor dimension of various flows", J. Mech. Theor. Appl., Special Issue (1984) 133-156.

[2] R. Badii and A. Politi, "Hausdorff dimension and uniformity factor of strange attractors", Phys. Rev. Lett. 52 (1984) 1661-1664.

[3] G. K. Batchelor, "Small-scale variation of convected quantities like temperature in turbulent fluid. Part 1. General discussion and the case of small conductivity", J. Fluid Mech. 5 (1959) 113-133.

[4] P. Berge', "Study of phase space diagrams through experimental Poincare' sections in prechaotic and chaotic regimes", Physica Scripta T1 (1982) 71.

[5] A. Brandstater, J. Swift, H. L. Swinney, A. Wolf, J. D. Farmer, E. Jen and P. J. Crutchfield, "Low-Dimensional Chaos in a Hydrodynamic System", Phys. Rev. Lett. 51 (1983) 1442-1445.

[6] D. S. Broomhead and G. P. King, "Extracting qualitative dynamics from experimental data", Physica 20D (1986) 217-236.

[7] S. Ciliberto and J. P. Gollub, "Pattern competition leads to chaos", Phys. Rev. Lett. 52 (1984) 922-925.

[8] S. Ciliberto and J. P. Gollub, "Chaotic mode competition in parametrically forced surface waves", J. Fluid Mech. 158 (1985) 381.

[9] J. P. Eckmann and D. Ruelle, "Ergodic theory of chaos and strange attractors", Rev. Mod. Phys. 57 (1985) 617-656.

[10] J. D. Farmer, E. Ott and J. A. Yorke, "The dimension of chaotic attractors", Physica 7D (1983) 153-180.

[11] A. Fote, J. McDonough and R. Egler, "Chaos theory and 1/f noise in HgCdTe photodiodes", Nucl. Phys. B (proc. suppl.) 2 (1987) 597.

[12] K. Fraedrich, "Estimating the dimensions of weather and climate attractors", J. Atmos. Sciences 43 (1986) 419-432.

[13] A. M. Fraser and H. L. Swinney, "Independent coordinates for strange attractors from mutual information", Phys. Rev. A33 (1986) 1134-1140.

[14] H. Froehling, J. P. Crutchfield, J. D. Farmer, N. H. Packard and R. Shaw, "On determining the dimension of chaotic flows", Physica 3D (1981) 605-617.

[15] C. J. R. Garrett and W. H. Munk, "Internal waves in the Ocean", Ann. Rev. Fluid Mech. 11 (1979) 339-369.

[16] P. Grassberger and I. Procaccia, "Measuring the strangeness of strange attractors", Physica 9D (1983) 189-208.

[17] P. Grassberger and I. Procaccia, "Estimation of the Kolmogorov entropy from a chaotic signal", Phys. Rev. A28 (1983) 2591-2593.

[18] H. S. Greenside, A. Wolf, J. Swift and T. Pignataro, "Impracticability of box counting algorithms for calculating the dimension of strange attractors", Phys. Rev. A25 (1982) 3453-3456.

[19] J. Guckheneimer and G. Buzyna, "Dimension measurements for geostrophic turbulence", Phys. Rev. Lett. 51 (1983) 1438-1441.

[20] A. Hense, "On the possible existence of a strange attractor for the southern oscillation", Beitr. Phys. Atmosph. 60 (1987) 34-47.

[21] C. Keppenne and C. Nicolis, "Toward a quantitative view of predictability of weather patterns", Annales Geophysicae, Special Issue "EGS General Assembly" (1988) 207.

[22] G. P. King, R. Jones and D. S. Broomhead, "Phase protraits from a time series: a singular system approach", Nucl. Phys. B (proc. suppl.) 2 (1987) 379-390.

[23] D. C. Leslie, "Developments in the Theory of Turbulence" (Clarendon Press, Oxford, 1973).

[24] B. Malraison, P. Atten, P. Berge and M. Dubuois, "Dimension of strange attractors: an experimental determination for the chaotic regime of two convective systems", J. Phys. Lettres 44 (1983) L897-L902.

[25] B. B. Mandelbrot, "Fractals: Form, Chance and Dimension" (Freeman, San Francisco, 1977).

[26] B.B. Mandelbrot, "The Fractal Geometry of Nature" (Freeman, San Francisco, 1982).

[27] R. Mane', "On the dimension of the compact invariant sets of certain nonlinear maps", in: Lecture Notes in Mathematics 898, D. A. Rand and L. S. Young, eds. (Springer, Berlin, 1981) 230-242.

[28] C. Nicolis and G. Nicolis, "Is there a climatic attractor?", Nature 311 (1984) 529-532.

[29] A. R. Osborne, A. D. Kirwan, A. Provenzale and L. Bergamasco, "A search for chaotic behavior in large and mesoscale motions in the Pacific Ocean", Physica 23D (1986) 75-83.

[30] A. R. Osborne and A. Provenzale, "Finite Correlation Dimension for Stochastic Systems with Power-Law Spectra", Physica 35D (1989) 357-381.

[31] A. R. Osborne, A. D. Kirwan, A. Provenzale and L. Bergamasco, "Fractal Drifter Trajectories in the Pacific Ocean", Tellus, in press.

[32] S. A. Orszag, "Lectures on the Statistical Theory of Turbulence", in: Fluid Dynamics, Proc. Les Houches, R. Balian and J.-L. Peube eds. (Gordon and Breach, London, 1977) 235-374.

[33] N. H. Packard, J. P. Crutchfield, J. D. Farmer and R. S. Shaw, "Geometry from a time series", Phys. Rev. Lett. 45 (1980) 712-716.

[34] S. Panchev, "Random Functions and Turbulence" (Pergamon Press, Oxford, 1971).

[35] I. Procaccia, "Exploring deterministic chaos via unstable periodic orbits", Nucl. Phys. B (proc. suppl.) 2 (1987) 527-538.

[36] A. Provenzale, A. R. Osborne, A. D. Kirwan and L. Bergamasco, "The Study of Fluid Parcel Trajectories in Large Scale Ocean Flows", in "Nonlinear Topics in Ocean Physics", A. R. Osborne Editor, Elsevier, in press.

[37] A. Provenzale, A. R. Osborne and R. Soy, "Convergence of the K_2 Entropy for Random Noises with Power Law Spectra", sub judice.

[38] G. Radnoti, "Characterizing the weather attractor", Annales Geophysicae, Special Issue "EGS General Assembly" (1988) 210.

[39] R. Salmon, "Geostrophic Turbulence", in: Topics in Ocean Physics, Proc. Int. School of Physics "E. Fermi", Course LXXX, A. R. Osborne and P. Malanotte Rizzoli eds. (North-Holland, Amsterdam, 1982) 30-78.

[40] H. L. Swinney and J. P. Gollub, "Characterization of hydrodynamic strange attractors", Physica, 18D (1986) 448-454.

[41] F. Takens, "Detecting strange attractors in turbulence", in: Lecture Notes in Mathematics 898, D. A. Rand and L. S. Young, eds. (Springer, Berlin, 1981) 366-381.

[42] A. A. Tsonis, "On the dimension of the weather attractor", Annales Geophysicae, Special Issue "EGS General Assembly" (1988) 208.

[43] L. S. Young, "Dimension, entropy, and Lyapunov exponents", Ergodic Theory and Dynamical Systems 2 (1982) 109.

[44] A. Wolf, J. B. Swift, H. L. Swinney and J. A. Vastano, "Determining Lyapunov exponents from a time series", Physica 16D (1985) 285-317

FRACTIONAL DIFFUSION

W. R. Schneider

Asea Brown Boveri Corporate Research
CH-5405 Baden, Switzerland

ABSTRACT: Loosely speaking, the fractional diffusion equation is obtained by replacing the first order time derivative in the diffusion equation by a derivative of order α with $0 < \alpha \leq 1$. A precise definition is obtained by using integrals instead of derivatives. This has the additional advantage of taking the initial condition automatically into account. The Green function for the fractional diffusion equation is obtained in closed form and its properties are exhibited. Finally, a stochastic process called grey Brownian motion is constructed based on a probability measure called grey noise. Its relation to fractional diffusion is the same as the one of ordinary Brownian motion to ordinary diffusion.

1. The Fractional Diffusion Equation

The fractional diffusion equation

$$u(\mathbf{x},t) = u_0(\mathbf{x}) + \frac{1}{\Gamma(\alpha)} \int_0^t ds\,(t-s)^{\alpha-1} \Delta u(\mathbf{x},s)\,, \quad 0 < \alpha \le 1\,, \quad (1.1)$$

has been introduced in [1]; other variants may be found in [2], [3]. The following discussion is an alternative to the one in [1], possibly simpler and hopefully more elegant.

Obviously, a solution $u\colon \mathbf{R}^n \times \mathbf{R}_+ \to \mathbf{R}$ of the integrodifferential equation (1.1) satisfies the initial condition

$$u(\mathbf{x},0) = u_0(\mathbf{x})\,, \quad \mathbf{x} \in \mathbf{R}^n\,. \quad (1.2)$$

Applying the Fourier transform

$$\mathcal{F}v(\mathbf{k}) = \int_{\mathbf{R}^n} d^n x\, v(\mathbf{x})\, e^{i\mathbf{k}\cdot\mathbf{x}}\,, \quad \mathbf{k} \in \mathbf{R}^n\,, \quad (1.3)$$

and the Laplace transform

$$\mathcal{L}f(p) = \int_0^\infty dt\, f(t)\, e^{-pt} \quad (1.4)$$

to (1.1) yields

$$\mathcal{L}\mathcal{F}u(\mathbf{k},p) = \frac{p^{\alpha-1}}{p^\alpha + k^2}\, \mathcal{F}u_0(\mathbf{k})\,, \quad k^2 = \mathbf{k}\cdot\mathbf{k}\,. \quad (1.5)$$

To invert the Laplace transform we introduce the function

$$F_\alpha(z) = \sum_{j=0}^\infty \frac{(-z)^j}{\Gamma(1+j\alpha)}\,, \quad z \in \mathbf{C}\,. \quad (1.6)$$

The function $E_\alpha(z) = F_\alpha(-z)$ is known as Mittag-Leffler function, an entire function of order $1/\alpha$ [4]. Now, by interchanging sum and integral we obtain

$$\int_0^\infty dt\, e^{-pt} \sum_{j=0}^\infty \frac{(-k^2 t^\alpha)^j}{\Gamma(1+j\alpha)} = p^{-1} \sum_{j=0}^\infty (-k^2 p^{-\alpha})^j = \frac{p^{\alpha-1}}{k^2 + p^\alpha} \quad (1.7)$$

for $\mathrm{Re}\,p > 0$ and $|k^2 p^{-\alpha}| < 1$. This leads to

$$\int_0^\infty dt\, e^{-pt}\, F_\alpha(k^2 t^\alpha) = \frac{p^{\alpha-1}}{k^2 + p^\alpha}\,, \quad \mathrm{Re}\,p > 0\,. \quad (1.8)$$

Consequently, the inverse Laplace transform of (1.5) is given by

$$\mathcal{F}u(\mathbf{k},t) = F_\alpha(k^2 t^\alpha)\,\mathcal{F}u_0(\mathbf{k})\ . \tag{1.9}$$

By the convolution theorem the solution of (1.1) may now be written as

$$u(\mathbf{x},t) = \int_{\mathbf{R}^n} d^n y\, G_\alpha(\mathbf{x} - \mathbf{y},t)\, u_0(\mathbf{y},t) \tag{1.10}$$

where the Green function $G_\alpha(\mathbf{x},t)$ has to satisfy

$$\int_{\mathbf{R}^n} d^n x\, G_\alpha(\mathbf{x},t)\, e^{i\mathbf{k}\cdot\mathbf{x}} = F_\alpha(k^2 t^\alpha)\ . \tag{1.11}$$

Anticipating the representation (2.13) from Section 2

$$F_\alpha(z) = \int_0^\infty d\lambda\, \zeta_\alpha(\lambda)\, e^{-\lambda z}\ , \quad \operatorname{Re} z > 0\ , \quad 0 < \alpha < 1\ , \tag{1.12}$$

where ζ_α is a probability density on $\mathbf{R}_+$, we obtain

$$G_\alpha(\mathbf{x},t) = \int_0^\infty d\lambda\, \zeta_\alpha(\lambda)\, G_1(\mathbf{x}, \lambda t^\alpha)\ . \tag{1.13}$$

Here $G_1(\mathbf{x},\tau)$ is the Green function of the ordinary diffusion equation, i.e., of (1.1) with $\alpha = 1$, explicitly

$$G_1(\mathbf{x},\tau) = (4\pi\tau)^{-n/2}\, e^{-r^2/4\tau}\ , \quad r^2 = \mathbf{x}\cdot\mathbf{x}\ . \tag{1.14}$$

Combining (1.13) and (1.14) yields

$$G_\alpha(\mathbf{x},t) = (4\pi t^\alpha)^{-n/2}\, g_\alpha(r^2 t^{-\alpha}/4) \tag{1.15}$$

with

$$g_\alpha(u) = \int_0^\infty d\lambda\, \zeta_\alpha(\lambda)\, e^{-u/\lambda}\, \lambda^{-n/2}\ , \quad u > 0\ . \tag{1.16}$$

Applying the Mellin transform

$$\mathcal{M}f(s) = \int_0^\infty du\, u^{s-1}\, f(u) \tag{1.17}$$

to (1.16) and taking

$$\zeta_\alpha(\lambda) = \frac{1}{\alpha}\, \lambda^{-1-1/\alpha}\, \rho_\alpha(\lambda^{-1/\alpha}) \tag{1.18}$$

into account leads to

$$\mathcal{M}g_\alpha(s) = \frac{\Gamma(s)\Gamma(s+1-n/2)}{\Gamma(\alpha s + 1 - \alpha n/2)} \ . \qquad (1.19)$$

Here we used

$$\mathcal{M}\rho_\alpha(s) = \frac{1}{\alpha}\frac{\Gamma(\frac{1-s}{\alpha})}{\Gamma(1-s)} \qquad (1.20)$$

which is easily obtained from the characterizing equation [5-7]

$$\mathcal{L}\rho_\alpha(p) = e^{-p^\alpha} \qquad (1.21)$$

of the one-sided stable probability density ρ_α. It is possible to invert the Mellin transform in (1.19) in terms of the so-called H-functions whose definition and basic properties are presented in Section 3. Inserting the result into (1.15) yields

$$G_\alpha(\mathbf{x},t) = (4\pi t^\alpha)^{-n/2} H_{12}^{20}\left(\frac{r^2}{4t^\alpha}\ \middle|\ \begin{matrix}(1-\alpha n/2, \alpha)\\ (0,1)\end{matrix}\ \ (1-n/2, 1)\right)\ . \qquad (1.22)$$

The behaviour of $g_\alpha(u)$ for small u is obtained from (3.5) leading to

$$g_\alpha(u) \sim \begin{cases} \frac{\Gamma(1/2)}{\Gamma(1-\alpha/2)}\, u^0\ , & n = 1 \\[2mm] \frac{-1}{\Gamma(1-\alpha)}\log u\ , & n = 2 \\[2mm] \frac{\Gamma(n/2-1)}{\Gamma(1-\alpha)}\, u^{1-n/2} & n \geq 3 \end{cases} \qquad (1.23)$$

Its asymptotic behaviour for large $u > 0$ is determined by (3.7) and reads

$$g_\alpha(u) \sim C\, u^{-\sigma} \exp(-c\, u^\tau) \qquad (1.24)$$

with

$$C = (2-\alpha)^{-1/2}\, \alpha^\nu\ , \quad c = (2-\alpha)\, \alpha^{\alpha/(2-\alpha)} \qquad (1.25)$$

and

$$\nu = \frac{\alpha(n+1)-2}{2(2-\alpha)}\ , \quad \sigma = \frac{n}{2}\frac{1-\alpha}{2-\alpha}\ , \quad \tau = \frac{1}{2-\alpha}\ . \qquad (1.26)$$

Combining (1.15) with (1.23) and (1.24) yields the behaviour of $G_\alpha(\mathbf{x},t)$ for small and large $|\mathbf{x}| = r$, respectively. Furthermore, $G_\alpha(\mathbf{x},t)$ is a probability density on $\mathbf{R}^n$ as is immediately evident from (1.13) and (1.14). A stronger result may be found in [1] (Theorem 3.1). It follows that in the case $n = 1$ (1.22) is a probability density in the extended

range $0 < \alpha \le 2$. For $\alpha > \alpha_c(n)$ with $\alpha_c(1) = 2$ and $\alpha_c(n) = 1$ for $n \ge 2$ (1.22) becomes indefinite but

$$\int_{\mathbf{R}^n} d^n x \, G_\alpha(\mathbf{x}, t) = 1 \;, \quad t > 0 \;, \tag{1.27}$$

remains valid.

The form and the name of (1.1) are based on the following elementary observations. Define the so-called fractional integral I_λ by

$$I_\lambda f(t) = \frac{1}{\Gamma(\lambda)} \int_0^t ds \, (t-s)^{\lambda-1} f(s) \;, \quad \lambda > 0 \;, \tag{1.28}$$

supplemented by $I_0 f = f$. It has the properties

$$I_\lambda I_\mu = I_{\lambda+\mu} \;, \quad \lambda, \mu \in \mathbf{R}_+, \tag{1.29}$$

and

$$D^k I_\lambda = I_{\lambda-k} \;, \quad k = 1, 2, ..., [\lambda] \;, \tag{1.30}$$

where D denotes differentiation and $[\lambda]$ is the largest integer not exceeding λ. For $\lambda = k \in \mathbf{N}$ we have

$$I_k f(t) = \int_0^t dt_1 \int_0^{t_1} dt_2 \, ... \int_0^{t_{k-1}} dt_k \, f(t_k) \;, \tag{1.31}$$

i.e., the fractional integral is a generalization of the k-fold integral.

In the same spirit we may introduce the fractional wave equation

$$u(\mathbf{x}, t) = u_0(\mathbf{x}) + t \, u_1(\mathbf{x}) + \frac{1}{\Gamma(\beta)} \int_0^t ds \, (t-s)^{\beta-1} \Delta u(\mathbf{x}, s) \;, \quad 1 < \beta < 2 \;, \tag{1.32}$$

which reduces to the integrated wave equation for $\beta = 2$. The solution $u(\mathbf{x}, t)$ of (1.32) may be expressed in terms of the initial conditions

$$u(\mathbf{x}, 0) = u_0(\mathbf{x}) \;, \quad u_t(\mathbf{x}, 0) = u_1(\mathbf{x}) \tag{1.33}$$

as follows

$$u(\mathbf{x}, t) = \sum_{k=0}^1 \int_{\mathbf{R}^n} d^n y \, _kG_\beta(\mathbf{x} - \mathbf{y}, t) \, u_k(\mathbf{y}) \;. \tag{1.34}$$

Here $_0G_\beta(\mathbf{x}, t)$ is given by (1.22) with β replacing α. Obviously, the second Green function $_1G_\beta(\mathbf{x}, t)$ has to satisfy

$$_1G_\beta(\mathbf{x}, t) = \int_0^t {}_0G_\beta(\mathbf{x}, s) \, ds \;, \tag{1.35}$$

leading to the H-function representation

$$_1G_\beta(\mathbf{x},t) = (4\pi t^{\beta/2})^{-n/2}\, t\, H^{20}_{12}\left(\frac{r^2}{4t^\beta}\,\middle|\,\begin{matrix}(2-\beta n/2,\beta)\\(0,1)\end{matrix}\quad (1-n/2,1)\right).$$

(1.36)

2. Grey Brownian Motion

Grey Brownian motion $B_\alpha(t)$, $t \in \mathbf{R}_+$, is a stochastic process whose characteristic function is given by

$$\mathrm{E}(e^{i\lambda(B_\alpha(t)-B_\alpha(s))}) = F_\alpha(\lambda^2(t-s)^\alpha)\,, \quad 0 \le s \le t\,, \quad 0 < \alpha < 1\,. \quad (2.1)$$

For $\alpha = 1$ ordinary Brownian motion is recovered. Before showing that such a process exists we exhibit the connection to fractional diffusion. We claim that $u: \mathbf{R} \times \mathbf{R}_+ \to \mathbf{R}$ given by

$$u(x,t) = \mathrm{E}(u_0(x + B_\alpha(t)) \quad (2.2)$$

solves the fractional diffusion equation (1.1) for n=1 if u_0 is sufficiently well behaved. To see this insert the representation

$$u_0(y) = \frac{1}{2\pi}\int_\mathbf{R} dk\, e^{-iky}\, \mathcal{F}u_0(k) \quad (2.3)$$

into (2.2) and interchange integration and expectation. Using (2.1) with $s = 0$ we obtain

$$u(x,t) = \frac{1}{2\pi}\int_\mathbf{R} dk\, e^{-ikx}\, F_\alpha(k^2 t^\alpha)\, \mathcal{F}u_0(k)\,. \quad (2.4)$$

This is nothing but the inverse Fourier transform of (1.9) thus proving our claim.

The next step is to define the underlying probability space for grey Brownian motion. We equip the Schwartz space $\mathcal{S}(\mathbf{R})$ of real-valued infinitely differentiable functions of rapid decrease with the scalar product

$$(\xi,\eta)_\alpha = \Gamma(1+\alpha)\sin\frac{\pi}{2}\alpha\int_\mathbf{R} d\omega\,|\omega|^{1-\alpha}\,\overline{\tilde\xi(\omega)}\,\tilde\eta(\omega) \quad (2.5)$$

where

$$\tilde\xi(\omega) = (2\pi)^{-1/2}\int_\mathbf{R} dt\, e^{i\omega t}\,\xi(t)\,, \quad \xi \in \mathcal{S}(\mathbf{R})\,. \quad (2.6)$$

We denote the Hilbert space obtained by completing $\mathcal{S}(\mathbf{R})$ with respect to (2.5) by H_α. The functions ϕ_k, $k \in \mathbf{Z}_+$, defined in terms of their transforms

$$\tilde{\phi}_k(\omega) = a(k,\alpha)\, e^{-\omega^2/2}\, L_{[k/2]}^{\sigma(k)-\alpha/2}(\omega^2)\, \omega^{\sigma(k)}\ , \quad k \in \mathbf{Z}_+\ , \qquad (2.7)$$

with

$$\sigma(k) = k - 2[k/2]\ , \qquad (2.8)$$

belong to $\mathcal{S}(\mathbf{R})$ and form an orthonormal basis of H_α for suitably chosen positive normalization constants $a(k,\alpha)$. Here

$$L_k^\epsilon(x) = (1/k!)\, e^x\, x^{-\epsilon}\, D^k\left(e^{-x} x^{k+\epsilon}\right)\ , \quad k \in \mathbf{Z}_+\ , \quad \epsilon > -1\ , \qquad (2.9)$$

are the Laguerre polynomials [8] normalized as follows

$$\int_0^\infty dx\, e^{-x}\, L_j^\epsilon(x)\, L_k^\epsilon(x) = \delta_{j,k}\, \frac{\Gamma(k+\epsilon+1)}{\Gamma(k+1)}\ . \qquad (2.10)$$

Two further Hilbert spaces $H_{\alpha\pm}$ are obtained by completing $\mathcal{S}(\mathbf{R})$ with respect to the scalar products

$$(\xi,\eta)_{\alpha\pm} = \sum_{k=0}^\infty (k+1)^{\pm 2}\, (\phi_k,\xi)_\alpha\, (\phi_k,\eta)_\alpha\ . \qquad (2.11)$$

The three Hilbert spaces satisfy $H_{\alpha+} \subset H_\alpha \subset H_{\alpha-}$ with proper dense inclusions and $H_{\alpha-}$ is the topological dual of $H_{\alpha+}$, i.e. the elements T of $H_{\alpha-}$ are the continuous linear functionals attributing the value $\langle T,\xi\rangle$ to $\xi \in H_{\alpha+}$.

We now define grey noise as the probability measure τ_α on $H_{\alpha-}$ by its characteristic functional

$$\int_{H_{\alpha-}} d\tau_\alpha(T)\, e^{i\langle T,\xi\rangle} = F_\alpha(\|\xi\|_\alpha^2)\ , \quad \xi \in H_{\alpha+}\ . \qquad (2.12)$$

In view of

$$F_\alpha(x) = \frac{1}{\alpha} \int_0^\infty d\lambda\, \lambda^{-1-1/\alpha}\, \rho_\alpha(\lambda^{-1/\alpha})\, e^{-\lambda x}\ , \quad x \in \mathbf{R}_+\ , \qquad (2.13)$$

the r.h.s. of (2.12) has the required properties (normalization, continuity and nonnegative definiteness) of a characteristic functional and according to a theorem of Yu. L. Daletzkii [9] (2.12) defines a unique σ-additive probability measure on $(H_{\alpha-}, \mathcal{B})$, $\mathcal{B}$ being the σ- algebra of the Borel sets of $H_{\alpha-}$. The term grey noise has been chosen to indicate that it generalizes (Gaussian) white noise. The latter is recovered for $\alpha = 1$.

In (2.13) ρ_α is the one-sided stable probability density characterized by its Laplace transform given in (1.21). Momentarily taking (2.13) as the definition of F_α we obtain for its Mellin transform

$$\mathcal{M}F_\alpha(s) = \frac{\Gamma(s)\Gamma(1-s)}{\Gamma(1-\alpha s)} \tag{2.14}$$

where (1.20) has been taken into account. From (3.5) we obtain the series expansion of F_α which coincides with (1.6).

Denoting the expectation with respect to grey noise by $E(\cdot)$ we consider the complex-valued functions F defined on $H_{\alpha-}$ with

$$E(|F|^2) < \infty . \tag{2.15}$$

They form a Hilbert space denoted by $\mathcal{H}_\alpha$ with scalar product $E(\overline{F}G)$. All polynomials P with

$$P(T) = p(\langle T, \xi_1\rangle, ..., \langle T, \xi_n\rangle) \tag{2.16}$$

belong to $\mathcal{H}_\alpha$ where p is a polynomial in n indeterminates. This follows from the fact that grey noise has moments of all orders which in turn is a consequence of F_α being entire. For the particular case $p(x) = x$ we denote P by ξ. From (2.12) and (1.6) we obtain

$$E(|\xi|^2) = \frac{2}{\Gamma(1+\alpha)} \|\xi\|_\alpha^2 . \tag{2.17}$$

Any $\xi \in H_\alpha$ is limit of a sequence in $H_{\alpha+}$. The associated sequence in $\mathcal{H}_\alpha$ is Cauchy in view of (2.17). Hence it has a limit conveniently denoted also by ξ and (2.17) extends to these limit elements. In particular, the indicator functions of half-open intervals $[a,b)$ belong to H_α with norm given by

$$\|\chi_{[a,b)}\|_\alpha^2 = (b-a)^\alpha , \quad 0 \le a < b . \tag{2.18}$$

The associated element in $\mathcal{H}_\alpha$ is denoted by $B_\alpha([a,b))$. Setting $B_\alpha(t) = B_\alpha([0,t)))$ for $t > 0$ defines grey Brownian motion which satisfies (2.1) in view of (2.18) and (2.12).

A more detailed exposition of the topics sketched in this section may be found in [10], where also additional material is presented. In particular, the extension to higher dimensions $n > 1$ is exhibited and the sample paths of grey Brownian motion are shown to be Hölder continuous with index arbitrarily close to $\alpha/2$.

3. The H-Functions

For the reader's convenience we reproduce the definition and the basic properties of the H-functions [11-13]. The general H-function is denoted by

$$H_{pq}^{mn}(z) = H_{pq}^{mn}\left(z \left| \begin{array}{c} (a_1, \alpha_1)...(a_p, \alpha_p) \\ (b_1, \beta_1)...(b_q, \beta_q) \end{array} \right. \right) . \tag{3.1}$$

It is defined by the contour integral

$$H_{pq}^{mn}(z) = \frac{1}{2\pi i} \int_L \frac{A(s)B(s)}{C(s)D(s)} z^{-s} ds \tag{3.2}$$

with

$$A(s) = \prod_{i=1}^{m} \Gamma(b_i + \beta_i s) , \qquad B(s) = \prod_{i=1}^{n} \Gamma(1 - a_i - \alpha_i s)$$

$$C(s) = \prod_{i=m+1}^{q} \Gamma(1 - b_i - \beta_i s) , \quad D(s) = \prod_{i=n+1}^{p} \Gamma(a_i + \alpha_i s) . \tag{3.3}$$

The integers m, n, p and q have to satisfy $0 \leq n \leq p$ and $1 \leq m \leq q$. Empty products in (3.3) are interpreted as one. The parameters a_i, $i = 1, ..., p$, and b_i, $i = 1, ..., q$, are arbitrary complex numbers, whereas α_i, $i = 1, ..., p$, and β_i, $i = 1, ..., q$, are positive. The set $P(A)$ of the poles of A is assumed to be disjoint of the set $P(B)$ of the poles of B. In addition, it is assumed that

$$\delta = \sum_{i=1}^{q} \beta_i - \sum_{i=1}^{p} \alpha_i > 0 ; \tag{3.4}$$

in [11] also the case $\delta = 0$ is treated. The contour L in (3.2) runs from $c - i\infty$ to $c + i\infty$ such that $P(A)$ lies to the left and $P(B)$ to the right of L. The integral (3.2) is independent of c. Under these conditions (3.2) defines the H-function (3.1) which is an analytic function for $z \neq 0$, in general multiple-valued and one-valued on the Riemann surface of $\log z$. It is given by

$$H_{pq}^{mn}(z) = \sum_{s \in P(A)} \text{res} \left\{ \frac{A(s)B(s)}{C(s)D(s)} z^{-s} \right\} \tag{3.5}$$

where res stands for residuum. Changing the sign in (3.5) and summing over $s \in P(B)$, which is non-empty for $n \neq 0$, yields an asymptotic expansion for large $|z|$, holding uniformly on closed subsectors of the domain

$$|\arg z| < \frac{\pi}{2} \{ \sum_1^m \beta_i - \sum_{m+1}^q \beta_i + \sum_1^n \alpha_i - \sum_{n+1}^p \alpha_i \} , \qquad (3.6)$$

provided the quantity in curly brackets is positive. For $n = 0$ there are cases where the H-function becomes exponentially small in certain sectors when $|z|$ becomes large. For $m = q$ we have

$$H_{pq}^{q0}(z) \sim F \, z^{\gamma/\delta} \, \exp(-E^{1/\delta} \delta \, z^{1/\delta}) , \qquad (3.7)$$

for large $|z|$, uniformly on every closed sector with vertex at the origin contained in $|\arg z| < \delta\pi/2$. The constants in (3.7) are given by (3.4) and

$$\gamma = \sum_1^q b_j - \sum_1^p a_j + (p - q + 1)/2$$

$$E = \prod_1^p \alpha_j^{\alpha_j} \prod_1^q \beta_j^{-\beta_j} \qquad (3.8)$$

$$F = (2\pi)^{(q-p-1)/2} \, E^{\gamma/\delta} \, \delta^{-1/2} \prod_1^p \alpha_j^{1/2-a_j} \prod_1^q \beta_j^{b_j-1/2} .$$

If the inequality

$$\max_{1 \le i \le m} \text{Re} \, \frac{-b_i}{\beta_i} < \min_{1 \le i \le n} \text{Re} \, \frac{1 - a_i}{\alpha_i} \qquad (3.9)$$

holds then the H-function (3.1) possesses a Mellin transform given by

$$\mathcal{M} H_{pq}^{mn}(s) = \frac{A(s)B(s)}{C(s)D(s)} , \quad c_- < \text{Re} \, s < c_+ , \qquad (3.10)$$

where c_- and c_+ are given by the l.h.s. and r.h.s. of (3.9), respectively. This follows from (3.2) and the fact that we may choose L parallel to the imaginary axis inside the strip $c_- < \text{Re} \, s < c_+$. By this choice (3.2) is nothing but the inverse Mellin transform of (3.10).

REFERENCES

[1] W. R. Schneider and W. Wyss, "Fractional diffusion and wave equations," J. Math. Phys. **30**, 134 (1989).

[2] W. Wyss, "The fractional diffusion equation," J. Math. Phys. **27**, 2782 (1986).

[3] R. R. Nigmatullin, "The realization of the generalized transfer equation in a medium with fractal geometry," Phys. Status Solidi: B **133**, 425 (1986).

[4] L. Bieberbach, *Lehrbuch der Funktionentheorie*, Band II (Teubner, Leipzig, Berlin, 1927).

[5] W. Feller, *An Introduction to Probability Theory and Its Applications*, Vol. II (John Wiley, New York, 1971).

[6] V. M. Zolotarev, *One-dimensional Stable Distributions* (American Mathematical Society, Providence, 1986).

[7] W. R. Schneider, "Stable distributions: Fox function representation and generalization," in *Stochastic Processes in Classical and Quantum Systems*, edited by S. Albeverio, G. Casati, and D. Merlini, Lecture Notes in Physics, Vol. 262 (Springer, Berlin, 1986).

[8] *Handbook of Mathematical Functions*, edited by M. Abramowitz and I. A. Stegun (Dover, New York, 1965).

[9] I. I. Gihman and A. V. Skorohod, *The Theory of Stochastic Processes I* (Springer, Berlin, Heidelberg, New York, 1974).

[10] W. R. Schneider, "Grey noise," in *Ideas and Methods in Mathematics and Physics, Memorial Volume dedicated to Raphael Høegh-Krohn (1938 - 1988)*, edited by S. Albeverio, J. E. Fenstad, H. Holden, and T. Lindstrøm (to appear).

[11] B. L. J. Braaksma, "Asymptotic expansions and analytic continuations for a class of Barnes-integrals," Compos. Math. **15**, 239 (1964).

[12] H. M. Srivastava, K. C. Gupta, and S. P. Goyal, *The H-Functions of One and Two Variables with Applications* (South Asian, New Delhi, Madras, 1982).

[13] A. M. Mathai and R. K. Saxena, *The H-function with Applications in Statistics and Other Disciplines* (Wiley Eastern, New Delhi, Bangalore, Bombay, 1978).

WHITE NOISE ANALYSIS AND
QUANTUM FIELD THEORY

L. Streit

BiBoS, Universität Bielefeld

D 4800 Bielefeld

and

UM - Area de Matematica

P 4700 Braga

HAMILTONIAN QUANTUM FIELD THEORY IN THE SCHRÖDINGER REPRESENTATION - A PRELUDE FROM THE SIXTIES

In 1960 when quantum field theory was in the doldrums because nobody could make solid sense of nontrivial field Hamiltonians and axiomatics had turned out to be not selective enough, there appeared almost simultaneously two papers [1], [2] which did not introduce dynamics perturbatively as everybody else had tried to before. Instead, they said, the whole dynamics is already encoded in the vacuum. They proposed different formalisms of how to extract these dynamics from the ground state, one on more solid mathematical ground than the other - we will sketch the other.

Starting point is a scalar field ϕ and its canonical conjugate π, obeying the canonical commutation relations (CCR):

$$[\phi(x),\pi(y)] = i\delta(x-y) \tag{1}$$

and commuting among themselves. These fields are operator valued distributions such that for test functions $f \in \mathcal{S}$ $\phi(f)$ and $\pi(f)$ are assumed self-adjoint. It is well known that there are unclassifiably many inequivalent representations of this algebra, and that because of Haag's theorem we cannot content ourselves with the the well studied Fock representation if we want to discuss nontrivial dynamics [3].

Dynamics will be a unitary group with generator H, the "Hamiltonian". It relates to the canonical fields (in analogy to the quantum mechanical $p = \dot{q}$) through the commutation relation

$$[\phi(x),H] = i\pi(x). \tag{2}$$

Semiboundedness of the Hamiltonian

$$H \geq 0 \qquad\qquad (3)$$

is a physical stability condition, it implies a lower bound for the energy of states of the system. There is a state naturally asssoiated with energy zero, we shall postulate the existence of a "vacuum state" Ψ_0 with

$$H\ \Psi_0\ =\ 0. \qquad\qquad (4)$$

As in Schrödinger theory where all wave functions are obtained through multiplication of the ground state wave functions by suitable functions of q, we shall assume that the vacuum vector Ψ_0 is cyclic with respect to the abelian algebra $\mathfrak{U}_\phi$ generated by the field ϕ: we obtain the full Hilbert space $\mathfrak{H}$ of states through

$$\mathfrak{U}_\phi \Psi_0\ =\ \mathfrak{H}. \qquad\qquad (5)$$

Again in analogy to Schrödinger theory we shall assume that the algebra of the field and its canonical conjugate is irreducible. As a consquence the Hamiltonian should be expressible in terms of ϕ and π. A candidate compatible with the commutation relation (2) is

$$H\ =\ \frac{1}{2}\int dx\ \pi^2(x)\ +\ H_1\ [\phi],$$

(in the spirit of this informal introduction we shall not be concerned here with the serious problems associated with the squaring of operator valued distributions).

The cyclicity (5) of the field ϕ can be used to replace the action of $\pi(x)$ on the vacuum by that of a function B_ϕ of the field:

$$\pi(x)\ \Psi_0\ =\ -iB_\phi(x)\ \Psi_0$$

and we rewrite

$$H\ =\ \frac{1}{2}\int dx\ (\pi(x)\ -\ i\ B_\phi(x))(\pi(x)\ +\ i\ B_\phi(x))\ +\ H_2\ [\phi].$$

Next we use the ground state property $H\Psi_0\ =\ 0$ to conclude $H_2\Psi_0\ =\ 0$. But then, by the cyclicity and abelianness of the algebra $\mathfrak{U}_\phi$, $H_2[\phi]$ must vanish identically, so that finally

$$H\ =\ \frac{1}{2}\int dx\ (\pi(x)\ -\ i\ B_\phi(x))(\pi(x)\ +\ i\ B_\phi(x)),$$

i.e. the Hamiltonian H is determined by the ground state. In strict analogy with Schrödinger theory one would try to realize the Hilbert space of states as an L^2-space

$$\mathcal{H} = \left\{ \Psi[\phi] \ \middle| \ \|\Psi\|^2 = \int \Psi^2[\phi] \ \mathcal{D}^{\infty}[\phi] < \infty \right\}$$

over the field, with the canonical conjugate represented by a (functional) derivative

$$\pi = -i\frac{\delta}{\delta\phi}.$$

As a consequence

$$B_{\phi}(x) \ \Psi_0 = i\pi(x) \ \Psi_0 = \frac{\delta}{\delta\phi(x)}\Psi_0[\phi]$$

$$B_{\phi}(x) = \Psi_0^{-1}[\phi] \ \frac{\delta}{\delta\phi(x)}\Psi_0[\phi].$$

However, one of the lessons that have since been understood is that there is no reasonable infinite-dimensional "flat" measure $\mathcal{D}^{\infty}[\phi]$. In Hilbert spaces $L^2(\ d\mu[\phi]\ )$ constructed from $\mathcal{J}$-quasiinvariant measures, i.e. measures for which there exists a Radon-Nikodym derivative

$$\frac{d\mu[\phi+f]}{d\mu[\phi]} \quad \text{for any test function } f \in \mathcal{J}$$

we can still construct a canonical conjugate field [4], using the definition

$$e^{i\pi(f)}\Psi[\phi] = \sqrt{\frac{d\mu[\phi+f]}{d\mu[\phi]}} \ \Psi[\phi+f] \,,$$

but we have lost the concept of the ground state wave function which after all was defined with respect to a flat measure and which was supposed to be the starting point for our non-perturbative construction of dynamics.

This is where White Noise will come in - the flattest thing there is in infinite dimensional measures. We shall develop White Noise Analysis in the following chapters and return to quantum fields in the final one.

WHAT IS WHITE NOISE?

Talking about White Noise we should specify right away that what we have in mind should more properly be called Gaussian, Continuous Parameter, Real Valued White Noise to distinguish it e.g. from its Poisson, or Discrete, or Complex counterparts. Depending on background and preference, we might also introduce White Noise as the velocity of (Wiener's) Brownian motion. Probably the most concise definition, and a very practical one indeed, is through its characteristic functional.
Consider

$$C(f) = e^{-\frac{1}{2}(f,f)}$$

where $f \in \mathcal{S}(\mathbb{R})$ and $(f,g) = \int f(t)g(t)dt$. Then by Minlos' Theorem [4] [5] C(f) is the Fourier transform of a probability measure μ

$$C(f) = \int_{\mathcal{S}^*} e^{i <x,f>} d\mu[x] = E(e^{i < \cdot ,f>})$$

on the space $\mathcal{S}^*$ of tempered distributions. In other words, x is a generalized (distribution valued) random process, it is this process that we shall call White Noise. The functional C plays a very useful role as generating functional. Note in particular that its correlation function, the expectation

$$E(x(s)x(t)) = \delta(s-t)$$

has a constant Fourier transform; it is this constant spectrum which prompted the name of the process.

Likewise it is easy to verify that

$$B(t) = \int_0^t x(s) \, ds$$

is a Gaussian process with zero mean and correlation

$$E(B(s)B(t)) = \min(s,t),$$

hence a version of Wiener's Brownian motion [6], in this sense functionals of Brownian motion can be considered as functionals of White Noise.

One of our central goals will be to obtain and to control large classes of nonlinear functionals of White Noise, in view of their many applications not only in physics but in various other branches of science and also for their intrinsic mathematical interest.

An obvious first choice is the space of square integrable (or finite variance, in probabilistic terminology) functionals which we shall denote as

$$(L^2) = L^2(\mathcal{S}^*, d\mu).$$

Admittedly, nonlinear functionals of generalized functions are not so easy to visualize, hence it is helpful to invoke the so-called $\mathfrak{S}$-transform [7] of functionals

F: $$(\mathfrak{S}F)(f) = \int F(x+f) \, d\mu[x] \quad \text{where } f \in \mathcal{S}(\mathbb{R}).$$

The $\mathfrak{S}$-transform is well defined for any test function f, and we are now dealing with functionals of *smooth* functions. Another useful formula is

$$\mathfrak{G}F(f) = C(f) \ E(\ F \ e^{< \cdot \, , f >}).$$

Decomposing this expression into its multilinear parts we obtain

$$(\mathfrak{G}F)(f) = \sum \int F^{(n)}(t_1, \ldots , t_n) \ f(t_1) \cdots f(t_n) \ d^n t$$

where the kernel functions $F^{(n)}$ are exactly the elements of the Boson Fock space

$$\mathfrak{F} = \bigoplus_n \ \text{Sym} \ L^2(\ \mathbb{R}^n, \ n! \ d^n t \ ).$$

In fact $\mathfrak{G}$ induces a unitary map between (L^2) and $\mathfrak{F}$, and we can recuperate the non-linear functionals F of White Noise, given the kernels $F^{(n)}$, through the formula

$$F(x) = \sum_n \int F^{(n)}(t_1, \ldots , t_n) : x(t_1) \cdots x(t_n) : d^n t$$

where Wick ordering is with respect to the expectation $E(x(s)x(t))$, so that e.g.

$$: x(s)x(t) : \ = \ x(s)x(t) - \delta(s\text{-}t).$$

From this series expansion in (L^2) it is clear that e.g. the $\mathfrak{G}$-transform of a smeared out Wick monomial such as

$$F = : x(g_1) \cdots x(g_n) :$$

is equal to

$$(\mathfrak{G}F)(f) = \prod_n < g_n, \ f >$$

As another example let us discuss the kinetic energy of Brownian motion. It is given by the square of the velocity i.e. of White Noise. Transition to the Wick square amounts to subtraction of an infinite constant. Hence the renormalized action integral is

$$F[x] = \int : x^2(t) : dt.$$

To study such an object it is useful to note that it is given by a kernel function

$$F^{(2)}(t_1, t_2) = \delta(t_1 - \ t_2)$$

which clearly is not in the Fock space $\mathfrak{F}$ of square integrable functions. This example thus indicates the direction for desirable generalizations, as does the following one. "Donsker's δ-function", which "lights up" each time Brownian motion passes through a point a, is given by the expression

$$F = \delta(\ B(t) - a\ )$$

which in the light of the above we can view as a functional of White Noise. Clearly we cannot expect such a functional to be square integrable. Invoking e.g. the Fourier representation of Dirac's δ-function it is not hard to calculate the $\mathfrak{S}$-transform and its kernel functions which turn out to be square integrable for each n. What makes them non-Fock is the fact that the summation over the "particle number" n inherent in the calculation of a Fock space norm turns out to be divergent [8] [9].

Now the stage is set for the construction of generealized White Noise functionals: to include the above and many more examples of functionals that arise naturally in applications we must relax both the square integrability of the kernels and the square summability of their norms. We shall do so in the next section.

WHITE NOISE ANALYSIS - FROM SMOOTH WHITE NOISE FUNCTIONALS TO HIDA DISTRIBUTIONS

We shall turn to the construction of distributions in finite dimensional analysis for inspiration on how to construct generalized functionals of White Noise. In the finite dimensional case one route to distribution theory is via the construction of a Gelfand triple

$$E \subset L^2 \subset E^*,$$

where E stands for a suitably chosen linear space of test functions and E^* then is the space of continuous linear functionals on E, synonymously the space of generalized functions or of distributions [4].

In White Noise Analysis we shall proceed in the same fashion, thus the first step will be that of constructing a space of test functionals inside (L^2). As in finite dimensional analysis there are many inequivalent ways to do so, a choice will depend on the type of problem that one has in mind. In the infinite dimensional case we cite the pioneering work of T. Hida [6], [10]. Here we mimick a construction of tempered distributions as it is given e.g. in [11]. Recall that the Schwartz space of test functions can be characterized as the intersection of the domains of arbitrary powers of the harmonic oscillator Hamiltonian H [12]. Analogously we shall consider functionals F with kernel functions that can see arbitrary powers of the second quantized harmonic oscillator Hamiltonian $\Gamma(H)$. Specifically we consider

$$H = 1 + s^2 - \frac{d^2}{ds^2}$$

and its 2nd quantization

$$\Gamma(H) \simeq \bigoplus_n H^{\otimes n}.$$

We note in passing that for self-adjoint operators A we have

$$\Gamma(A^p) = \Gamma(A)^p$$

and for example if

$$F[x] = e^{i(x,f)}$$

we find

$$\Gamma(A^p)F[x] = \text{const. } e^{i(x,A^p f)},$$

so that in particular these functionals are in the domain of $\Gamma(H^p)$ for all positive p and any test function f. We define

$$\mathfrak{D}(\Gamma(H^p)) = (\mathfrak{I}_p) \subset (L^2)$$

and have

$$\ldots (\mathfrak{I}_p) \subset (\mathfrak{I}_{p-1}) \subset \ldots \subset (L^2) \subset \ldots \subset (\mathfrak{I}_{-p}) \subset \ldots .$$

In analogy to the definition of Schwartz space in finite dimensional analysis we define the space of test functionals by

$$(\mathfrak{I}) = \bigcap_p (\mathfrak{I}_p)$$

and endowing $(\mathfrak{I})$ with the projective limit topology, we consider the triple

$$(\mathfrak{I}) \subset (L^2) \subset (\mathfrak{I})^*.$$

Note that

$$(\mathfrak{I})^* = \bigcup_p (\mathfrak{I}_{-p}).$$

Furthermore we have the following results

LEMMA 1: $(\mathfrak{I})$ is dense in (L^2).
Proof: Recall that for any $f \in \mathfrak{I}$, $F[x] = e^{i(x,f)} \in (\mathfrak{I})$. The span of such F is dense in (L^2).

LEMMA 2: $(\mathfrak{I})$ is a *-algebra.

The proof amounts to verification of an estimate

$$\|F\,G\|_{2,p} \leq \text{const. } \|F\|_{2,p+1} \|G\|_{2,p+1} ,$$

where the norm with index 2,p refers to the Hilbert space $\mathfrak{D}(\Gamma(H^p)) = (\mathfrak{J}_p)$. Essential for the argument is the product formula for normal ordered monomials and the estimate

$$\Gamma(H^p) \geq 2^{pN}.$$

LEMMA 3: (Kubo, Yokoi [13]): Any test functional $F \in (\mathfrak{J})$ has a version

$$\tilde{F}[x] = \sum_n \left\langle :x^{\otimes n}:, F_n \right\rangle \quad \text{with} \quad F_n \in \text{Sym } \mathfrak{J}(\mathbb{R}^n),$$

well-defined for any $x \in \mathfrak{J}^*(\mathbb{R})$.

Conversely, $F \in (\mathfrak{J})$ iff

$$\sum_n n! \left\| (H^p)^{\otimes n} F_n \right\|_2^2 < \infty \quad \text{for all } p \geq 0.$$

We shall use this version to extend functionals beyond the support of μ, usually omitting the indication $\tilde{}$ to simplify notation.

Let us recall that Constructive Quantum Field Theory can be viewed as the construction of suitable non-Gaussian measures on infinite dimensional spaces. Now in finite dimensional analysis we obtain measures from distributions by the very useful theorem which states that positive distributions are indeed measures [4]. Clearly it is extremely desirable to demonstrate such a theorem also for the Hida distributions $\Phi \in (\mathfrak{J})$. This has been done by Yokoi [14]. The first thing to do is to introduce positive cones in the triple of spaces as follows.

DEFINITION: $(\mathfrak{J})_+ = \{ F \geq 0 \} \in (\mathfrak{J})$

$$(\mathfrak{J})_+^* = \{ \Phi \in (\mathfrak{J})^* : (\Phi, F) \geq 0 \text{ if } F \in (\mathfrak{J})_+ \}.$$

For positive generalized functionals $\Phi \in (\mathfrak{J})_+^*$ we have, as in the finite dimensional case, the alternate characterization that the expression

$$C_\Phi(f) = < \Phi, e^{i<\cdot, f>} >$$

is positive definite. But more can be shown.

THEOREM: To any positive generalized functional $\Phi \in (\mathfrak{J})^*$ there corresponds a unique measure ν on the space of tempered distributions with

$$(\Phi, F) = \int \tilde{F} \, d\nu.$$

In other words the positive Hida distributions offer a natural framework for the construction of measures. They play the role of generalized densities for such measures. Conversely we may ask for conditions which assure that the expectations of a given random field may be expressed in terms of White Noise and such a density functional. An answer to this question is given by the following

THEOREM [15]: If the correlation functions

$$G^{(n)}(t_1 \cdots t_n) = E(\phi(t_1) \cdots \phi(t_n))$$

are distributions of a finite order, in the sense of being in the dual of $\mathfrak{D}((H^{\otimes n})^p) \subset L^2(\mathbb{R}^n)$ for all n and some fixed p, then

$$E(e^{i<\phi,f>}) = < \Phi, e^{i<x,f>}>$$

for some positive Hida distribution Φ. An analogous statement holds for random fields φ if we replace the parameter t and the harmonic oscillator Hamiltonian by their multidimensional analogues.

We note in passing that on the basis of this theorem Euclidean fields, not only free ones, but also highly non-trivial constructs such as the Sine-Gordon, Hoegh-Krohn and $P(\varphi)$ models in two space-time dimensions, can thus be expressed in terms of Hida distributions [15], [16].

It is important to note that the partial derivatives of finite dimensional analysis generalize naturally in White Noise Analysis to a directional derivative

$$\partial_f F[x] = \frac{d}{d\lambda}F[x+\lambda f]\big|_{\lambda=0}.$$

LEMMA 4: For any $f \in \mathcal{S}'(\mathbb{R})$ the derivative ∂_f defines a continuous linear map $(\mathcal{S}) \to (\mathcal{S})$. It obeys the product rule, and, for differentiable $g\circ F$, the chain rule.

Central to the proof is an estimate of the form

$$\big| \partial_f F \big|_{2,p} \leq \text{const.} \big\| H^{-q}f \big\|_2 \big| F \big|_{2,q}$$

for a q depending on the order of the distribution f.

To illustrate the calculation of these derivatives let us consider as an example the functional

$$F[x] = e^{(x,g)}$$

for which one obtains

$$\partial_f F[x] = \frac{d}{d\lambda} F[x+\lambda f]\big|_{\lambda=0} = (f,g) \cdot F[x].$$

In particular the lemma applies to the sharp time derivatives that one obtains from the choice of the Dirac distribution δ_t for f. With a slight abuse of notation we shall denote them by ∂_t. It is interesting to note that their images in the Fock space $\mathfrak{F}$ are canonical annihilation operators so that together with their adjoints they furnish a representation of the canonical commutation relations in (L^2) [7], [17]:

$$[\partial_t, \partial_s^*] = \delta(t - s) \ .$$

ENERGY FORMS - HAMILTONIAN QUANTUM FIELD THEORY IN THE SCHROEDINGER REPRESENTATION

Let us now return to the problem of a correct formulation for the dynamics of interacting quantum fields. The heuristic discussion of the first section suggests that in canonical theories it should be possible to extract the time development, i.e. the Hamiltonian, from the knowledge of the vacuum. Actually this program has turned out to be mathematically feasible and rather successful in the setting of non-relativistic quantum mechanics [18-21] where the vacuum vector turns into the ground state wave function ψ and the Hamiltonian in the "Ground State Representation"

$$\mathfrak{H} = L^2(\mathbb{R}^n, \ \psi^2 d^n x)$$

is given by the "energy form"

$$\epsilon(f) = \ <f, H f> \ = \ \int (\nabla f)^2(x) \ \psi^2(x) \ d^n x,$$

a definition - in terms of the ground state density ψ^2 - which is non-perturbative and very effective as a means to define extemely singular interactions that defy a perturbative treatment.

With the tools that we have acquired in the previous section it is now obvious how we should generalize this formula from quantum mechanics with n degrees of freedom to quantum field theories.

$$\text{For} \ (\nabla f)^2 = \sum_1^n \left(\frac{\partial f}{\partial x_j}\right)^2 \quad \text{we substitute} \quad (\nabla F)^2 = \int ds \ (\partial_s F)^2$$

which is a nonlinear map of $(\mathfrak{F})$ into itself [8], [22] and for the vacuum densities we invoke positive Hida distributions, so that for quantum field theories energy forms are simply given by

$$\epsilon(F) = <\Phi, (\nabla F)^2> = \int d\nu[x] \int ds \ (\partial_s F)^2[x]$$

as a quadratic form defined for all test functionals $F \in (\mathfrak{f}) \subset L^2(d\nu)$ for suitable positive Hida distributions Φ resp. their associated measures ν. As in the finite dimensional case we must make sure that this form defines a self-adjoint Hamiltonian by imposing an "admissibility condition".

DEFINITION: Consider $\Phi \in (\mathfrak{f})_+^*$ and the measure ν corresponding to it. If

$$\epsilon(F) = \int d\nu \ (\nabla F)^2 \quad \text{with } \mathfrak{D}(\epsilon) = (\mathfrak{f})$$

is closable on $L^2(d\nu)$ we shall call ϵ an "energy form" and Φ "admissible".

As is well-known we obtain a self-adjoint, positive "Hamiltonian" operator in $L^2(\mathfrak{f}^*, d\nu)$ with

$$< F, H \ F > = \epsilon(F) \text{ for } F \in (\mathfrak{f})$$

canonically whenever we construct an energy form:

THEOREM (Kato [23]): To every energy form ϵ there corresponds a unique self-adjoint positive operator H in $L^2(\mathfrak{f}^*, d\nu)$ such that

$$\epsilon(F) = \| H^{1/2}F \|^2 \text{ and } \mathfrak{D}(H^{1/2}) = \mathfrak{D}(\bar{\epsilon})$$

Admissibility of Hida distributions follows e.g. if

$$\partial_s \Phi = B(s) \cdot \Phi$$

for some $(\mathfrak{f})$-valued tempered distribution B. For the Fock vacuum of a massive free relativistic scalar field it is straight forward to caculate

$$B(s) = - ((\sqrt{-\Delta + m^2} -1)x)(s)$$

which clearly is in $(\mathfrak{f})$ after smearing out with a test function.

Another sufficient condition for the admissibility of positive Hida distributions is the quasiinvariance of the associated measure ν mentioned in section I [24]. Quasiinvariance has been established in particular for the vacuum measures of bosonic quantum field theories such as the Hoegh-Krohn and the sine-Gordon model which at the same time permit a representation of their vacuum in terms of a Hida distribution [15], [16]. This shows that energy forms in terms of White Noise are suitable for the discussion of non-trivial quantum field theories.

References:

[1] H. Araki: "Hamiltonian Formalism and the Canonical Commutation Relations in Quantum Field Theory". J. Math. Phys. 1, 492 (1960).

[2] F. Coester, R. Haag: "Representation of States in a Field Theory with Canonical Variables". Phys. Rev. 117, 1137 (1960).

[3] see e.g. G. Emch: "Algebraic Methods in Statistical Mechanics and Quantum Field Theory". Wiley, New York, 1971.

[4] I.M.Gelfand, N.J.Vilenkin: "Generalized Functions" vol. 4, Academic Press, New York, 1964.

[5] T. Hida: "Stationary Stochastic Processes". Princeton University Press, Princeton, 1970.

[6] T. Hida: "Brownian Motion". Springer, Berlin, 1980.

[7] I. Kubo, S. Takenaka: "Calculus on Gaussian White Noise I-IV" Proc. Japan Acad. Sci. 56, 376, 411 (1980); 57, 433 (1981); 58, 186 (1982).

[8] T. Hida, J. Potthoff, L. Streit: "White Noise Analysis and Applications" in "*Mathematics + Physics. Lectures on recent results*". vol.3, L. Potthoff and L. Streit, eds.. World Scientific, Singapore, 1989.

[9] T. Hida: "Generalized Brownian Functionals". *In "Theory and Application of Random Fields"*. G. Kallianpur, ed. . Springer, Berlin 1983.

[10] T. Hida: "Analysis of Brownian Functionals". Carleton Math. Lecture Notes no. 13. Carleton, 1975.

[11] K. Ito: "Foundations of Sochastic Differential Equations in Infinite Dimensional Space". Soc.Industr..Appl.Math. Philadelphia, 1984. Ch.1

[12] M. Reed, B. Simon: "Functional Analysis", vol. I. Academic Press, New York, 1972.

[13] I. Kubo, Y. Yokoi: "A Remark on the Space of Testing Random Variables in the White Noise Calculus". Preprint 1987.

[14] Y. Yokoi: "Positive Generalized Brownian Functionals". Kumamoto preprint, 1987.

[15] S. Albeverio, T. Hida, J. Potthoff, L. Streit: "The Vacuum of the Hoegh-Krohn Model as a Generalized White Noise Functional". Phys. Lett. B. 217, 511 (1989).

[16] S. Albeverio, T. Hida, M. Roeckner, J. Potthoff, L. Streit: "Dirichlet Forms in Terms of White Noise Analysis I". Preprint, 1989.

[17] T. Hida: "Brownian Functionals and the Rotation Group". In "*Mathematics + Physics. Lectures on recent results.*" Vol.1, L. Streit, ed.. World Scientific, Singapore, 1985.

[18] M. Fukushima: "Dirichlet Forms and Markov Processes". North Holland - Kodansha 1980.

[19] S. Albeverio, R. Hoegh-Krohn, L. Streit: "Energy Forms, Hamiltonians, and lm +5 Distorted Brownian Paths ". J. Math. Phys. 18, 907 (1977).

[20] S. Albeverio, R. Hoegh-Krohn, L. Streit: "Regularization of Hamiltonians and Processes". J. Math. Phys. 21, 1636 (1980).

[21] S. Albeverio, M. Fukushima, W. Karwowski, R. Hoegh-Krohn, L. Streit: "Capacity and Quantum Mechanical Tunneling". Comm. Math. Phys. 81, 501 (1981).

[22] T. Hida, J. Potthoff, L. Streit: "Dirichlet Forms and White Noise Analysis". Comm. Math. Phys. 116, 235 (1988).

[23] T. Kato: "Perturbation Theory for Linear Operators". Springer, Berlin, 1966.

[24] S. Albeverio, R. Hoegh-Krohn: "Quasiinvariant masures, symmetric diffusion processes and quantum fields." In *Proc. Int. Colloq. Math. Methods Quantum Field Theory*. CNRS 1976.